MEDIUM AND HIGH SPEED DIESEL ENGINES FOR MARINE USE

MEDIUM AND HIGH SPEED DIESEL ENGINES FOR MARINE USE

by

S. H. HENSHALL
B.Sc.(Eng.), C.Eng., M.I.Mar.E., F.I.Mech.E.

THE INSTITUTE OF MARINE ENGINEERS

MEDIUM AND HIGH SPEED DIESEL ENGINES
FOR MARINE USE

I S BN 0 900976 01 2

Printed in England by St. Stephen's Bristol Press, Bristol.

Contents

PREFACE

During my working life I have been fortunate in the many good friends I have made in the ship-operating, ship-building and marine engineering industries and in the universities also, throughout the world. The benefits I have derived from their patient instruction and their sound judgements in the experiences we have shared are my only real qualifications for writing this book.

Some of them have given direct practical help. Particularly, I wish to thank the directors of Mirrlees Blackstone Limited for placing secretarial and typing facilities at my disposal and my colleagues in that firm for their helpful advice and criticism, the directors of Turnbull Marine Design Co. Ltd. for carrying out drawing office work involved in constructing many of the figures, the directors of Bryce Berger Ltd. and the Joseph Lucas Group for supplying nearly all the illustrations that appear in Chapter 5 and the directors of Holset Engineering Co. Ltd. for providing material on torsional vibration dampers, as well as turbochargers. My gratitude is due also to the engine and turbocharger manufacturers who so willingly sent me information concerning their products; their names appear in the text.

An explanation of the variety of units used is needed as I can foresee the critics bouncing up and down like a row of valve springs in their eagerness to proclaim that they have never before come across a book containing such a motley, incoherent assortment. When the book was started it seemed a good idea to use S.I. units as the industry was intent on turning over to them. The British Standards Institution publication PD6430:1969 "The adoption of the metric system in the Marine Industry" was taken as a guide. The new system did not have (and still does not have) the stability that comes from long usage and just as printing of the book had started a revision was issued changing certain recommendations; notably proposing the use of MN/m^2 instead of bar for the measurement of pressure. This came too late and throughout the book the bar is the unit used.

One or two friends who read the earlier chapters said they felt all at sea with the S.I. system so some of the well known quantities were stated in established units as well, to serve as navigation marks until the waters became more familiar. As more chapters were written it was realized that it was misleading to give the impression that S.I. units were commonly used in the practical measurement of certain properties so I reverted to the units of the trade in covering these subjects. The chapters containing descriptions of engines and turbochargers retain the units which each manufacturer used when giving the information as it would not be honest to say that an engine was designed to develop 372·6 kilowatts per cylinder at 8·333 r/s when the figures the designer had in mind were clearly 500 h.p. per cylinder at 500 revolutions per minute.

The result of all this is that S.I., imperial and metric system units occur throughout the book. Although this may fill the purists with dismay I can be sure that the marine engineers for whom the work is intended will feel quite at home as it is just like the familiar world in which they live.

SHH. 1 . 10 . 71

INTRODUCTION

The diesel engine, or compression ignition engine, in its various sizes, covers a wider range of rotational speed than most, if not all, other prime movers. It is used in all sizes and types of vessel, from small launches to large ocean going ships. When we speak of high speed engines or low speed engines in connexion with this type of machine we are referring to the rotational speed. High and low are terms which have a meaning only relative to each other so it is worthwhile examining just exactly what is meant.

In the marine world the low speed engines are a well defined group distinguished by their revolutions per minute being of the same order as those required by the propellers of ocean going ships. This type of engine is usually directly coupled to the propeller and rotates at about 100 to 150 revolutions per minute. Its development has been limited by this propeller speed. At the upper end of the scale the high speed engine has been developed for use in boats and small craft. An engine with a speed of over 1,000 rev/min is usually looked upon as being high speed. The medium speed engines fall between these two and cover the speed range from about 300 to about 1,000 rev/min.

Both medium and high speed engines have been familiar to marine engineers for many years as auxiliary generator engines and in the propulsion of fishing vessels, tramps, coasters as well as tugs, dredgers and other service craft. If one considers engines having similar brake mean pressures and piston speeds it can be shown that power per unit swept volume, is greater for the smaller engines. That is to say the smaller engines give higher values of power per unit swept volume. Thus the most compact engines can be designed with the smallest cylinders but in order to obtain large powers the number of cylinders must increase and for high powers a great number of cylinders are needed.

Engines having cylinder bores of less than say 500 mm, can have their cylinders arranged in V form, a form which adds greatly to the possibility of providing compact high powered engines. Engines having sixteen cylinders are quite common and engines of eighteen and even twenty cylinders have been built in this way. Because of their small bulk two such engines or three or even four can be geared together and with their gearing occupy a smaller space than the large bore engine of similar total power. They are also much lighter and the combined saving in bulk and weight makes them very attractive for ship propulsion. There have been practical difficulties in the construction of reliable and inexpensive gearing and couplings to connect together the groups of these medium speed engines and in equipment to control them satisfactorily but it is now quite apparent that these difficulties have been completely overcome.

The early history of the development of the medium and high speed engine has been connected with the burning of light distillate fuels because such fuels burn sufficiently rapidly for the cycle to be carried out in the short space of time that occurs when the crankshaft is run at high rev/min. It has also been a story of the demand for compact power. In consequence of these factors this type of engine is almost invariably of trunk piston design. A characteristic of this design is the very much reduced headroom and overhauling height which is required when compared with the large bore slow speed engine of cross head construction. This feature has resulted in an understandable demand for this type of engine in ferries where headroom is of paramount importance.

The extraordinary reliability of medium speed engines has been proved by their history of operation in trawlers and fishing vessels which are away from port for much longer periods than practically any other type of vessel and continue to operate under the most arduous conditions with the minimum of attention. The more recent development of the medium speed engine has proceeded parallel with that of lubricating oils and the two developments together have enabled this type of engine to run reliably and satisfactorily on heavy fuels. Lubricants can be maintained sufficiently alkaline at all times to ensure that liner wear is not excessive and that crankcase contamination with the possibility of corrosion is entirely non-existent. The only limiting factor at present to the burning of heavy fuel is the speed at which it can be burnt in the cylinder and currently this restricts engines to a maximum speed of 750 rev/min or thereabouts if they are to run on fuel containing residual constituents. Because of this, the high speed engine at present cannot be run on heavy fuels and is confined to the use of distillates.

The development of turbocharging, whilst helping compression ignition engines of all sizes to increase their specific power has been of most benefit in the medium speed range. Compared with high speed engines, turbochargers dealing with the flows required by medium speed engines have been more efficient, although lately radial flow turbochargers are having their effect in the high speed range of engine. Compared with large bore engines the size of the medium speed engine is conducive to containing higher pressures in the cylinder and coping with the increased thermal loading far better than can be done in the large bore cylinders. In consequence the greatest objection to the use of multi-engine geared machinery in ships is gradually disappearing, namely the large number of cylinders that must be employed. Although the numbers of cylinders are obviously greater than when large bore engines are used they are now numbers that are quite manageable from the maintenance point of view.

Multi-engined medium speed plant is eminently suitable for installation near to the stern of a ship where it can be suited to a space less satisfactory for other purposes and where its light weight does not affect the trim unduly. A medium speed engine is often light enough to be handled as a complete unit from the factory and can be installed without dismantling and rebuilding and often without involving heavy lifts down a long engine room casing.

It is true that there is some supplementary loss in power arising from the use of gearing, but this can, in most cases, be more than offset by the facility of choosing a propeller speed more suited to the ship, the gain in propeller efficiency being greater than the loss occasioned by the gearing efficiency. Another feature is that the reliability of the ship is no longer restricted to the reliability of one engine only and in a two-engined ship with a fixed pitch propeller a speed of 70% of full speed can be obtained on one engine alone. If a controllable pitch propeller is used then the speed of vessel obtainable on one engine alone is usually of the order of 80% or more of full speed.

The multi-engined medium speed installation is well suited to forming a power source from which nearly all the diverse energy requirements of the ship can be taken. In conjunction with a controllable pitch propeller and suitable alternators, electric power can be provided whilst on passage and in port. The output shaft speed is suitable for electrical power generation and because there are a number of engines it is possible to run only one of them which may amount to only a half, a third or a quarter of the whole power installation in order to provide electrical power when propulsion is not required. In other words, the multi-engine medium speed geared installation is a very flexible power source.

Smaller engines with comparatively larger numbers of cylinders have other advantages, one of which is that the provision of spare gear can form a comparatively small part of the capital outlay as spares items are suitable for a larger number of cylinders than in the case of an engine room which has only a few cylinders in it. Furthermore a fleet may consist of vessels of different sizes and it is possible to cover a wider range of vessels of various powers with basically the same engine cylinder size and an owner may, in consequence, stock a very suitable range of spares at a small capital outlay.

Whilst the high speed engine continues to serve its traditional function of propulsion of small craft it is also becoming widely used as a power generator in large vessels. The high speed results in electrical equipment which is cheaper in first cost so that the outlay for a given generator capacity is more economic when high speed engines are used. As the power requirements of vessels are increasing rapidly this is a matter of the greatest importance.

Both medium speed and high speed engines are regarded as making rather more noise than slow speed engines and there is some truth in this although a large bore engine when turbocharged can, in parts of the engine room, have noise levels as high and as uncomfortable as those in compact engine rooms full of medium and high speed machinery. In all cases, there is a continuing tendency in modern ships to provide engine control rooms from which the operators can see the machinery whilst remaining in an acoustically shielded zone. With the increased complexity of modern machinery, not only the propulsion and auxiliary engines but all the other machines that are in engine rooms, there is increasingly the necessity to group the instruments and controls together where all can be seen rapidly. The control room is, therefore, a natural development and the noise problem is being solved without a great deal of worry in this way.

The chief aim of this book is to assist those who are responsible for the operation and maintenance of medium and high speed engines in marine use where the popularity of this type of machinery will undoubtedly continue to increase. Developments are proceeding rapidly in both total and specific output and a book consisting largely of detailed descriptions of current engines would quickly become out-dated. The approach adopted has been to explain the principles on which design and operation are based and the examples described in the later chapters are confined to a few representative types selected to illustrate the practical application of these principles.

An understanding of diesel engines demands a thorough acquaintance with a wide range of disciplines and in a work of this size is not possible to do more than effect an introduction to some of the many branches of the art, and attempt to promote an appreciation of both the capabilities and the limitations within which designers, manufacturers and operators all must work. The references and bibliographies which are listed at the end of each chapter are intended to provide some indication of where more detailed information may be found and it is hoped that this feature will extend the usefulness of the book throughout the industry.

CHAPTER ONE

Fundamental Thermodynamics

1.1. The modern turbocharged and intercooled diesel engine operates on a somewhat complex thermodynamic cycle. For a full appreciation of its capabilities and its limitations an understanding of certain basic thermodynamic relationships is essential. The treatment given in this chapter, whilst adequate for the purposes of this book, does not claim to be either complete or rigorous and readers requiring more precise analysis are referred to the bibliography given at the end of the chapter.

1.2. *Laws of perfect gases*

Boyle's law states that at constant temperature, the volume V of a gas varies inversely as the pressure P. Charles' law states that at constant pressure the volume V varies directly as the temperature T.

These laws are usually combined as

$$\frac{PV}{T} = \text{Constant}$$

If unit mass of gas is considered, then V is the specific volume V_{sp} (m^3/kg or ft^3/lb) ($= 1/\rho$ where ρ is the density) and the constant is termed the gas constant denoted by R. Its value for air is 0·287 kJ/kg K or 53·3 ft.lb/lb/°F.

This gives what is known as the characteristic equation for gases

$$PV_{sp} = RT \tag{1.1}$$

or

$$\frac{P}{\rho} = RT \tag{1.1a}$$

or in differential form

$$PdV + VdP = RdT \tag{1.1b}$$

For a given mass of gas, m, equation (1.1) becomes

$$PV_m = mRT$$

where V_m is the actual volume, not the specific volume.

For a quantity of 1 mol. (molecular wt. in mass units) of a gas

$$PV\text{mol.} = m_{mol}RT$$

From Avogadro's law it is known that at any given pressure and temperature a mol. of a gas occupies the same volume as a mol. of any other gas, hence $m_{mol}R$ is a fixed quantity. It is known as the universal gas constant denoted by $\bar{R}$. Its value is 8·3143 kJ/k mol. K or 1,547 ft.lb/mol./°F.

Thus,

$$PV_{sp} = \frac{\bar{R}}{m_{mol}} T \qquad (1.2)$$

1.3. *First law of thermodynamics*

The first law of thermodynamics is a form of the principle of conservation of energy. When heat is added to a gas it may appear as (i) a rise in temperature, (ii) as internal work, (iii) as external work, (iv) as an increase in kinetic energy.

A rise in temperature increases the sensible heat or internal energy of the gas. It is the product of mass, specific heat and rise in temperature of the gas. Internal work signifies a change in the molecular state of the gas, such as dissociation. (Whilst we are concerned with perfect gases this will not enter our calculations but it becomes important when dealing with actual gases.) External work is performed by the gas on its surroundings and for a gas enclosed in a cylinder is measured in terms of pressure and change of volume. Kinetic energy is measured by velocity of flow of the gas, this is negligibly small in cylinder processes but often substantial in flow processes.

If we consider unit mass of a perfect gas in a cylinder, or flowing with uniform velocity, then the first law can be stated as "the heat supplied is equal to the change of internal energy plus the work done". Using the symbols Q for heat supplied, U for internal energy and W for work done, we have

$$Q = U + W \qquad (1.3)$$

in the differential form

$$dQ = dU + dW \qquad (1.3a)$$

Substituting $c\,.\,dT$ for dU where c is specific heat and PdV for dW, we have

$$dQ = c\,.\,dT + \frac{PdV}{J} \qquad (1.3b)$$

the constant J being necessary if dQ and $c\,.\,dT$ are in heat units and PdV in mechanical work units. If the coherent units of the SI system are used, this constant becomes unity (as 1 Nm = 1 J) and may be omitted.

Equation (1.3) in its various forms is known as the energy equation of gases.

1.4. *Specific heats*

Specific heat is usually defined as the heat required to raise unit mass of a substance through unit temperature, but for a gas it is necessary to specify the conditions under which the heat is added. Two cases are of special importance, constant volume and constant pressure.

Using the energy equation, when volume is constant, $dV = 0$, so there is no work done and

$$dQ = C_v dT$$

And when pressure is constant, the work done $= PdV/J$ as in (1.3b) and

$$dQ = C_v dT + \frac{PdV}{J}$$

using equation (1.1b)

$$dQ = C_v dT + \frac{R}{J}dt \quad \text{(since } dP = 0\text{)}$$

$$= \left(C_v + \frac{R}{J}\right) dT$$

$$= C_p dT$$

Therefore, C_p is greater than C_v and is equal to $C_v + R/J$

or
$$C_p - C_v = \frac{R}{J} \tag{1.4}$$

The energy equation thus becomes

$$dQ = C_v dT + \frac{Pdv}{J} \tag{1.5}$$

for all cases. Note that the internal energy is dependent only on the temperature.

The ratio C_p/C_v is denoted by γ.

1.5. *Enthalpy*

A thermodynamic system has what are known as properties, the criterion of a property being that it is a quantity whose change is determined by the end states alone and is not affected by the process. Pressure and temperature are properties but heat and work are not.

Enthalpy or total heat is denoted by H and defined as

$$H = U + \frac{PV}{J} \tag{1.6}$$

i.e. enthalpy is equivalent to the sum of the internal energy and the external work of expanding the gas from zero volume to its actual volume whilst at its actual pressure. As enthalpy is composed entirely of properties it must itself be a property and is independent of the process by which a given state is reached.

The differential form of (1.6) is

$$dH = dU + \frac{d\,(PV)}{J}$$

$$= C_v dT + \frac{PdV + VdP}{J} \tag{1.6a}$$

and from (1.1b)

$$dH = C_v dT + \frac{RdT}{J} \tag{1.6b}$$

and from (1.4)

$$dH = C_p dT \tag{1.7}$$

Note that under constant pressure conditions

$$dH = dQ$$

i.e. change of enthalpy is equal to the heat received or given out.

1.6. *Entropy and reversible processes*

A process is said to be reversible if all the elements in the system can be restored to their initial states by reversing the process. In practice, no operation is completely reversible but the concept provides a criterion by which actual processes may be assessed. Friction is the outstanding example of a factor limiting the attainment of a reversible process. In flow processes friction and eddying of the fluid are transformed into heat and render actual compression and expansion processes irreversible.

Entropy is a property but is not physically measurable and therefore difficult to define satisfactorily. For our purpose it can be regarded as the extensive factor of heat, temperature being the intensive factor. Just as

$$\text{Force} \times \text{change of distance} = \text{work done}$$

and

$$\text{Pressure} \times \text{change of volume} = \text{work done}$$

so

$$\text{Temperature} \times \text{change of entropy} = \text{heat supplied}$$

In symbol form:

$$Td\phi = dQ \tag{1.8}$$

or

$$d\phi = \frac{dQ}{T} \tag{1.8a}$$

Note that change of entropy can only be calculated by $\int dQ/T$ when the process is reversible. Irreversible processes always result in an increase in entropy.

As entropy is a property, changes may be evaluated from a knowledge of the end states alone, values at any state from a datum being found by $\int dQ/T$ following any reversible path.

1.7. *Cylinder processes*

Consider a cylinder fitted with a piston as shown in Figure 1–1. If the cylinder is connected to a supply of gas at a constant pressure, P, by means of a valve and the gas displaces the piston from position 0 to position 1. The work done is given by

$$W_{01} = \int_0^1 PdV = P(V_1 - V_0) \tag{1.9}$$

This can be regarded as positive if it is performed by the gas on the piston (outward movement of the piston increasing the volume) and as negative if it is performed by the piston on the gas (inward movement of the piston decreasing the volume and displacing the gas out of the cylinder).

In an actual engine there is a pressure close to atmospheric pressure on the outside of the piston which may have to be taken into account when applying equation (1.9) in practice.

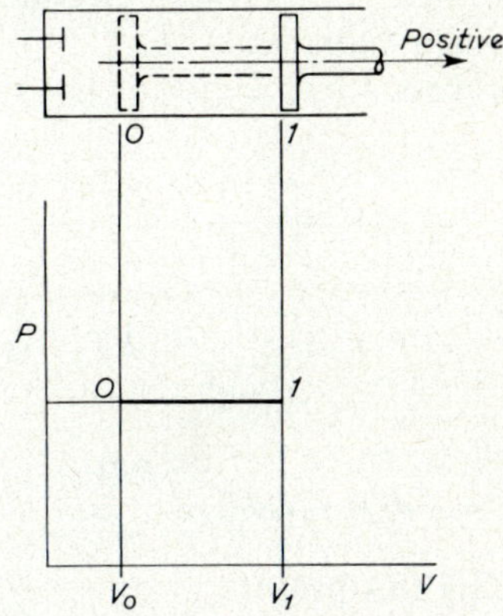

FIG. 1–2—*Work done during compression or expansion of a gas in a cylinder.*

If gas is enclosed in the cylinder, perhaps by closing the valves with the piston in position 2, then it will be compressed if the piston is moved inwards or expanded if it moves outwards, see Figure 1–2. The relationship between pressure and volume is given by

$$PV^n = \text{constant}$$

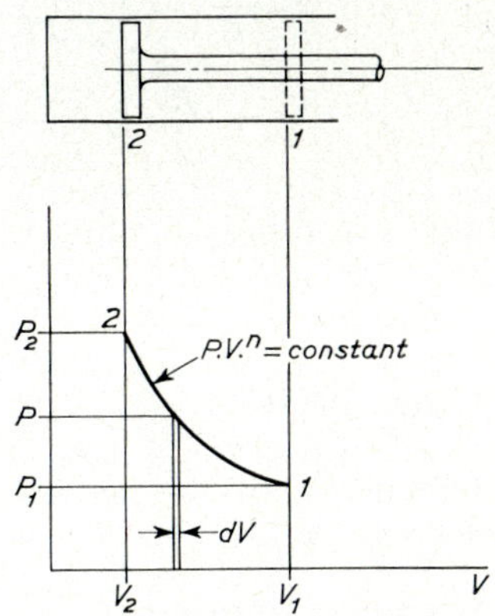

FIG. 1–2—*Work done during compression or expansion of a gas in a cylinder.*

The work done is given by

$$W_{12} = \int_1^2 PdV = P_1V_1^n \int_1^2 \frac{dV}{V^n} \tag{1.10}$$

i.e.

$$W_{12} = P_1V_1^n \left[\frac{V_2^{1-n} - V_1^{1-n}}{1-n}\right]$$

$$= \frac{P_1V_1 - P_2V_2}{n - 1} \qquad (1.10a)$$

which will be negative for compression and positive for expansion.

Two special cases are of interest. First, a process at constant temperature following Boyle's law, termed "isothermal". Second, a process with neither gain nor loss of heat, termed "adiabatic".

In an isothermal process PV = constant (Boyle's law) *i.e.* $n = 1$. The work done is given by

$$W_{12} = P_1V_1 \int_1^2 \frac{dV}{V}$$

$$= P_1V_1 \log \frac{V_2}{V_1} = RT_1 \log \frac{V_2}{V_1} \qquad (1.11)$$

From the first law

$$dQ = c_v dT + \frac{PdV}{J}$$

and as T is constant during this process

$$dQ = \frac{PdV}{J}$$

i.e. the heat to be removed from the gas = work done on it. (In the case of compression.)

The change of entropy can be calculated from

$$\int_1^2 d\phi = \int_1^2 \frac{dQ}{T} = \int_1^2 \frac{PdV}{JT}$$

i.e.
$$\phi_2 - \phi_1 = \frac{R}{J} \log \frac{V_2}{V_1} \qquad (1.12)$$

In an adiabatic process n will have a value which is determined in the following way.

Using the energy equation

$$dQ = C_v dT + \frac{PdV}{J} = 0$$

as there is no exchange of heat.

Substituting $(PdV + VdP)/R$ for dT

$$\frac{C_v}{R}(PdV + VdP) + \frac{PdV}{J} = 0$$

Putting $R/J = C_p - C_v$ and dividing by PV

$$\frac{C_v}{C_p - C_v}\left[\frac{dV}{V} + \frac{dP}{P}\right] + \frac{dV}{V} = 0$$

i.e.
$$\frac{dV}{V}\left[\frac{C_v}{C_p - C_v} + 1\right] + \frac{dP}{P}\left[\frac{C_v}{C_p - C_v}\right] = 0$$

$$\frac{dV}{V}\left[\frac{C_p}{C_p - C_v}\right] + \frac{dP}{P}\left[\frac{C_v}{C_p - C_v}\right] = 0$$

$$\frac{dV}{V}\left[\frac{C_p}{C_v}\right] + \frac{dP}{P} = 0$$

Denoting the ratio C_p/C_v by γ and integrating:

$$\log V^{\gamma} + \log P = \text{Constant}$$

i.e. $$\log PV^{\gamma} = \text{Constant}$$

and $$PV^{\gamma} = \text{Constant}$$

Thus, in an adiabatic process $n = \gamma = C_p/C_v$.

Referring again to Figure 1–2 and equation (1.10a), the work done

$$W_{12} = \frac{P_1V_1 - P_2V_2}{\gamma - 1} = \frac{R}{J}\left(\frac{T_2 - T_1}{\gamma - 1}\right) \tag{1.13}$$

Writing $$\frac{R}{J} = C_p - C_v$$

and $$\gamma - 1 = \frac{C_p - C_v}{C_v}$$

then $$W_{12} = C_v(T_2 - T_1) = U_2 - U_1 \tag{1.13a}$$

the work done = change in internal energy as would be expected from the energy equation.

For a reversible adiabatic process as $dQ = 0$

then $$d\phi = \frac{dQ}{T} = 0$$

i.e. it is a process at constant entropy.

The term "isentropic" is used to describe ideal adiabatic processes.

Equation (1.13) gives the "absolute" work for the process. If the work of taking in and expelling the gas is taken into account, the "total" work for the process is obtained. Referring to Figure 1–2 and considering compression the total work is given by:

$$dW = \int_1^2 VdP = \int_1^2 PdV + P_2V_2 - P_1V_1$$

If the process is isentropic and $n = \gamma$ then total work

$$dW = \frac{1}{\gamma - 1}(P_2V_2 - P_1V_1) + P_2V_2 - P_1V_1$$

$$= \frac{\gamma}{\gamma - 1}(P_2V_2 - P_1V_1) \tag{1.14}$$

$$= \frac{\gamma}{\gamma - 1}\frac{R}{J}(T_2 - T_1)$$

$$dW = C_p(T_2 - T_1) = H_2 - H_1 \tag{1.14a}$$

the total work done = change in enthalpy (total heat).

A similar expression will be obtained in the case of expansion.

1.8. *Flow processes*

When a gas undergoes compression or expansion in turbo-machinery it is necessary to take into account the kinetic energy as mentioned in Section 1.3. Figure 1–3 represents a thermodynamic machine with gas entering at velocity v_1, its pressure, specific volume, temperature and entropy being P_1, V_1, T_1 and ϕ_1 respectively, and leaving with velocity v_2 and corresponding properties P_2, V_2, T_2 and ϕ_2. During passage through the machine energy may be added or taken away in the form of heat or work.

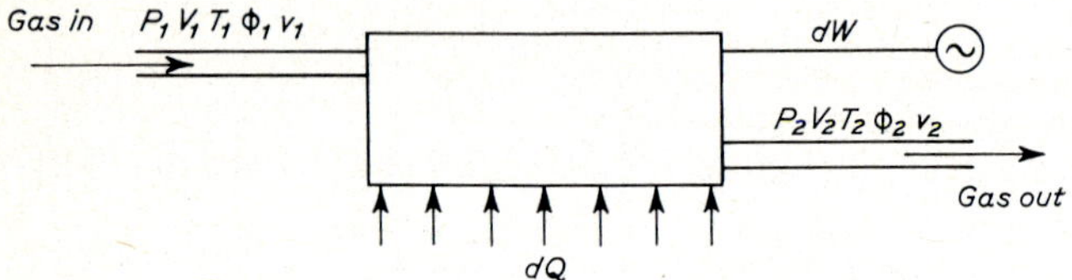

FIG. 1–3—*Energy balance.*

The general form of the energy equation to suit this case may be written:

$$H_1 + \frac{v_1^2}{2g} + dQ + dW = H_2 + \frac{v_2^2}{2g} \qquad (1.15)$$

If the machine is the compressor or the turbine of a turbocharger, there will be no heat exchange and v_1 may be assumed equal to v_2 as both will be negligibly small at the inlet to and outlet from these machines. Equation (1.15) then becomes

$$dW = H_2 - H_1 = C_p (T_2 - T_1) \qquad (1.16)$$

This is the same equation as (1.14a) and it should be noted that it applies in all cases and not only for isentropic processes.

Practical processes of this kind may be adiabatic in the sense that no heat is lost or gained but because of the effects of friction and eddies the entropy will increase during the process so that they are not isentropic.

Figure 1–4 shows a comparison of isothermal, isentropic and practical adiabatic compression processes in PV and $T\phi$ diagram form and Figure 1–5 illustrates in a similar way expansion processes. The lines 1 to 2 on these

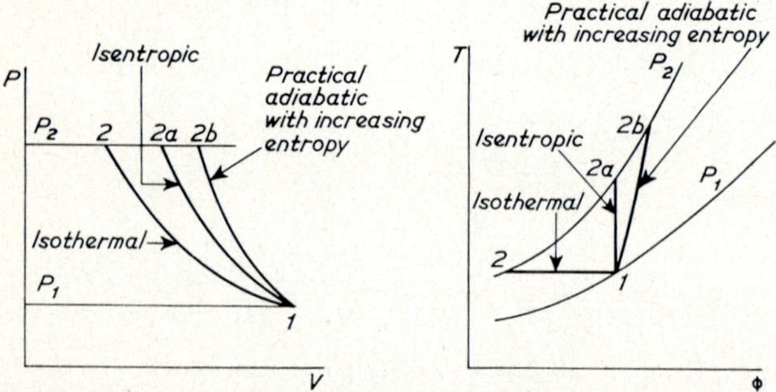

FIG. 1–4—*PV and Tφ diagrams of compression processes.*

diagrams denote an isothermal process as is evident from the $T\phi$ diagram. The lines 1 to 2a denote an isentropic process as is also evident from the $T\phi$ diagram. Both these are ideal reversible processes. The lines 1 to 2b denote the practical adiabatic processes and these are not reversible but result in an increase of entropy in both compression and expansion as can be seen by comparing the $T\phi$ diagrams of Figure 1–4 and Figure 1–5.

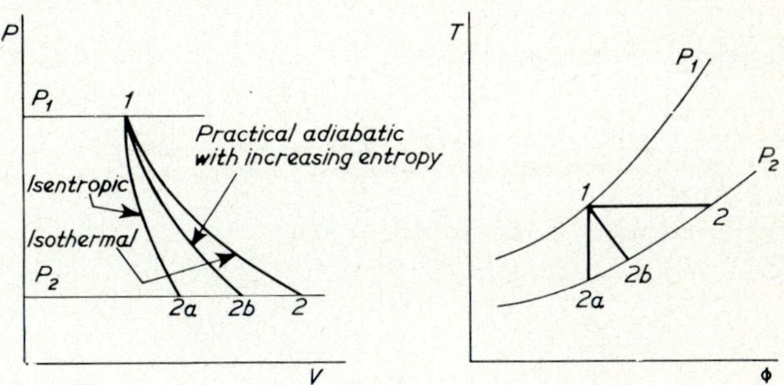

FIG. 1–5—*PV and Tφ diagrams of expansion processes.*

It is most important to realize that when considering flow processes, lines such as those drawn in Figures 1–4 and 1–5 merely link initial and final states of the gas; they no longer represent processes as in the case of gas enclosed in a cylinder and areas on the PV diagram ARE NOT representative of work done. Generally speaking PV diagrams are of greatest use in analyzing cylinder processes and $T\phi$ diagrams in analyzing flow processes.

1.9. *Flow through passages of varying cross-section*

During its progress through turbomachinery gas often has to proceed through passages of changing cross-section such that the velocity of the gas changes to an extent that is highly important when applying the energy equation. Figure 1–6 shows a passage which, at first, converges and then diverges. The plane at which the cross-section is a minimum is termed the throat.

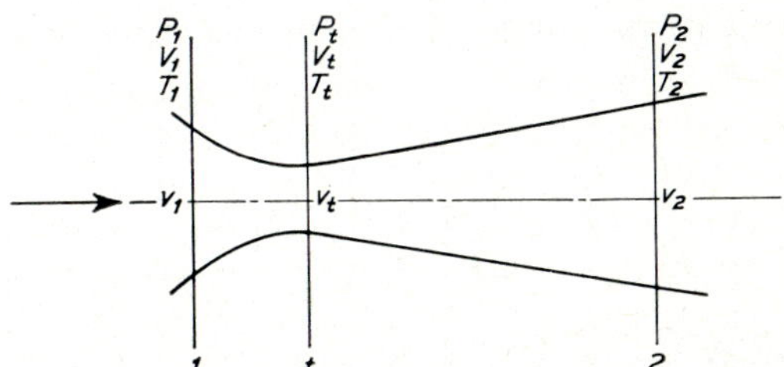

FIG. 1–6—*Flow in passage of varying cross section.*

The mass flow of gas at all sections is constant, thus

$$m = A_1 v_1 \rho_1 = A_t v_t \rho_t = A_2 v_2 \rho_2 \tag{1.17}$$

This is termed the continuity equation in which A is the cross-section area.

$$dm = d\,(Av\rho) = 0$$

Differentiating ($Av\rho$) and dividing through by $Av\rho$ gives

$$\frac{dA}{A} + \frac{d\rho}{\rho} + \frac{dv}{v} = 0 \qquad (1.18)$$

Applying the energy equation (1.15)

$$H_1 + \frac{v_1^2}{2g} + dQ + dW = H_2 + \frac{v_2^2}{2g}$$

There is no external work done, so $dW = 0$

Thus, $$H_2 - H_1 + \frac{v_2^2 - v_1^2}{2g} = dQ$$

or, $$dH + \frac{vdv}{g} = dQ$$

Substituting for dH from equation (1.6a)

$$C_v dT + PdV + VdP + \frac{vdv}{g} = dQ$$

For a reversible change, $dQ = C_v dT + PdV$

$$\therefore \qquad VdP + \frac{vdv}{g} = 0$$

Replacing V by $1/\rho$

$$\frac{dP}{\rho} = -\frac{vdv}{g}$$

i.e. $$\frac{dv}{v} = -\frac{dP}{\rho}\,\frac{g}{v^2} \qquad (1.19)$$

From the adiabatic relationship PV^γ = constant $= P/\rho^\gamma$

Differentiating

$$d\left(\frac{P}{\rho^\gamma}\right) = 0$$

$$= dP\,\rho^{-\gamma} - \gamma\rho^{-\gamma-1}\,.\,P\,.\,d\rho$$

Dividing through by $P\rho^{-\gamma}$

$$0 = \frac{dP}{P} - \gamma\frac{d\rho}{\rho}$$

or, $$\frac{d\rho}{\rho} = \frac{1}{\gamma}\frac{dP}{P} \qquad (1.20)$$

Substituting for dv/v (equation 1.19) and for $d\rho/\rho$ (equation 1.20) in equation (1.18).

$$\frac{dA}{A} = dP\left(\frac{g}{\rho v^2} - \frac{1}{\gamma P}\right)$$

or,

$$dP = \frac{dA}{A}\left(\frac{\gamma P v^2}{g\gamma\, P/\rho - v^2}\right) \tag{1.21}$$

This general expression relates the change of pressure in a gas flowing along a passage of changing area.

1.10. *Convergent flow*

Consider flow between planes 1 and t in Figure 1–6 where the flow is convergent. dA is negative (decreasing area), therefore dP is negative (decreasing pressure) and v increases if the term in brackets in the denominator is positive.

i.e. if

$$v^2 < g\gamma\frac{P}{\rho}$$

or

$$v < \sqrt{g\gamma\frac{P}{\rho}}$$

or

$$v < \sqrt{g\gamma RT} \tag{1.22}$$

The expression $\sqrt{g\gamma RT}$ is the velocity of sound in a gas. Therefore, in a convergent passage, starting with a subsonic velocity the velocity of the gas cannot exceed the sonic velocity.

Thus, the maximum velocity at the throat $v_t = \sqrt{g\gamma RT}$.

1.11. *Convergent nozzle*

The turbine nozzle is an example of a convergent passage. From the energy equation (1.15)

$$H_1 - H_t = \frac{v_t^2 - v_1^2}{2g}$$

as there is no external work done and no heat added or subtracted. $dQ = 0$ and $dW = 0$.

Putting $H = C_pT$ and, if necessary because of the units, using J to convert to mechanical energy.

$$v_t^2 - v_1^2 = 2gJC_\rho\,(T_1 - T_t) \tag{1.23}$$

Usually in the case of a nozzle v_1 is very small in comparison with v_t and equation (1.23) may be written

$$v_t^2 = 2gJC_\rho\,(T_1 - T_t) \tag{1.23a}$$

If V_t is the maximum value

$$g\gamma RT_t = 2gJC_\rho\,(T_1 - T_t)$$

Substituting $C_p - C_v$ for R/J this reduces to

$$\frac{T_1}{T_t} = \frac{\gamma + 1}{2} \tag{1.24}$$

or

$$\left(\frac{P_1}{P_t}\right)^{(\gamma-1)/\gamma} = \frac{\gamma + 1}{2} \tag{1.25}$$

Thus, the pressure ratio from entry to throat has a maximum value dependent only on the specific heats of the gas. This is known as the critical pressure ratio. For pressure ratios below the critical the velocity at the throat will be subsonic and can be calculated from (1.23a). As the pressure ratio is increased the velocity at the throat will rise reaching sonic velocity at the critical pressure ratio. At pressure ratios higher than the critical, the throat velocity will remain at sonic; it will not be influenced by the downstream conditions and the nozzle is said to be choking.

1.12. *Divergent flow*

When the cross-section is increasing in the direction of flow, dA is positive, and, therefore, dP is positive, *i.e.* the pressure increases, and v decreases if $v^2 < gRT$, *i.e.* the velocity decreases if it is less than the velocity of sound.

This is the case for diffusion in which velocity is converted into static pressure.

1.13. *Convergent-divergent flow*

If a divergent passage follows a convergent one as in Figure 1–6 and the velocity at the throat has reached the sonic velocity, P_2 being less than P_t, then dA is positive and dP is negative and $v^2 > gRT$, *i.e.* the velocity becomes supersonic. The mass flow, however, does not increase as it is limited by the choked condition at the throat. Convergent-divergent nozzles are used in steam turbines but not normally in gas turbines for turbochargers.

1.14. *Fluid energy and mechanical energy*

Figure 1–7 shows the rotor of a turbo-machine rotating at a speed of ω radians per second and with fluid entering with velocity v_i at one particular angle and radius and leaving with velocity v_o at another angle and radius. As shown in the Figure the velocity at inlet and outlet can be resolved into components which are axial, radial and tangential with reference to the rotor. The tangential component is often referred to as the velocity of whirl.

By Newton's Law of Motion the torque given to or by the rotor is equal to the rate of change of angular momentum of the fluid. For unit mass flow of fluid

$$\text{Torque} = \frac{1}{g}\,(v_{wi}r_i - v_{wo}r_o) \tag{1.26}$$

where v_{wi} is velocity of whirl at inlet and v_{wo} is velocity of whirl at outlet.

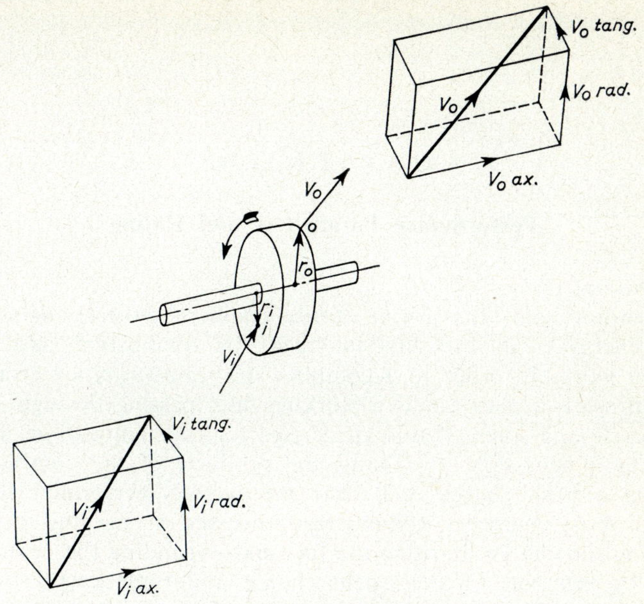

FIG. 1–7—*Rotor with fluid velocities at inlet and outlet.*

The work done per unit time for unit mass flow, or the rate of energy transfer for unit mass flow, W, equals torque times angular velocity.

i.e.
$$W = \frac{\omega}{g}(v_{wi}r_i - v_{wo}r_o)$$

and since $\omega r = U$, the rotor peripheral velocity at the radius r,

$$W = \frac{1}{g}(U_i v_{wi} - U_o v_{wo}) \qquad (1.27)$$

This is the energy equation for a fluid and a rotor.

BIBLIOGRAPHY

BENSON, R. S., *Advanced Engineering Thermodynamics*, Pergamon Press.
KIEFER, P. J., KINNEY, G. F. and STUART, M. C., *Principles of Engineering Thermodynamics*, Chapman and Hall, London.
OBERT, E. F. and GAGGIOLI, R.A., *Thermodynamics*, McGraw-Hill.
PHILLIPS, J. and OWEN JAMES, J. B., *Concise Applied Thermodynamics*, Van Nostrand.
VAN WYLEN, G. J., *Thermodynamics*, Chapman and Hall, London.

CHAPTER TWO

Performance Parameters and Rating

2.1. *Operating cycles*

In common with other forms of heat engine, the diesel engine converts chemical energy in its fuel first into heat, by combustion, and then into mechanical work. In order to accomplish the transformation from heat to mechanical work it uses air as a working fluid passing through a thermodynamic cycle and some of this air is used for the combustion of the fuel. This makes it necessary to exhaust the products of the combustion and to draw in a fresh charge each time the cycle is performed. The cycle, familiar to most engineers, follows the processes of charging the cylinder, compressing the charge, burning the fuel and expanding the heated charge, followed by removal of this spent charge and replacement by a fresh charge ready for the next cycle. The essence of the diesel cycle is that each charge of air is compressed to such a degree that its temperature at the end of the compression stroke is high enough for it to ignite the fuel which is introduced into the cylinder by injecting it in the form of a fine spray. The diesel engine is also known as the compression ignition engine because of this essential feature.

Practical engines work on either the four stroke cycle or the two stroke cycle. In a four stroke cycle engine the exhaust gases after completing the expansion stroke are expelled from the cylinder by the piston completing a full stroke and the fresh charge is drawn in by the piston making a full stroke in the opposite direction. The flow of gases is controlled by valves in the cylinder head, Figure 2–1(a). In the two stroke engine the exhaust gases at the end of expansion are allowed to escape by opening valves or ports, the residue being displaced by the incoming fresh charge which is arranged to be at a somewhat higher pressure. The flow of gases can be controlled by ports only, as in the loop scavenge, Figure 2–1(b) or opposed piston engines, Figure 2–1(c) or by a combination of piston controlled ports and valves in the cylinder head, Figure 2–1(d).

It is clearly desirable to obtain the maximum amount of power from a given cylinder providing this can be done reliably. The work done per cycle can be increased by using a larger proportion of the working fluid, air, to burn more fuel, but there are limits to this, both in the amount of air that can be successfully brought into intimate contact with the fuel in the short time available and also in the higher temperatures reached during the cycle which affect, adversely, the components forming the combustion chamber and exhaust system of the engine. Alternatively, the frequency with which the chosen cycle is performed can be raised by increasing the speed of the engine but obviously a number of limits will be reached. These are discussed in later chapters. These problems can be overcome by increasing the density of the charge thus providing more air for combustion and at the same time a greater mass of working fluid per cycle. The increased density is achieved by raising the pressure of the charge air prior to its entering the cylinder. If the pressure ratio is more than 1·2 or 1·3

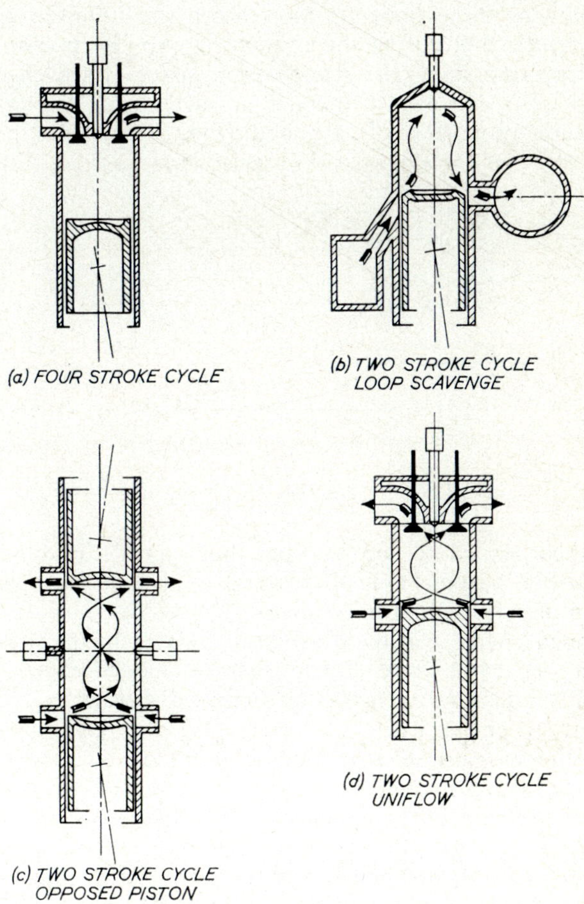

FIG. 2–1—*Diesel engine cylinder types.*

it is usually beneficial and sometimes necessary to cool the air in order to keep it at a reasonably high density, and fortunately this is easily done as the temperature reached after compression provides a temperature difference for the cooling process. The increased pressure is provided by a blower or pressure charger. This may be mechanically driven, using power take-off from the engine crankshaft, or it may be driven by an exhaust gas turbine using energy available in the exhaust gases after they have left the cylinder.

2.2. *The four stroke cycle*

Figure 2–2 shows an indicator diagram, *i.e.* a real PV diagram, for a four stroke engine. In naturally aspirated four stroke cycle engines the suction stroke 0–1 draws in the charge at atmospheric pressure, it is then compressed 1–2 and heated by burning fuel in the charge air itself 2–3–4, and then expanded, doing work on the piston until the exhaust valves open at 5, allowing the cylinder to blow down 5–6 followed by expulsion of the burnt gases, 6–0 during the exhaust stroke. By pressure charging, even at low pressure level, the compression space is scavenged of combustion products so that even if there is no increase in density the charge contains a greater proportion of oxygen allowing more fuel to be burnt and

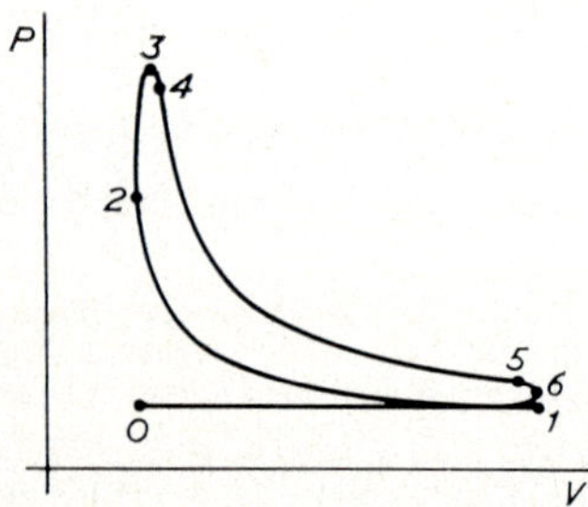

FIG. 2–2—*Indicator diagram for four-stroke engine.*

hence more heat added to the working fluid per cycle with a resultant increase in power output. This scavenging is assisted by a valve timing giving overlap of exhaust and inlet valves.

The trapped charge will be at a density related to the pressure of the air supply and its temperature. For a pressure charged engine this will be a higher density than for a naturally aspirated engine enabling a larger amount of fuel to be burnt and also increasing the mass of working fluid per cycle. If the pressurized air supply is provided by an exhaust driven turbocharger the exhaust pressure will also rise but of course will remain well below the pressure in the cylinder at the exhaust valve opening, point 5 Figure 2–2. In general, the exhaust pressure and air supply pressure are about the same level when compared with other (higher) pressures throughout the cycle, so that a turbocharged engine has a PV diagram of much the same appearance as a naturally aspirated engine but with pressures increased proportionately to the air supply (or boost) pressure, except where limited by the maximum pressure. Actually, at all but low loads an efficient turbocharging cycle will result in the air supply pressure being slightly higher than the exhaust pressure. The pumping loop 6–0–1 then becomes positive (*i.e.* of the same sign as the power part of the diagram) instead of negative as in the naturally aspirated engine. This effect, coupled with the large increase in power output compared with only small increases in friction losses results in significant improvement in mechanical efficiency; values of 91% or 92% being quite usual in modern engines.

Engines which are pressure charged by means of mechanical blowers have an air supply pressure considerably in excess of the exhaust pressure which remains close to atmospheric. This results in a comparatively large positive contribution of power from the pumping loop but the mechanical efficiency of this type of engine is low as the power used for providing the air is taken directly from the crankshaft and not obtained from energy available in the exhaust gas.

2.3. *The two stroke cycle*

The naturally aspirated two stroke cycle follows a similar pattern of events to that of the four stroke, the charge being trapped at point 1 is on the indicator diagram shown in Figure 2–3 compressed 1–2 and heated

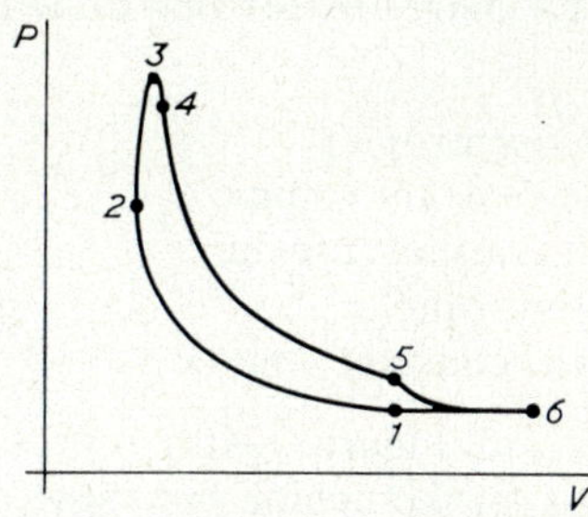

FIG. 2–3—*Indicator diagram for two-stroke engine.*

2–3–4. It is then expanded until the exhaust ports open at 5 giving blow down of the cylinder 5–6, followed by scavenging and charging 6–1. The scavenging and charging process requires the air supply in the manifold to be at a higher pressure than in the exhaust manifold, and in naturally aspirated two stroke engines this is achieved by a mechanically driven scavenge pump. The pumping losses of the four stroke engine in carrying out the exhaust and induction strokes are eliminated. It should be noted that in the two stroke engine compression starts at point 1 when the charge is trapped by the ports or valves closing and this point is part way along the piston stroke. The volume, V_1, is the effective stroke volume. The pressure at this point of the cycle is usually higher than that at point 1 of the four stroke cycle, because of the nature of the charging process, so that pressures at point 2 and pressures at points 3, 4 and 5 are of the same order in both two stroke and four stroke cycle engines. As the two stroke cycle engine does not have a pumping loop the mechanical efficiency when turbocharged cannot improve with positive output from this source but only from the smaller proportion of power lost in friction. As will be seen in Chapter 3, the trapped charge pressure of the turbocharged two stroke engine is as likely to depend on the exhaust manifold pressure as on the air manifold pressure, according to the design of the particular engine considered.

2.4. *Idealized cycle (cylinder processes)*

For thermodynamic analysis purposes it is convenient to separate the processes in the cylinder from those outside the cylinder, the former being cylinder type processes and the latter flow processes. The cylinder cycle follows the well known dual combustion cycle whilst the flow processes follow a gas turbine type cycle. An air standard cycle is the usual basis against which to judge the thermodynamic cycles of actual internal combustion engines, and it is the same for both two stroke cycle and four stroke cycle engines from trapping of the charge to blow down of the cylinder. It is based on the assumptions that the working fluid is air behaving as a perfect gas, having constant specific heats and that addition and rejection of heat do not involve chemical reactions or changes in mass. In the case of the compression ignition engine the appropriate air standard or ideal cycle is the dual combustion cycle, so called because the addition of heat is supposed to take place partly at constant volume and partly at constant pressure. The PV and Tϕ diagrams for this cycle are shown in Figure 2–4. The processes are:

1–2 Compression (isentropic)
2–3 Heat added at constant volume
3–4 Heat added at constant pressure
4–5 Expansion (isentropic)
5–1 Heat rejected at constant volume

The dotted line 0–1 in Figure 2–4(a) represents the exhaust and induction strokes of a four stroke engine but these do not affect the thermodynamic analysis and the ideal diagrams are otherwise identical for two or

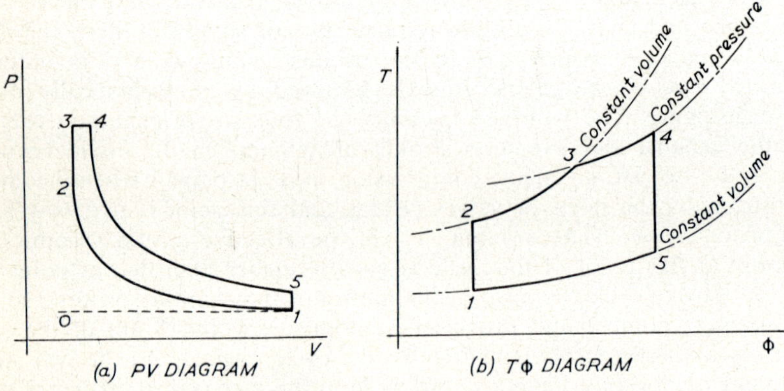

FIG. 2–4—*Cylinder cycle air standard diagrams.*

four stroke engines. If naturally aspirated engines are being considered the pressure at point 1 is taken as atmospheric. In pressure charged engines this pressure is above atmospheric according to the success of the charging process.

The work done during one cycle is equivalent to the difference between the energy supplied as heat in the processes 2–3 and 3–4 and that rejected in the process 5–1. It is also the difference between the work done by the gas in expanding 3–4–5, and the work done on the gas by compressing 1–2.

Thus:

$$\text{Work done} = (P_4V_4 - P_3V_3) + \left[\frac{P_4V_4 - P_5V_5}{\gamma - 1}\right] - \left[\frac{P_2V_2 - P_1V_1}{\gamma - 1}\right]$$

$$= P_4V_4 - P_3V_3 + \frac{1}{\gamma - 1}(P_4V_4 - P_5V_5 - P_2V_2 + P_1V_1)$$

$$= R\left[T_4 - T_3 + \frac{C_v}{C_p - C_v}(T_4 - T_5 - T_2 + T_1)\right]$$

$$= J(C_p - C_v)\left[T_4 - T_3 + \frac{C_v}{C_p - C_v}(T_4 - T_5 - T_2 + T_1)\right]$$

$$= J[C_pT_4 - C_vT_4 - C_pT_3 + C_vT_3 + C_vT_4 - C_vT_5 - C_vT_2 + C_vT_1]$$

$$= J[C_v(T_3 - T_2) + C_p(T_4 - T_3) - C_v(T_5 - T_1)]$$

i.e. Work done = J (Heat supplied − heat rejected)

$$\text{The efficiency} = \frac{\text{work done}}{J\,(\text{Heat supplied})} = \frac{\text{Heat supplied} - \text{heat rejected}}{\text{Heat supplied}}$$

$$= 1 - \frac{C_v(T_5 - T_1)}{C_v(T_3 - T_2) + C_p(T_4 - T_3)}$$

$$= 1 - \frac{T_5 - T_1}{(T_3 - T_2) + \gamma(T_4 - T_3)}$$

Let $r = \dfrac{V_1}{V_2}$, the compression ratio.

Let $\alpha = \dfrac{P_3}{P_2}$, the ratio $\dfrac{\text{maximum pressure}}{\text{compression pressure}}$

sometimes called the explosion ratio.

Let $\beta = V_4/V_3$, sometimes called the cut-off ratio.

Then
$$T_2 = T_1 r^{\gamma-1}$$
$$T_3 = \alpha T_2 = T_1 \alpha r^{\gamma-1}$$
$$T_4 = \beta T_3 = T_1 \alpha\beta r^{\gamma-1}$$

$$T_5 = T_4\left(\frac{\beta}{r}\right)^{\gamma-1} = T_1\alpha\beta r^{\gamma-1}\left(\frac{\beta}{r}\right)^{\gamma-1} = T_1\alpha\beta^{\gamma}$$

Substituting these values into the expression for efficiency:

$$\text{Efficiency, } \eta = 1 - \frac{T_1(\alpha\beta^{\gamma-1})}{T_1(\alpha r^{\gamma-1} - r^{\gamma-1}) + T_1\gamma(\alpha\beta r^{\gamma-1} - \alpha r^{\gamma-1})}$$

$$\eta = 1 - \frac{1}{r^{\gamma-1}}\frac{(\alpha\beta^{\gamma-1})}{(\alpha - 1) + \alpha\gamma(\beta - 1)} \qquad (2.1)$$

If all the heat is supplied at constant volume, then $\beta = 1$ and equation (2.1) reduces to

$$\eta = 1 - \frac{1}{r^{\gamma - 1}}$$

which is well known as the efficiency of the constant volume cycle.

In the same way the work done can be expressed in terms of T_1, and the ratios r, α and β. Thus:

$$\begin{aligned}\text{Work done} &= C_v\,[T_1\alpha r^{\gamma-1} - T_1 r^{\gamma-1}] + C_p\,[T_1\alpha\beta r^{\gamma-1} - T_1\alpha r^{\gamma-1}] - \\ &\quad C_v\,[T_1\alpha\beta^{\gamma} - T_1] \\ &= C_v T_1\,[r^{\gamma-1}\,\{(\alpha - 1) + \alpha\gamma\,(\beta - 1)\} - (\alpha\beta^{\gamma} - 1)]\end{aligned}$$

In order to relate the work done per cycle to the size of engine cylinder needed to produce it we can divide this work done by the volume displaced per cycle.

The volume displaced per cycle is the swept volume,

$$\begin{aligned}V_s &= V_1 - V_2 = V_1 - \frac{V_1}{r} \\ &= V_1\left(\frac{r-1}{r}\right)\end{aligned}$$

The ratio $\dfrac{\text{work done per cycle}}{\text{swept volume}}$ is the same thing as the mean effective pressure.

$$\text{Thus m.e.p.} = \frac{C_v T_1}{V_1\left(\dfrac{r-1}{r}\right)}\,[r^{\gamma-1}\{(\alpha - 1) + \alpha\gamma\,(\beta - 1)\} - (\alpha\beta^{\gamma} - 1)]$$

P_1/R can be substituted for T_1/V_1, and the expression multiplied by J to convert from heat units to mechanical units of energy, giving:

$$\text{M.e.p.} = \frac{JC_v P_1}{R}\,\frac{r}{r-1}\,[r^{\gamma-1}\,\{(\alpha - 1) + \alpha\gamma\,(\beta - 1)\} - (\alpha\beta^{\gamma} - 1)] \quad (2.2)$$

Clearly, it is desirable to choose a cycle that will give both a high output and a high thermal efficiency. In considering idealized cycles it is reasonable to assume that the air to fuel ratio will be limited by combustion factors to the same value whatever cycle is chosen; in other words that the heat supplied per unit mass of air is a fixed value. Assuming such a value and taking a given compression ratio, r, and explosion ratio, α, the m.e.p. and maximum pressure, P_3, can be calculated for a series of charge pressures, P_1, using equation (2.2). Figure 2–5 has been constructed in this way using a value for the heat supplied corresponding to 100 m.e.p. for $r = 12:1$ and $\alpha = 1{\cdot}6:1$, so that m.e.p. is shown as a percentage of this condition. It has also been assumed that the charge temperature, T_1, would be constant for all cycles. These compression ratios are shown with α values ranging from 1·2 to 2·0 and charge pressure ratios from 1:1 to 4:1. The figure illustrates some important considerations.

For a given charge pressure ratio (*e.g.* 1 : 1 or naturally aspirated) the cycle efficiency and therefore the output, increases with compression ratio but only slightly whereas maximum pressure increases considerably.

For fixed values of compression ratio and explosion ratio the m.e.p. can be increased by raising the charge pressure ratio proportionally but the maximum pressure will also increase proportionally.

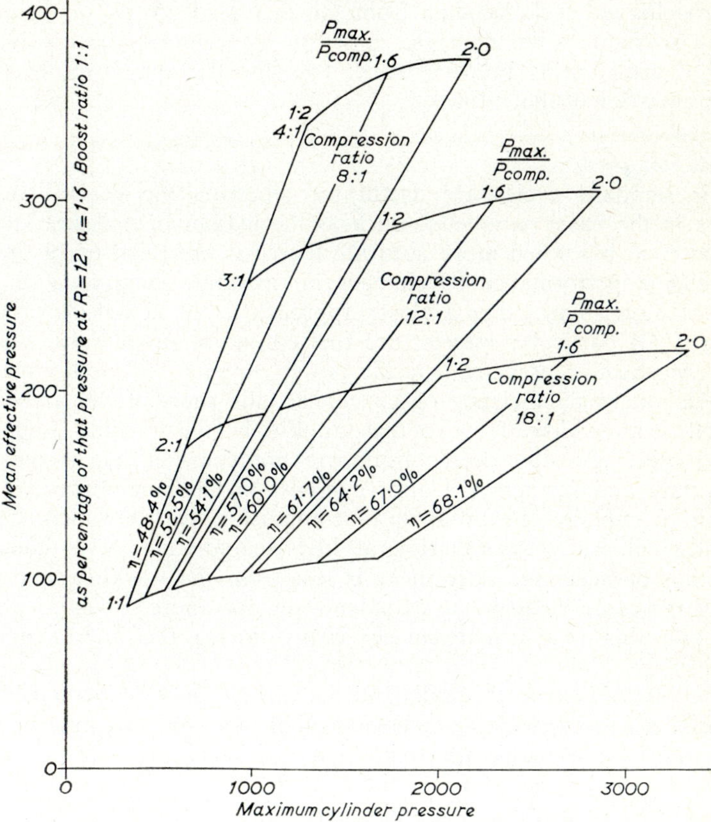

FIG. 2–5—*Variation of m.e.p. with maximum pressure, boost pressure, charge pressure and explosion ratio.*

The maximum pressure is a fundamental consideration in the mechanical design of an engine and usually there will be a limiting value of the maximum pressure. An increased m.e.p. can be obtained without exceeding a given maximum pressure only if the compression ratio or explosion ratio or both are reduced, but this will result in a reduction in thermal efficiency. The loss of efficiency incurred in reducing the compression ratio down to about 12 : 1 is tolerable but below this ratio the losses are increasingly heavier. A practical limitation also comes into consideration in that if the engine is to be designed for cold starting (and most medium and high speed engines are) then the compression ratio cannot be allowed to fall much below 12:1. In a similar way reducing the explosion ratio is tolerable down to a value around 1·6 but reduction below this figure incurs increasing losses in efficiency and also leads to higher cycle temperatures.

Fortunately, brake thermal efficiency does not always suffer so much as might be supposed from a study of the cycle thermal efficiency, this is because the mechanical efficiency improves as a result of two factors: one, the turbocharger cycle benefits from the increased energy in the exhaust gas and so reduces the pumping losses in the cylinder, frequently to the extent of contributing positive work; two, the friction power becomes a lower proportion of the output.

2.5. *Actual cycles*

Air standard cycles are useful for assessing the general result of changes in the basic parameters such as the maximum pressure and compression ratio but when more accurate forecasts are required of the effect upon engine performance of changes in operating conditions or of the choice of design limits then a closer approach to the actual cycle becomes necessary. Of particular interest are fuel consumption (thermal efficiency) and mean effective pressure (specific power output). It is well known that the actual thermal efficiency of an engine falls short of the air standard cycle efficiency, chiefly for two reasons. Firstly, the working fluid (air) is not a perfect gas. Its specific heats are not constant but increase with temperature and furthermore after combustion has commenced, it is not even air. It contains the products of combustion, two of which, CO_2 and H_2O, are subject to dissociation at high temperatures with consequent absorption of heat. Secondly, heat is lost from the working fluid during most of the cycle through the cylinder walls and the combustion chamber walls, and this loss is particularly heavy during the high temperature period.

The loss of heat to the walls and the variable specific heats modify the compression and expansion processes so that they can no longer be treated as adiabatic. A reasonable approximation to the real processes can be made by assuming they follow laws of the form $PV^n = C$, where n has a value of about 1·35 for compression and about 1·27 for expansion. These values vary a little with different types of engine and individual manufacturers no doubt have their own values which apply to their own engines.

Variable specific heats and dissociation play important roles in the rate of heat release during combustion. The combustion processes are examined in more detail in Chapter 4 where it will be seen that the temperatures attained at the high pressure part of the cycle are only about half of what might commonly be supposed from a consideration of the idealized cycle.

Nowadays a powerful tool, the electronic computer, is available and by its use more satisfactory approaches to calculating the condition of the cylinder contents at every point round the cycle can be made using step by step methods. Information on the properties of gases under the range of different states occurring in internal combustion engine cycles has been fully documented by Keenan and Kay (Ref. 1). The heat loss during operation of the cycle has been given attention by Eichelberg (Ref. 2) and, more recently, by Annand (Ref. 3). The work of these authors leads to the formation of equations relating heat transfer coefficients to the pressure and temperature of the gas and the engine geometry. It is still necessary, however, to choose values for the constants in these equations by experience with known engine cycles. The relation between fuel injection and rate of heat release during combustion has been explored by Lyn (Ref. 4) although again, the final choice of heat release pattern will rest a good deal upon previous experience with engines similar to the one under study.

However, by using information of this kind and applying it to the energy equation the following relationship is obtained representing any small change during the cycle:

$$\delta (Q_f - Q_s) = \delta V + \frac{P\delta V}{J} \tag{2.3}$$

in which $\delta V = V_{n+1} - V_n$

where V_n = the volume determined from the engine geometry for a particular step and hence V_{n+1} is the volume for the next step. Knowing T_n for the step then T_{n+1} can be calculated and so on, step by step, throughout the cycle, starting at the trapped charge temperature. During compression and during the latter part of expansion there will, of course, be no heat from the fuel, that is δQ_f will be zero, but the equation nevertheless holds right throughout the trapped cycle.

The trapped charge temperature is calculated from a knowledge of the turbocharger cycle, the effectiveness of intercooling and the gas exchange processes. As both the turbocharger cycle and the gas exchange processes are influenced by the exhaust gas temperature the calculation is commenced by assuming a value for this which is then compared with the value resulting from the calculation. If necessary the initial value is corrected and the cycle calculated again, iteration continuing in this way until satisfactory agreement is reached. As the calculation concerns a cycle, the conditions at any convenient point may be chosen as the starting point and iterative calculations carried out until agreement is reached. The method is fully described by Whitehouse *et al* (Ref. 5).

2.6. *Turbocharger cycle*

A turbocharger is a form of gas turbine, the compressor and turbine are flow machines continuously passing the working fluid and as it is almost always a free running unit, not mechanically connected to the engine or to any power supply or take off, the whole of the turbine power is absorbed in driving the compressor.

Figure 2–6 shows the cycle of operation on PV and Tϕ diagrams. Starting at atmospheric pressure, P_A, compression takes place ideally at constant entropy but in practice with increasing entropy up to the boost pressure, P_B. Heat is added by passage through the cylinder cycle, *B* to *L*, followed by expansion to atmospheric pressure, P_L to P_M, again ideally with constant entropy but in practice with increasing entropy. If we assume a perfect gas with constant specific heats the work done during compression will be given by the change in total heat (enthalpy) = $C_p\ (T_B - T_A)$. This must be equal to the work done during expansion = $C_p\ (T_L - T_M)$.

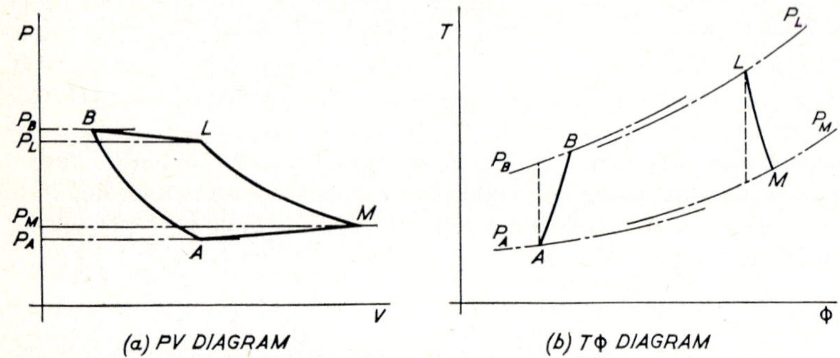

FIG. 2-6—*Turbocharger cycle diagrams.*

The isentropic temperature rise during ideal compression can be calculated:

$$T_{B\phi} - T_A = T_A\left(\frac{T_{B\phi}}{T_A} - 1\right)$$
$$= T_A\left(\left[\frac{P_B}{P_A}\right]^{(\gamma-1)/\gamma} - 1\right)$$

The actual temperature rise will be greater than this, *i.e.*

$$T_B - T_A > T_{B\phi} - T_A$$

The ratio $(T_{B\phi} - T_A)/(T_B - T_A)$ is known as the adiabatic efficiency of the compressor η_c.

The isentropic temperature drop during ideal expansion:

$$T_L - T_{M\phi} = T_L\left(1 - \frac{T_{M\phi}}{T_L}\right)$$
$$= T_L\left(1 - \left[\frac{P_M}{P_L}\right]^{(\gamma-1)/\gamma}\right)$$

The actual temperature drop will be less than this and the ratio $(T_L - T_M)/(T_L - T_{M\phi})$ is known as the adiabatic efficiency of the turbine η_T.

Thus, as $C_p\ (T_B - T_A) = C_p\ (T_L - T_M)$

$$\frac{1}{\eta_c} \,.\, (T_{B\phi} - T_A) = \eta_T \,.\, (T_L - T_{M\phi})$$

i.e.
$$\frac{1}{\eta_c} T_A\left(\left[\frac{P_B}{P_A}\right]^{(\gamma-1)/\gamma} - 1\right) = \eta_T T_L\left(1 - \left[\frac{P_M}{P_L}\right]^{(\gamma-1)/\gamma}\right) \qquad (2.4)$$

The product of $\eta_c\ \eta_T$ is the overall efficiency of the turbocharger. As the overall efficiency approaches unity, that is as compression and expansion processes approach the ideal of isentropic, P_L falls below P_B providing a pressure drop across the engine which can be used for scavenging the cylinder and adding to the cylinder output by doing work during the pumping strokes. As the overall efficiency falls away this pressure drop diminishes and may become a rise in pressure making unassisted scavenging impossible and detracting from the output of the cylinder cycle. Figure 2–7 has been constructed from equation 2.4 and using values for overall

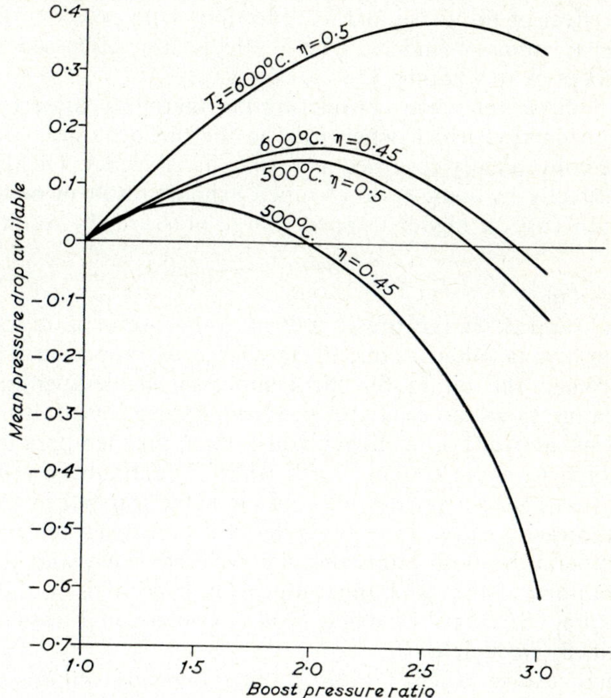

FIG. 2–7—*Pressure drop available for scavenging at various boost pressures, exhaust temperatures and turbocharger overall efficiencies.*

efficiency of 45% and 50%. The curves show also the effect of turbine inlet temperature T_L, a higher pressure drop being available from a higher gas temperature. In a practical case the pressure drop available, the corresponding air flow and the resulting turbine inlet temperature will reach an equilibrium operating point for the engine and turbocharger combination providing the machines are sufficiently well matched.

When dealing with an actual cycle greater accuracy is achieved by taking into account the increased mass flow through the turbine compared with that through the compressor due to the fuel which has been added during passage through the cylinder, and by taking account of changes in specific heats. Also P_L does not equal P_A precisely as there is a pressure drop between turbine outlet and the outlet from the stack as well as one on the air inlet side through the silencer to the compressor.

2.7. *Exhaust pulses*

When the exhaust valve or port is opened to allow the gases in the cylinder at the end of the power stroke to escape the pressure of these gases is higher than the mean pressure in the exhaust manifold. The blow down process of the cylinder therefore creates a pulse of pressure in the exhaust manifold and this pulse travels along the exhaust pipe with the velocity of sound. It may be reflected as a negative pulse on reaching the open end of the pipe or reflected as a positive pulse on reaching the turbocharger nozzle. The effect of the pulse and its reflections both positive and negative can be felt at other cylinders. These effects are discussed under gas exchange processes in Chapter 3.

What is important when considering the thermodynamic cycle is that pulses contain energy, which when used in the turbocharger, can for cycle purposes, be conveniently regarded as raising its efficiency. Or alternatively, when used directly by some form of tuned exhaust result in better cylinder filling. In both cases a higher trapped charge pressure, P_1 results.

2.8. *Intercooling*

A high trapped charge pressure P_1, is only useful if it results in an increased density, in other words, if the charge is trapped at a sufficiently cool state. When the air leaves the compressor its temperature will be higher than at entry as can easily be seen from Figure 2–6(b). If the pressure ratio P_2/P_1 is above 1·3 or thereabouts, then this temperature will be sufficiently high to make cooling worth while. The air density is increased by cooling so enabling a greater mass of air to be trapped in the cylinder, also the compression curve in the cylinder will be carried out over a lower range of temperature, both improving the cycle efficiency and reducing the mean temperature of the cycle thus improving conditions of thermal loading. Intercooling necessarily carries a small penalty of pressure drop but this is practically negligible.

The temperature reached after cooling depends on the size of the intercooler used (*i.e.* the surface area available for cooling) and the temperature of the cooling medium. Invariably sea water is used as being the coolest and most readily available. The lowest temperature to which the intercooler can possibly reduce the air is clearly the temperature of the cooling water as it enters the cooler. In practice, the air will leave the cooler at a temperature a few degrees higher than this. The comparison of the temperature range through which the cooler cools the air compared with the maximum range possible is a measure of a kind of efficiency of the cooler. It is known as its effectiveness or thermal ratio.

$$\text{Effectiveness} = \frac{\text{Temperature of air entering cooler} - \text{temperature of air leaving cooler}}{\text{Temperature of air entering cooler} - \text{temperature of water entering cooler}}$$

A cooler designed for a high effectiveness will have a larger surface area and will occupy a larger space than one designed for a lower effectiveness. Values of 95% thermal effectiveness can be obtained but the space restrictions of most marine installations result in the use of coolers having effectiveness value of 80% to 85%.

2.9. *Brake mean effective pressure*

In constructing the pressure-volume diagram and in refining it from the idealized air standard to a more realistic cycle an attempt is being made to predict an indicator diagram for design purposes or to provide a standard with which actual engine performance may be compared. The mean effective pressure calculated from this P–V diagram is therefore an indicated mean effective pressure.

An actual indicator diagram in PV form and an indicated mean effective pressure derived from it is only rarely used in analysing the performance of high speed and medium speed engines. The rapidity of their operation is such that mechanical indicators do not reproduce events in the cylinder with sufficient accuracy. Electronic instruments can give much more precise information but by their nature they present it on a time or crank angle basis, *i.e.* a P–θ diagram. Academically, this information can be translated into a P–V diagram but the mean effective pressure obtained by measuring the area of such a diagram is open to considerable error as very small changes in the positioning of the dead centres in relation to the diagram can produce large changes in its area.

Because of these difficulties it has become accepted practice to use "brake mean effective pressure" as the usual measure of the specific output of medium and high speed engines. Brake mean effective pressure, b.m.e.p., is analogous to indicated mean effective pressure and is determined from the measured brake horsepower, speed and dimensions of the engine. Thus:

Indicated power per cylinder $= P_iLAN_c$ Watts, or $\dfrac{P_iLAN_c}{550}$ horsepower

Brake power per cylinder $= P_bLAN_c$ Watts, or $\dfrac{P_bLAN_c}{550}$ horsepower

where			
P_i	= indicated m.e.p.	N/m²	lb./in²
P_b	= brake m.e.p.	N/m²	lb./in²
N_c	= No. of cycles per second	Hz	
L	= Stroke	m	ft.
A	= Piston area	m²	in²

Also b.m.e.p. $= \eta_m \times$ i.m.e.p.
where η_m = mechanical efficiency.

Development testing of medium and high speed engines is carried out with the engine coupled to a dynamometer which provides accurate measurement and control of the b.h.p. and b.m.e.p. For a production engine, it is customary to conduct an acceptance test at the maker's works using a dynamometer, and at this time it is important to record carefully the readings of all instruments relating to engine performance as, after installation in the ship, it is difficult to obtain as accurate a measurement of horsepower and m.e.p.

2.10. *Piston speed*

The power that an engine can develop will obviously be increased if the working fluid can be passed through the cycle more often; in other words if the rev/min of the engine can be increased. Speed, however, is limited by mechanical considerations relating to reliability and economic life as well as by performance considerations. In both cases the commonly used criterion for judging the limit is that of mean piston speed $= 2 \times$ stroke $\times$ revolutions per unit time. Usual units are ft per min or metres per sec.

There is a good deal of evidence to suggest that high speeds are associated with short life but it would be quite wrong to assume that speed alone causes rapid wear, in fact, this is manifestly untrue in the cylinder where the maximum wear occurs at the top where the ring pressure is greatest and not in the middle where the rubbing speed is highest. The wear which occurs at high speeds is rather the result of the greatly increased inertia loads.

The force set up by the reciprocating or rotary movement of a part in the running gear will generally be of the form:

$F \propto m\,\omega^2\,r$, at least in first order terms

Where $m =$ mass

$\omega =$ angular velocity $\propto N =$ rev/min

$r =$ radius: centre of rotation to centre of gravity.

The mass, m, will be proportional generally to the cube of a linear dimension, *i.e.* $m \propto r^3$.

Thus, $F \propto \omega^2 r^4 \propto N^2 r^4$.

The area resisting this inertia force, a stressed area, or a bearing area is proportional to the square of a linear dimension, *i.e.* $A \propto r^2$.

Thus, stresses or bearing pressure, $F/A \propto N^2 r^2$.

If we choose as a typical dimension for r the crank radius, then Nr is proportional to $2NS$, *i.e.* the piston speed, and the stresses or bearing pressures are seen to be proportional to:

$(Nr)^2 \propto (2NS)^2$, the (piston speed)^2

This reasoning applies strictly to geometrically similar engines only and if a comparison is made between engines having different stroke/bore ratios this should be taken into account thus:

The mass, m, of the piston may be assumed proportional to bore^3 (B^3) and, r, is proportional to stroke (S).

Hence $F \propto \omega^2 B^3 S \propto N^2 B^3 S$

The projected bearing area of the small end, for instance, or the cross sectional area of the connecting rod, will be limited by the bore alone and so can be assumed $\propto B^2$.

Thus:

$$\frac{F}{A} \propto \frac{B^3}{B^2}\,\omega^2 S \propto BS\omega^2$$

$$\propto \frac{B}{S}\,S^2\omega^2$$

$$i.e. \propto (\text{piston speed}\,\sqrt{B/S})^2$$

So that for a more precise analysis

$$\frac{\text{piston speed}}{\sqrt{\text{stroke/bore ratio}}}$$

is a better criterion.

Nevertheless, piston speed alone is more commonly used.

The geometry of any engine design is such that the cross sectional area of the inlet and exhaust passages related to the cross sectional area of the cylinder can be altered only within fairly narrow limits. The gas velocity in these passages is, therefore, closely related to the piston speed to the extent that at a limiting piston speed any gain in horsepower that might be obtained by a further increase in speed is offset by a reduction in b.m.e.p. as a result of reduced trapped charged density.

Piston speed is thus regarded as the limiting criterion for both mechanical reliability and for output, but different design approaches may result in different features reaching the limit first. Because larger engines have tended to be a little more troublesome and costly in the matter of replacing worn parts they have tended to be restricted in piston speed more than has been done with smaller engines. Generally speaking small bore high rev/min engines for marine use are limited in piston speed to 10 to 13 m/s, or say 2,000 to 2,500 ft/min, whilst larger medium speed engines are restricted to 7 to 10 m/s, or say 1,500 to 2,000 ft/min.

2.11. *Thermal loading*

In most modern highly rated engines the b.m.e.p. which it would be possible to obtain by burning the maximum amount of fuel in the charge that could be trapped at the rated rev/min would result in cycle temperatures, exhaust gas temperatures and hence component temperatures too high for reliable operation over a reasonable life. For example, exhaust valves could attain temperatures which would shorten their life to a matter of a few hundred hours. Turbine nozzles and blades would also suffer in the same way. Cylinder heads and pistons, where they form the walls of the combustion chamber, would also be subjected to high heat flows and high temperatures. These points are dealt with in Chapter 5, where the problems of thermal loading are discussed.

A thermal loading problem on an engine is usually concerned with conditions at a particular hot spot but a comparison between the degree of thermal load which one engine is called upon to withstand and that imposed upon another of similar design may be made in a general way by relating the rate of heat release in the cylinder to the surface area upon which it impinges. An often used criterion based on this approach is that of brake power/unit of piston area. Care has to be taken when using this as it clearly does not apply between engines of dissimilar construction. In the present state of the art, the b.m.e.p. of many engines is limited because of thermal loading considerations.

2.12. *Rating*

It is the responsibility of the engine builder to ensure that his products are adequately rated for the duties they are to perform. A reputable manufacturer is able to do this by virtue of his intimate knowledge of the design, by extensive development testing and by collected experience in service over the years and is able to guarantee their performance and reliability.

Many limitations, such as the compression ratio and the maximum cylinder pressure are fixed by the design but within these parameters the b.m.e.p. can be increased by burning more fuel. Near the designed load the specific fuel consumption curve will be more or less flat depending on the cycle chosen but as the fuel injected per cycle is increased a condition will be reached when the specific fuel consumption takes a marked upward turn. This is often accompanied by the onset of smoke in the exhaust, and the exhaust temperatures, which will have been climbing steadily as more b.m.e.p. is applied, will curve upwards at an ever increasing rate. All these are indications that the limits have been reached at which the fuel can be burnt efficiently in the oxygen supply, and although beyond this point the engine may produce more power with no obvious signs of immediate distress, it will only be at the expense of the rapid deterioration of its various components.

The speed at which it may be run will be limited, perhaps by its breathing, which will restrict the b.m.e.p. that can be applied at higher speeds or perhaps by the rate of wear of the wearing parts being too high or by the moving parts being stressed too highly. Again, the thermal stressing of parts will be too high if the rate per unit time at which fuel is consumed is excessive. In this regard, it is interesting to note that the criterion of brake horsepower per unit piston area happens to be proportional to the b.m.e.p. multiplied by the mean piston speed, and is thus an indication of all three of the major limiting factors in the output of an engine.

The load on any article is only meaningful when compared to its strength, and the same is true of an internal combustion engine. With naturally aspirated engines the b.m.e.p. which can be achieved is limited by the mass of air which can be drawn into each cylinder during each cycle and this naturally tends towards a limiting b.m.e.p. for all designs. By virtue of careful design, some engines are able to approach this limit a little more closely than others, but nevertheless b.m.e.p. can be regarded as having a limit. With turbocharged engines the situation is completely different as the amount of air available for the production of power can be varied as is apparent in considering the thermodynamic cycle. A b.m.e.p. therefore, cannot be regarded as high unless it is considered in relation to all the cycle parameters. What is high for one design may be well within the capacity of another.

The b.m.e.p. which an engine can carry will probably be limited by different criteria at different rev/min over the permissible speed range. A good design making full use of all components will approach the upper limiting features together at the maximum b.m.e.p. at the maximum speed.

Figure 2–8 displays a "map" of lines of equal fuel consumption over a range of b.m.e.p. and rev/min for a typical medium speed engine and also indicates some of the limiting features. The dotted line is the operating line when coupled to a well-matched fixed pitch propeller. In Figure 2–9 is shown the performance curves over a speed range for the engine when operating on the same propeller law.

Figure 2–10 shows performance curves for an auxiliary engine at constant rev/min on a basis of b.m.e.p.

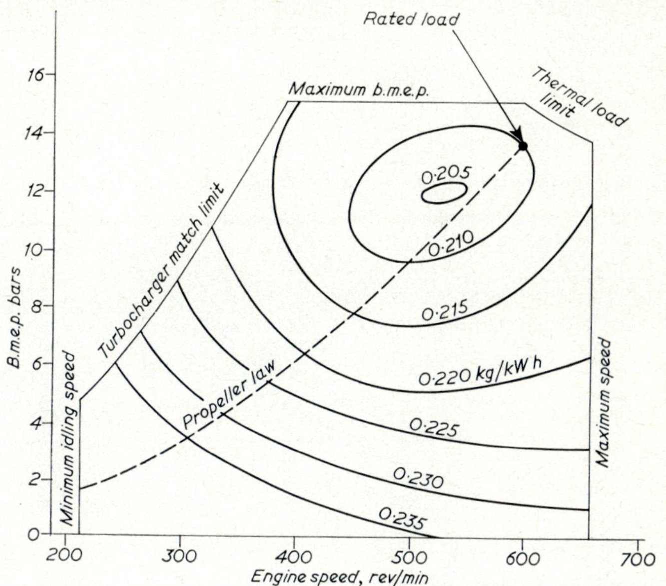

FIG. 2–8—*Lines of constant fuel consumption and limits of performance.*

The maximum practical safe b.m.e.p. which the engine will carry continuously at the highest safe rev/min is termed the maximum continuous rating. For operation for short periods an overload rating is permissible, for example, one hour in twelve or two hours in twenty four. Conversely by operating the engine at a load lower than the maximum continuous rating the need for overhauling certain components will occur at less frequent intervals. Other benefits obviously accrue from running with a safety margin of this kind. It is usual practice to operate engines in ocean-going ships at a rating that is at least ten per cent below the maximum continuous rating. This rating is known as the service rating. An engine builder may have a recommended service rating which he puts forward to his clients and some ship owners have, from past experience, settled their own margin by which they determine the service rating.

2.13. *Ratings in service*

The majority of engine builders' factories are situated in temperate climates and the development testing of engines is carried out in these locations. When an engine operates in a climate where the barometric pressure, the temperature or the humidity are substantially different from that in which its rating was determined, consideration will have to be given to the effect of these conditions on its performance. Any reduction of barometric pressure is that due to operation at altitude. This is a case which does not affect the vast majority of marine engines, but it does apply to any which operate on inland lakes in the higher parts of the world.

A reduction in barometric pressure, an increase in ambient temperature or an increase in the humidity of the air will obviously result in less

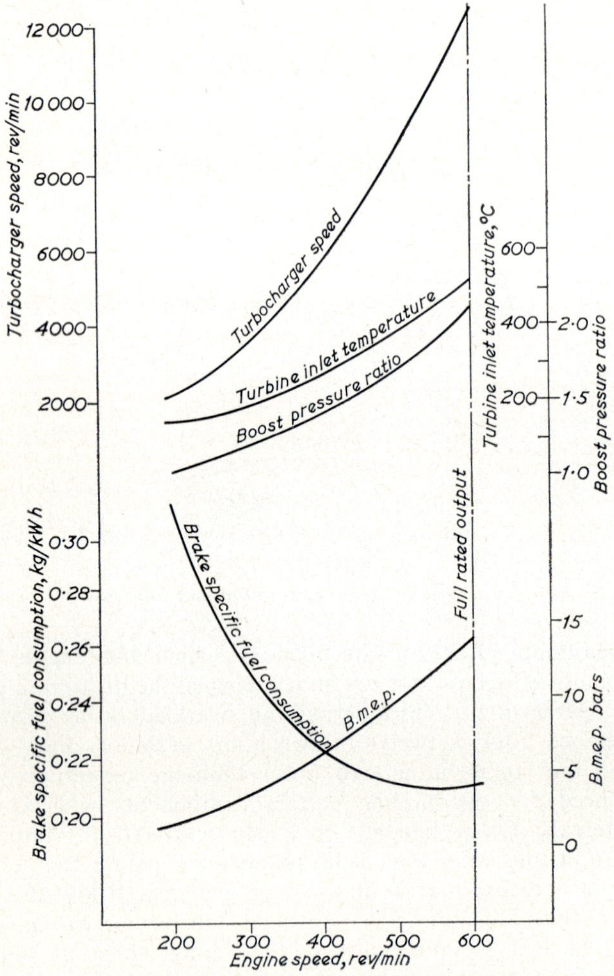

FIG. 2–9—*Performance curves propeller law.*

oxygen entering the cylinders of a naturally aspirated engine and which, therefore, will not be available for use in burning the fuel. With naturally aspirated engines the variation in the mass of air taken into the cylinder for different conditions of pressure and temperature is straightforward:

$$m \propto P/T$$

The mass of water vapour in unit volume of the air can also be ascertained from the observed relative humidity. Given the ambient conditions at which the engine is rated the effects of these factors can be reduced to simple rules such as those tabled in B.S.649.

The situation is a little different with turbocharged engines as it is the conditions of the air in the inlet manifold which matter and these conditions depend on the turbocharger and intercooler. Humidity does not have to be taken into account as the maximum output is no longer limited by the amount of oxygen available for combustion.

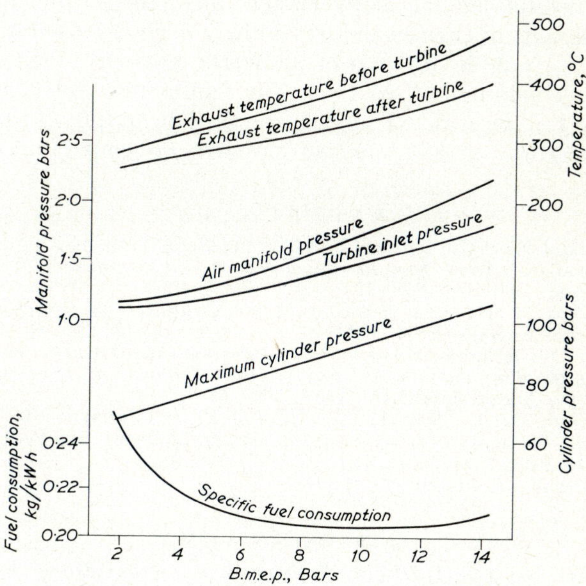

FIG. 2–10—*Performance curves at constant speed.*

The effect of reduced barometric pressure on the turbocharger is to cause it to speed up and in doing so it, to some extent, restores the performance of the engine by increasing the boost pressure towards that which is obtained at sea level. Whether such increased speed can be permitted depends upon the aerodynamic match of the turbocharger and its mechanical design. The closeness with which the sea level performance is approached will also depend upon the effectiveness of the intercooler, supposing that one is used. For an engine permanently operating at a high altitude most makers prefer to rematch the turbocharger and by this means the sea level performance of the engine can sometimes be completely restored.

The temperature of the ambient air obviously affects its density but the density of the air in the inlet manifold of the engine does not necessarily depend wholly on this, as intercooling when applied plays a very large part, and the sea water temperature is often a greater factor in the charge temperature than is the ambient temperature. Once again a great deal depends on the design of the system, the cycle chosen and particularly the thermal effectiveness of the intercooler.

Two stoke engines and four stroke engines can react differently to ambient conditions. The four stroke engine to some extent always meters the air supply, whereas the two stroke engine, unless fitted with a positive displacement scavenge pump, can allow airflow depending wholly on aerodynamic considerations. All these factors make it difficult to lay down standard rules for the derating of internal combustion engines for site conditions. An attempt has been made to do this by bodies such as the British Standards Institution, but many makers have their own rules which fit their own designs of engines and which there is good reason to apply. Ships which have to trade in any part of the world must have their engines rated to meet any conditions at sea level. The requirements of Lloyd's Register of Shipping for unrestricted service are that the rating is based on a sea water temperature of 30°C and an ambient air temperature of 45°C.

REFERENCES

1. KEENAN and KAY, *Gas Tables,* J. Wiley and Sons, 1950.
2. EICHELBERG, G., *Some New Investigations on Old Combustion Engine Problems,* Engineering, London, 1939.
3. ANNAND, W. J. D., *Heat Transfer in the Cylinders of Reciprocating Internal Combustion Engines,* Proc. I. Mech. E., 1963, Vol. 177 No. 36.
4. LYN, W. T., *Relation Between Fuel Injection and Heat Release in a Direct Injection Engine and the Nature of the Combustion Process,* I. Mech. E. Proc. of the Automobile Divn. No. 1, 1960/61.
5. WHITEHOUSE, N. D., STOTTER, A., GOUDIE, G. O., PRENTICE, B. W., *Method of Predicting Some Aspects of Performance of a Diesel Engine Using a Digital Computer,* Proc. I. Mech. E., 1962, Vol. 176 No. 9.
6. *British Standards Institution,* B.S.649.

BIBLIOGRAPHY

KÜHN, K. and GALLOIS, J., *Highly Supercharged 4-stroke Diesel Engine, Thermal Loads, Mechanical Loads, and Running Zones,* C.I.M.A.C. Congress, Copenhagen, 1962—Paper A.7.

LOWE, W., *The effect of Ambient and Environmental Conditions,* Proc. I. Mech. E., 1969–70, Vol. 184, Part 3P, Paper 3.

POPE, J. A. and LOWE, W., *The Development of a Highly-Rated Medium Speed Diesel Engine,* Trans. I. Mar. E. 1966—Vol. 78 No. 8.

CHAPTER THREE

Gas Exchange Processes and Scavenging

3.1. *Naturally aspirated four stroke cycle engines*

In the simple conception of the four stroke cycle the exhaust gas is pushed out of the cylinder by the piston during one complete stroke, and the fresh charge is drawn in by the next stroke, the valves being opened and closed at the dead centres corresponding to each end of the stroke. In order to accomplish this in reality, the valves have to commence opening before the dead centre and do not finally close until after the next dead centre. This is because a finite time is required to open the valve from a position tight on its seat, to a position in which it is fully open and passing the maximum flow of gas and similarly when closing. The need for early opening and late closing can be more readily seen when it is considered that the amount by which the valve is open at the inner centre is only a fraction of the full opening. Figure 3–1 shows a valve timing diagram for a four stroke engine and Figure 3–2 shows the valve displacement plotted on a basis of crank angle.

The gas exchange process commences with the opening of the exhaust valve, which causes the cylinder to "blow down". From the pressure

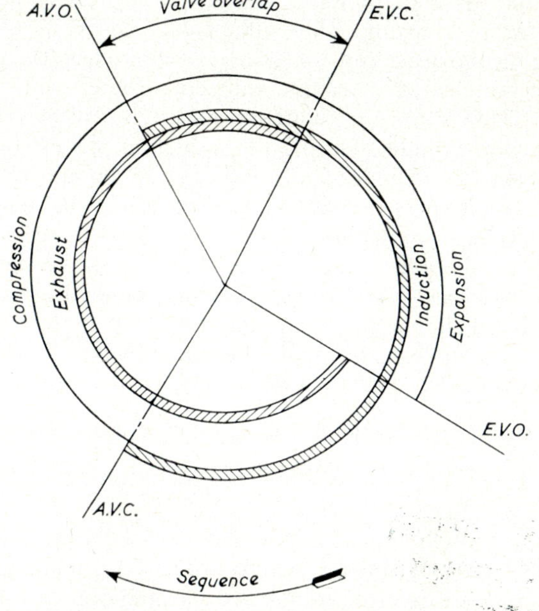

FIG. 3–1—*Valve timing diagram.*

volume diagram, Figure 2–4(a), the pressure at the end of the power stroke is seen to be well above that during the exhaust stroke, so that when the exhaust valve is opened, the cylinder commences to discharge the exhaust gases which are in it by blowing through the valve irrespective of any piston motion. Blow down to the exhaust manifold pressure must occur before the piston commences the exhaust stroke if extra work done by the piston is to be avoided. Because of this feature the exhaust valve opens quite early compared with the inlet valve.

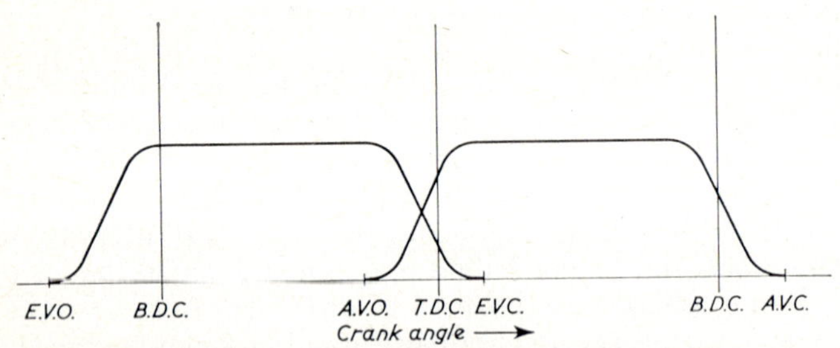

FIG. 3–2—*Valve displacement diagram.*

The exhaust gas remaining in the cylinder after blow down is displaced by the movement of the piston from outer to inner dead centre during the exhaust stroke, and as this stroke nears completion the inlet valve is opened.

In between the exhaust and admission strokes there is a period when both exhaust valve and inlet valve are, at least partially, open. This period is known as the "overlap period". With naturally aspirated engines there is little need for overlap and only sufficient to ensure large enough valve openings to give the least loss of energy is provided. At the end of the exhaust stroke the clearance volume will remain filled with exhaust gas; at this point the piston starts the induction stroke, and draws air through the inlet valve into the cylinder. During the induction process the incoming air receives heat from the cylinder walls and a small amount from the residual combustion gases. Its pressure will vary also, falling slightly as the velocity of the piston reaches a maximum at mid-stroke and increasing again as the piston comes to the outer dead centre. The closing of the inlet valve is timed to trap the largest possible mass of unburnt air. In general the "free" volume of the fresh charge of air can only be as big as the displaced volume of the piston, and unless the engine is fitted with a "tuned" induction system it will be less. The ratio

$$\frac{\text{volume of free air per cycle}}{\text{swept volume}}$$

is known as the volumetric efficiency.

In flowing through the passages and valves the air and exhaust gas streams incur pressure losses which affect the volumetric efficiency adversely. These losses can be minimized by attention to the shapes of the passages and the valve heads as described by Wallace (Ref. 1).

3.2. *Turbocharged four stroke cycle engines*

In turbocharged engines the blow down of the cylinder is very important because the pulse created in the exhaust manifold by the blow down is a valuable source of energy for the turbine, and in many cycles this energy is all important to the correct operation of the turbocharger. By using an early opening for the exhaust valve there is a greater amount of energy in the blow down pulse which can be used in the turbine. This energy is, of course, taken from the cylinder cycle but it represents only a small part of the energy in the cylinder, and can be a large part of the energy for the turbine. The best use of the pulse must be made by timing it in a fashion to give optimum energy in both turbine and cylinder cycles in the overall consideration.

Figure 3–3 shows typical pressure—crank angle diagrams taken in the

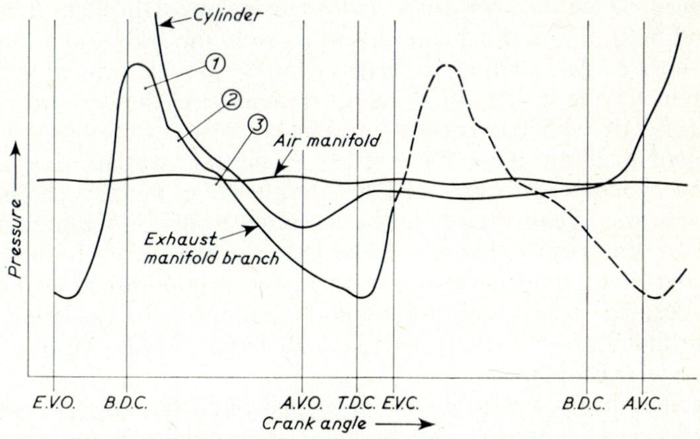

FIG. 3–3—*Light spring indicator diagram.*

cylinder and in the exhaust manifold branch of a turbocharged four stroke cycle engine. The formation of the blow down pulse is seen at (1).

When a pressure pulse is initiated in a pipe it spreads throughout the pipe system with the velocity of sound in the gas with which the system is filled. On reaching an open end or sudden enlargement of the section a negative reflection will take place. If a closed end or sudden contraction is reached a positive reflection results. The turbocharger nozzle acts as a partially closed end and a positive reflection of the blow down pulse can be traced at (2).

The positive pulse (3) is caused by the piston velocity increasing at mid-stroke, followed by decreasing velocity as inner dead centre is approached.

The valve overlap period of a turbocharged engine is often quite large in order to take advantage of the ability of this type of engine to scavenge the compression space.

In the turbocharged engine the pressure of air in the inlet manifold is usually greater than the mean pressure in the exhaust manifold, and must certainly be greater than the pressure in the exhaust manifold when the valves are open together during the overlap period. Any exhaust gas in the compression space at the end of the exhaust stroke can therefore be displaced by air flowing through the inlet valve across the compression space or combustion chamber displacing the exhaust gas that remains, out through the exhaust valve. In this way the fresh charge that is drawn in is greater than the displacement of the piston by the amount of the compression space or combustion chamber.

There are one or two exceptions to this, for instance engines that have pre-combustion chambers are obviously restricted as pre-combustion chamber shapes are very difficult to scavenge by air flow.

There is also a further need for large overlap on highly rated turbocharged engines. When such engines are operating at or near their full power the exhaust gases at the end of the expansion stroke are at a temperature which is very high and their passage through the exhaust valve and into the turbocharger would cause the valve and the turbocharger nozzle and blading to reach temperatures at which they would have a not very long life. In order to decrease the temperatures of these components, air which is present in abundance when the boost levels are high is used to dilute the exhaust gases during the overlap period, flowing across the cylinder and over the hot exhaust valve, thereby cooling it and mixing with the exhaust gases in the manifold to give a temperature more suitable for the metals that are currently in use in turbochargers for the nozzles and blading. This feature is of a special importance in the case of engines burning heavy fuel, containing vanadium salts, as the ash from these constituents is very corrosive at high temperatures. More about this is said in a later chapter.

In order that scavenging can be carried out efficiently, or even at all, during the overlap period the pressure in the inlet manifold must be higher than that in the exhaust manifold. Now, as we have already seen, the exhaust manifold mean pressure has imposed upon it pulses of pressure which occur when exhaust valves open allowing a cylinder to blow down. If a number of cylinders exhaust into the same manifold, this manifold will experience cyclic waves of pressure, as is shown in Figure 3–4. If one of these positive pressure pulses were to occur during the overlap period of any one of the cylinders, then the position would be that the high pressure in the exhaust manifold at that instant would cause a flow of exhaust gas through to the inlet manifold which is the very opposite of what is desired. The solution to this problem lies in dividing the manifolds of turbocharged engines so that no more than three cylinders exhaust into the same manifold. This matter is discussed further in section 3.7.

During the exhaust stroke, when the piston reaches a condition of maximum velocity it displaces exhaust gas through the valve at quite a high rate, often sufficient to cause a distinguishable positive pulse in the exhaust diagram. The opening of the inlet valve is therefore chosen not only with regard to the need for sufficient opening when the piston commences the admission stroke, but also with reference to the relative pressures between the exhaust manifold and the inlet manifold and the instantaneous pressure in the cylinder. Once the exhaust valve has closed, the admission stroke proceeds smoothly with the maximum opening of the air valve that can be accommodated to avoid loss of pressure between the inlet manifold and the cylinder.

As the inlet air is at a higher pressure than the mean exhaust manifold pressure, the gas exchange will add a small amount of work to the engine cycle output.

The greater the mass of unburnt air that is trapped in the cylinder when the inlet valve closes the greater is the potential output of the engine.

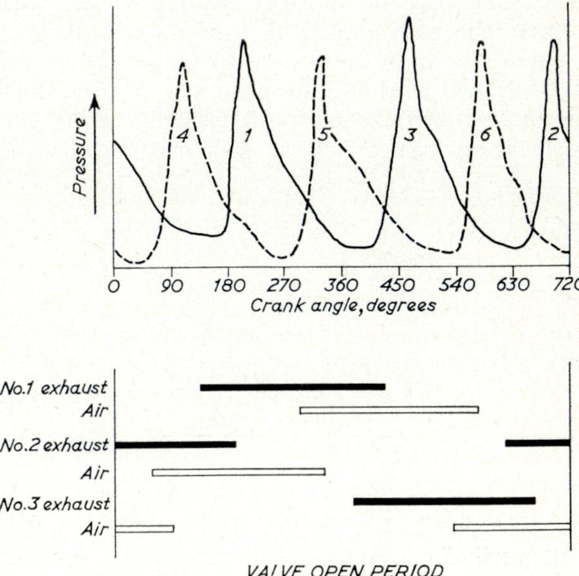

FIG. 3–4—*Exhaust manifold pressure diagram for the two groups of three cylinders of a six-cylinder turbocharged four-stroke engine.*

It is obviously desirable for purposes of design and development to be able to calculate the density of the trapped charge and to assess the effect upon it of variations in valve dimensions, rate of opening, timing and the like. The use of computers has enabled this to be done by considering small steps throughout the gas exchange process and using energy balance and flow equations. (Ref. 2).

3.3. *Two stroke cycle engines*

Compared with a four stroke cycle engine a two stroke cycle engine has a relatively short period during which to get the exhaust gases out of the cylinder and get the air in. Also these processes must be accomplished without the positive displacement effects of the piston and in fact are carried out whilst the piston is around the bottom dead centre, although, as will be seen, the process begins quite early and can finish fairly late. During these processes both air and exhaust ports are open to the cylinder during most of the gas exchange period. Frequently use is made of pressure waves and reflections to improve scavenging and charging as discussed in section 3.7.

There are three basic forms of the two stroke engine: the opposed piston engine in which the gas flow is controlled by the pistons uncovering air ports at one end of the cylinder and exhaust ports at the other, the uniflow two stroke engine which has exhaust valves in the cylinder head whilst the air ports are controlled by the piston at the other end of the cylinder and the loop scavenge engine, in which a single piston controls both inlet and exhaust ports at one end of the cylinder. This latter type has subdivisions each named according to the pattern of gas taken during the scavenge process. The path may be a loop in which the gas enters by air ports going across the top of the piston, up the side of the cylinder wall to the cylinder head and returning down the side of the cylinder containing the ports and out through exhaust ports virtually on the same side of the cylinder, as shown in Figure 3–5(a). Or it may attempt to form a loop

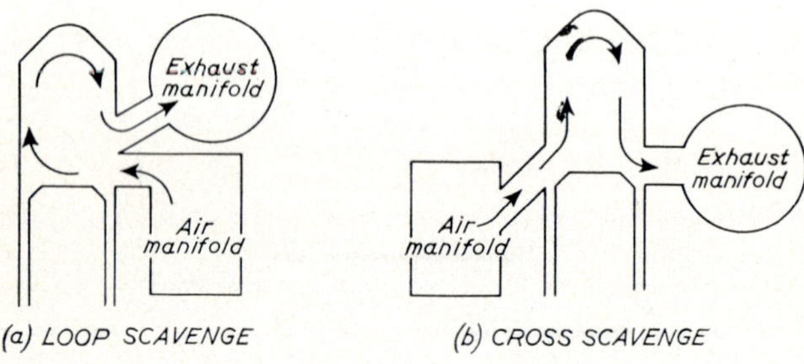

FIG. 3–5—*Cross and loop scavenge paths.*

going across the cylinder as shown in Figure 3–5(b). Because the cylinder is a three dimensional shape and it is necessary to get as much port area as possible, the actual paths followed are not so simple as those shown in the figures and are often a combination of cross and loop patterns.

3.4. *Opposed piston and uniflow engines*

With the uniflow types of engine, both opposed piston and valve in head, it is usual to cause the entering air to swirl around the axis of the cylinder by directing the inlet ports. In this way a better scavenging can be obtained and the air continues to swirl after the ports are closed and helps in better distribution and mixing of the fuel with the air during the combustion process.

Compared with valves, ports have one big advantage and one big disadvantage. The advantage is that the rate of opening and the rate of closing is much more rapid. The disadvantage is that the timing is linked directly to the crankshaft position and if an opening timing is chosen then, with reference to the crank that operates that piston, the closing timing is the same figure on the opposite side of the dead centre. The rapid opening of exhaust ports enables the cylinder to blow down much more quickly and for a steep fronted pulse containing a lot of energy to be formed in the exhaust manifold. When the exhaust and air ports are closed they close equally rapidly, cutting off sharply the flow of gas. This rapid opening and closing is particularly of value in two stroke engines where the absolute time for the gas exchange process is a minimum. The difference between ports and valves occurs because the piston is moving rapidly at the instant of port opening or port closing, whereas in the case of the valve, the valve is stationary at the start of the operation and cannot be accelerated off its seat or retarded on to it with any other than a gentle motion if hammer is to be avoided. Some attempt to improve the valve opening rate can be made by arranging for the valve head to be in a pocket so that whilst the movement of the valve off the seat breaks the seal, the real commencement of opening is when the valve passes out of the pocket.

The ideal sequence of the gas exchange processes in the two cycle engine is first to open the cylinder to the exhaust passages rapidly and allow the blow down of the cylinder to take place with all possible speed; when the cylinder pressure has fallen below the air manifold pressure, then the inlet ports can be opened and the air allowed to pass into the cylinder displacing before it the residual exhaust gas which remains inside; when this scavenging process has taken place the exhaust ports should be closed to prevent loss of further air, the inlet ports remaining open long enough for the cylinder to charge to the highest pressure and coolest temperature, that is the highest density. In the uniflow engine the air is controlled by the piston passing over the inlet ports which thereby open at a certain angle before bottom dead centre and close at the same angle after bottom dead centre, this period being timed to be long enough to secure good scavenging of the cylinder. Before the air ports are opened by the piston the cylinder must have blown down so the exhaust valves have to open sufficiently early to give a period of, say, 25° before the air ports open. The exhaust valves can then be arranged to close at the same time or slightly before the inlet ports are covered by the piston. In this way an ideal timing is approached. The exhaust valves, however, when compared with piston controlled ports have the disadvantage that the blow down period for the cylinder is usually longer owing to the slow start and because the two stroke cycle is repeated every revolution of the crankshaft the valves have to be operated twice as fast as on a four stroke running at the same crankshaft speed, and being very hard worked in this way they present a design problem.

The opposed piston engine attempts to get round this problem by using exhaust ports which are controlled by an exhaust piston. The majority of high speed opposed piston engines have the same stroke for each piston and the pistons are of the same diameter working in the same cylinder bore. If the two crankshafts, which are usually termed the "air crank" and the "exhaust crank", are exactly in phase, then if the exhaust ports are arranged to open some period before the inlet ports in order to achieve blow down of the cylinder, they will close after the inlet ports have closed by the same amount. This is a disadvantage in that the cylinder is connected to the exhaust after scavenging is completed, instead of to the air manifold, a circumstance which allows the charge in the cylinder to leak away down the exhaust or to become contaminated with exhaust gases. The usual solution to this problem is to give the exhaust crank a lead of 10° or 12° in advance of the air crank as well as allowing a slightly longer period of opening for the exhaust ports. In this way an exhaust port lead of 25° over the air port opening can be achieved, whilst the two ports will close at roughly the same time, although the exhaust port is likely to be the later of the two to close unless very high lead is given to the exhaust crank.

There are two objections to using high exhaust crank leads; one applies only to direct reversing engines and is obvious in that when the engine is running in the astern direction not only will there be little or no exhaust lead but there will be a long exhaust lag at the port closing position. The other objection is that the power given out by the cylinder is divided unequally between the exhaust crank and the air crank and the crank which is leading gives out the greater proportion of power. This can be realized if one considers the PV diagram in which V is the volume contained between the two pistons. If one piston is leading the other then during the expansion stroke at the high pressure end of the diagram the leading crank will have turned through a greater angle from inner dead centre, thus presenting a larger torque arm radius, so giving out a higher torque. At the low pressure end of the diagram the leading crank will have the smaller torque arm radius, but at the lower pressures; during the compression stroke the reverse will occur, thus not only does the leading crank give out greater power but it also suffers larger fluctuations in torque. The difference in power between the two cranks is quite substantial and for a 12° lead the exhaust crank will deliver almost twice as much power as the inlet crank. In spite of this limitation the opposed piston engine has many great attractions for its gas exchange processes and has been a deservedly popular type for large bore slow speed engines, taking in its stride the direct reversing required in this size. It is also widely used in medium and high speed engines, although high speed engines are rarely direct reversing.

In some forms the stroke of the exhaust piston is shortened and in certain of these it also operates in a smaller bore. Designs of this form are aimed at reducing the unbalance of power that the exhaust shaft transmits in an attempt to ease the conditions of the exhaust piston and so relieve the design problems posed by this component.

The orthodox opposed piston engine has the easiest shape of cylinder to scavenge together with an admirable combustion chamber shape. It has been used widely for modern high speed multifuel engines.

3.5. *Loop scavenge engines*

The loop scavenge engine is the one which presents the most difficulty with the gas exchange processes, for as both exhaust and inlet ports are tied to the same piston any blow down lead given to the exhaust must appear as lag at port closing. A number of devices are used to overcome this fundamental objection. One of these is to use light, quick acting non-return valves, in the transfer passages between the air manifold and the inlet ports. The use of these non-return valves enables the inlet port opening, to be timed just as early as the exhaust port opening, the valves preventing back flow into the air manifold during the blow down period. When the pressure in the cylinder falls to that in the air manifold the valves open and allow the scavenging air to pass through. After the scavenging period both sets of ports are closed at the same time.

Another approach is to use an additional valve, usually a rotary valve, in the exhaust branch. Such a valve can have a uniform rotary motion and is used in conjunction with exhaust and air port timings that are relatively widely spaced so that as the piston descends the exhaust port opens and an adequate blow down period is allowed before the air ports open. After the scavenging period is over and the exhaust ports are about to close, the rotary valve is arranged to seal off the exhaust branch so that compression then takes place in the cylinder and the exhaust branch is not opened until the piston has covered the exhaust ports. This arrangement is an additional mechanical complication introducing a valve which has to work under very arduous circumstances. The complexity is an unfortunate addition to an engine which is mechanically simple in all other respects and especially as simplicity is the chief appeal of this type of engine.

The path to be followed by the incoming air in scavenging the cylinder and sweeping out the products of combustion is clearly more devious and difficult to achieve without mixing of the outgoing and incoming charges than it is in either of the uniflow forms of two stroke engine.

3.6. *The trapped charge of the two stroke cycle engine*

Because of the limitations of ports controlled by the piston all loop scavenge and many opposed piston engines cannot be pressure charged simply by raising the inlet manifold pressure, they must be turbocharged in a manner which raises the exhaust manifold pressure as well as the inlet pressure. The effect as far as the gas exchange processes are concerned is much the same as putting the engine into a denser atmosphere in which all pressures are raised above the atmospheric. Compare Figure 3–8 and Figure 3–9 in which the various events take place at similar times but at a higher pressure level. The higher this pressure level is, the higher the density of the trapped charge. However, the level is limited by the maximum pressure that can be tolerated in the cylinder and the lowest compression ratio of the engine for starting, and other cycle limitations as already explained in Chapter 2.

A charge trapped at a high density is not necessarily one that contains a lot of oxygen and in consequence it is necessary to consider the purity of the charge during the gas exchange. Whereas in the four stroke cycle engine the great bulk of exhaust gas leaving the cylinder and fresh charge going into the cylinder is displaced positively by the piston, in the two stroke cycle engine the whole process has to be done by the incoming charge alone. Clearly if each cycle is to commence with a complete charge of fresh air, then the volume of the fresh air entering the cylinder each cycle must be at least equal to the volume of the cylinder.

In naturally aspirated four stroke engines an ideal condition can be regarded as that in which the piston draws in for each cycle a volume of fresh air equal to its displacement. In other words the ideal volume of the incoming charge equals the swept volume of the engine. The same criterion can be used for two stroke cycle engines and naturally aspirated two stroke cycle engines using a scavenge pump to supply the air are designed so that the volumetric displacement of the scavenge pump exceeds that of the cylinder. The ratio of the volume of the incoming air to the volume of the cylinder is known as the scavenge ratio. For a naturally aspirated two stroke using a positive displacement scavenge pump the ratio of the volume of the scavenge pump to the swept volume of the cylinder is the "geometric" scavenge ratio. If the volumetric efficiency of the scavenge pump is known, then the ratio of the volume of air at atmospheric conditions to the swept volume of the cylinder can be calculated giving a more realistic scavenge ratio. If for design purpose we assume an atmosphere at normal temperature and pressure then this becomes

$$\mathrm{R}_o = \frac{\text{(volume of air/rev/cyl) at N.T.P.}}{\text{swept volume of cyl (Vdis)}}$$

Even this, however, is not the best criterion with which to judge the purity of the charge and a better approach is to consider a scavenge ratio consisting of the volume of air entering the cylinder each cycle under air manifold conditions of pressure and temperature, compared with the displaced volume of the piston. This form of scavenge ratio can also be applied to turbocharged engines and is much more useful.

$$\mathrm{R}_{JJ} = \frac{\text{volume of air/rev/cyl at air manifold temperature and pressure}}{\text{swept volume of cylinder (Vdis)}}$$

Another form is to compare the volume of air entering the cylinder each cycle at manifold conditions with the trapped charge volume rather than the piston displaced volume. This form of scavenge ratio is more useful still.

$$\mathrm{R}^1_{JJ} = \frac{\text{volume of air/rev/cyl at air manifold pressure and temperature}}{\text{cylinder volume at trapping (inlet port closure) } (V_z 1)}$$

For some engines, particularly loop scavenge and opposed piston engines, in which the exhaust ports are the last openings in the cylinders to close, the trapped charge may owe more to the pressure of the exhaust manifold than it does to the inlet manifold, although obviously, being fresh charge, its temperature will be more influenced by that of the inlet manifold. The form of scavenge ratio used for estimating trapped charge conditions in these circumstances is:

$$\mathrm{R}^1_{JL} = \frac{\text{volume of air/rev/cyl at air manifold temp and exhaust pressure}}{\text{cylinder volume at exhaust port closure } (V_z 1)}$$

All these are different forms of scavenge ratio and in calculating charge purity it is essential to understand at any one time which is being used.

Typical values for R^1_{JJ} and R^1_{JL} range from 1·2 to 1·5.

When the fresh charge enters the cylinder to displace the combustion gases which are left in the cylinder it would perform the scavenging process in an ideal fashion if there was no mixing of the incoming gases with those which are to go out, that is, if perfect displacement occurred. If this were the case then a volume of incoming air at the appropriate pressure and temperature conditions equal to that of the trapped volume of the cylinder would be sufficient to fill the cylinder with a completely pure charge of fresh air. However, it is most unlikely that perfect displacement takes place, and there is bound to be some diffusion or mixing. If we assume perfect diffusion then the higher the scavenge ratio the nearer we get to a completely pure charge, but only when the scavenge ratio is infinite do we arrive in theory at the completely pure charge. Figure 3–6 shows charge

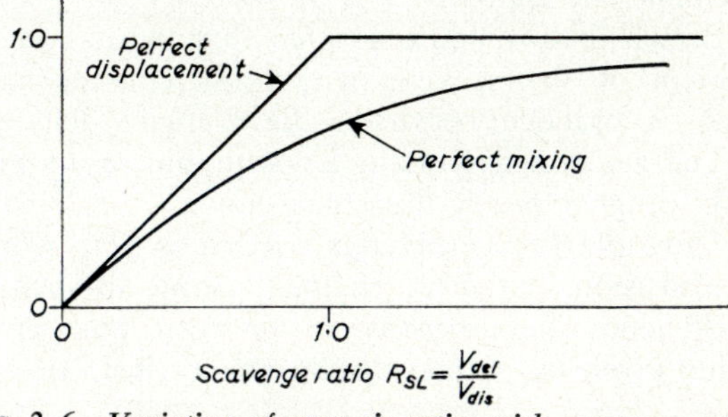

FIG. 3–6—*Variation of pure air ratio, with scavenge ratio.*

quality (trapping efficiency) plotted against the scavenge ratio. As can be seen, with perfect displacement the quality rises for scavenge ratios less than unity until at a scavenge ratio of unity a completely pure charge is attained and at scavenge ratios higher than one there will be always a pure charge together with some of the unmixed air blowing through to exhaust.

Also plotted in this figure is the line assuming perfect diffusion.

If, at any instant

y = the volume of air from the new charge trapped in the cylinder at port closure, the old charge being regarded as combustion products containing no air.

and

V_{Z1} = cylinder volume at port closure,

then when a small volume, dV_{del}, of air is introduced the outgoing volume of exhaust gas, also dV_{del}, will contain $y \,.\, dV_{del}/V_{z1}$ volume of air.

∴ the increase of air in the cylinder

$$dy = dV_{del} - y\,\frac{dV_{del}}{V_{z1}}$$

or

$$dV_{del} = \frac{dy}{1 - y/V_{z1}}$$

Integrating

$$[dV_{del}]_0^1 = [-V_{z1}\log_e(1 - y/V_{z1})]_0^1$$

At the beginning of the period $V_{del} = 0$, $y = 0$

At the end of the period $V_{del} = V_{del}$, $y = P \,.\, V_{z1}$

where

P = ratio of pure air in trapped charge.

i.e. $$V_{de} = -V_{z1} \log_e (1 - P)$$

i.e. $$\frac{V_{del}}{V_{z1}} = R^1{}_{JL} = -\log_e (1 - P)$$

$$e^{-R^1{}_{JL}} = 1 - P$$

$$P = 1 - e^{-R^1{}_{JL}} \tag{3.1}$$

The trapped charge assuming perfect mixing is not so pure as that of perfect displacement but approaches it as scavenge ratio increases.

The lines drawn in this figure assume that the gas in the cylinder at the commencement of the scavenging process consists entirely of combustion products. In practice, because a diesel engine does not burn all the oxygen in the charge, it will contain air with the result that the trapped charge will have a higher purity than that shown.

It is quite possible for a scavenge pattern to give worse results than that of perfect diffusion, but most engines operate in such a manner that the trapping efficiency lies somewhere between that given by perfect diffusion and that given by perfect displacement. (Ref. 3).

3.7. *Exhaust systems*

The primary purpose of the exhaust system is to convey the products of combustion from the cylinder to the atmosphere, or to the turbocharger and then to the atmosphere. The movement of gas in the system is started by the opening of the exhaust valve or port allowing the gases under pressure in the cylinder to blow down along the pipe. The same process initiates a pressure pulse which is propagated throughout the pipe system with the velocity of sound in the gas. The speed and direction of the pulse is not to be confused with the speed and direction of the gas particles, as the two are separate and largely independent, although arising from the same release of energy.

When the wave front of the pulse reaches an open end of the pipe, it experiences a negative reflection. When the wave front reaches a closed end it experiences a positive reflection. The pulse itself may become attenuated as it travels along the pipe and as it undergoes reflections.

A sudden enlargement of the pipe, or entry into an expansion box or silencer, causes a partial negative reflection, and a sudden contraction of the pipe causes a partial positive reflection. A turbocharger nozzle, for instance, causes a partial positive reflection.

The exhaust pulses contain energy which is part of the energy released into the exhaust system from the blow down of the cylinder. Exhaust systems are designed either to minimize the amount of energy contained in the pulses and carried out of the exhaust systems by them or to channel the energy into a useful purpose.

The most common form of utilization of energy in pulse form is with turbocharged engines where the energy in the pulse combined with the energy in the exhaust gas is converted into mechanical energy in passing through the turbine. Exhaust systems and turbochargers designed with this end in mind are known as "pulse systems".

With a pulse system, attempts are made to conserve the energy in the exhaust pulses by keeping the manifold and piping to as small a volume as possible, short in length, and without unnecessary bends. It is also important to have a uniform bore of pipe to avoid energy losses in reflections. The bore of the pipe must be large enough to avoid restricting the flow of gas and yet small enough to avoid attenuating the pulses. In a 4-stroke cycle turbocharged engine, the bore is usually of the same order as the maximum valve area. However, considerations of space and engine layout often take precedence over the ideal design for energy conservation.

As the pulses travel in all directions throughout the system, they enter the exhaust manifold branches of all cylinders and can give rise to an undesirable condition if they raise the pressure at the exhaust valve of a cylinder during its scavenge or overlap period (Figure 3–4). To avoid interference of this kind, cylinders are grouped so that usually not more than three cylinders exhaust into the same manifold. This results in the pressure pulses being phased at intervals of no less than 240° and since the duration of a pulse is only of the order of 40° there is a period of about 200° before the end of one pulse and the beginning of the next. Sometimes, however, the pulse is lengthened beyond 40° by a positive reflection of itself from the turbocharger nozzle, but by careful arrangement it is usually possible to arrange for a low pressure period in the exhaust manifold during the time that the valve openings overlap. The pulsed exhaust gas from each group of cylinders is kept separate right up to and through the turbine nozzle, the gas streams being allowed to mix only after the energy has been transformed into mechanical energy in the rotor.

Turbochargers are made with their turbine entry casings to receive a number of entries, thus a six cylinder engine usually has a turbocharger with two entries, receiving the manifolds of two groups of three cylinders. An eight cylinder engine often has one turbocharger with four entries each from two cylinders. With larger numbers of cylinders various groupings are possible and the choice depends on the best compromise which the designer has to make between 1) the crankshaft arrangement for mechanical balance of the engine, 2) even turning moment, 3) reasonable bearing loads, and 4) the conditions in the manifolds to suit turbocharging.

The turbocharger cycle is more efficient if the grouping avoids periods when there is no flow into any of the turbochargers; for instance, it is undesirable to have only one cylinder feeding one turbocharger. A practical consideration in the grouping of cylinders is to divide the gas flow so that all turbochargers are of the same size. Figure 3–7 shows some of the more common arrangements.

An alternative to the pulse system is the attempt to convert the energy released from the cylinder into a steady pressure by connecting the cylinders into one large manifold, which has damping capacity to absorb the pulse energy. Inevitably, a lot of the pulse energy is down-graded, appearing as low temperature heat in the exhaust but on the other hand, turbines designed for the steady flow conditions of constant pressure can be more efficient than those that have to cope with pulsating flow and at high boost levels the "constant pressure" system works quite successfully.

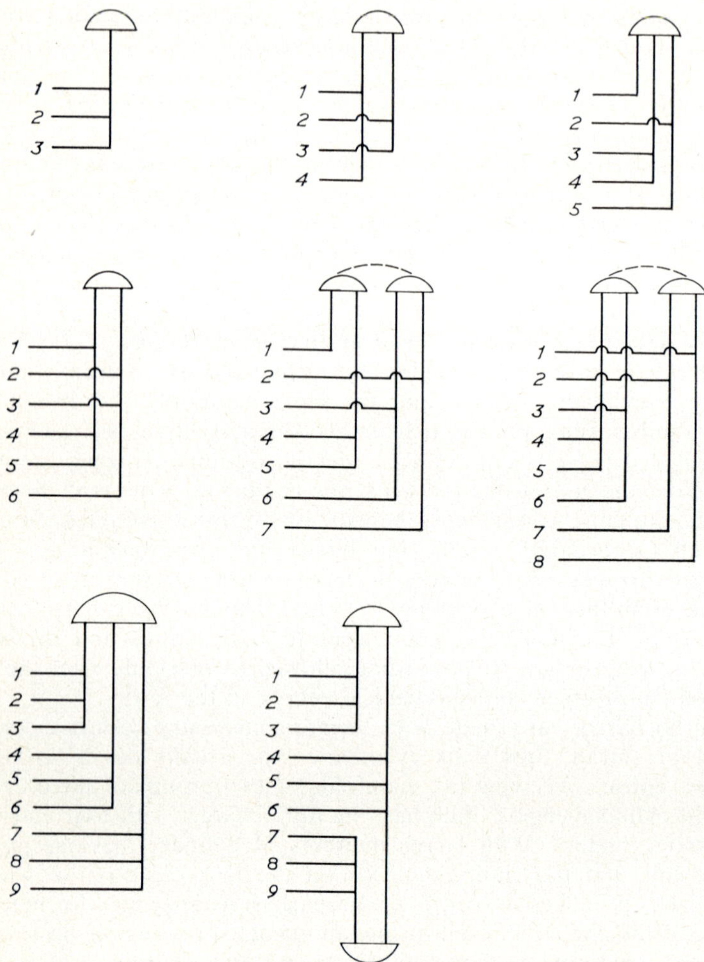

FIG. 3–7—*Some of the more popular exhaust grouping arrangements for four-stroke turbocharged engines.*

Although early two stroke cycle engines used large capacity exhaust manifolds, pulses have played a large part in the development of this type of engine. Naturally aspirated two stroke engines can make good use of the negative reflections in the exhaust system by tuning the length of exhaust pipes so that the negative reflection causes a depression at the exhaust ports during the scavenge period and thus assists scavenging. This system bears the name of its inventor, Kadenacy.

An interesting development known as exhaust pulse pressure charging is due to Carter (Ref. 4) and is illustrated in Figure 3–8(b). Here, the blow down pulse of one cylinder is timed by port opening angle and exhaust system design to raise the pressure in the exhaust branch of another cylinder at the moment of exhaust port closing. The system was developed in connexion with loop scavenge engines and it enabled a larger lead of exhaust to air port opening to be employed without loss of charging air

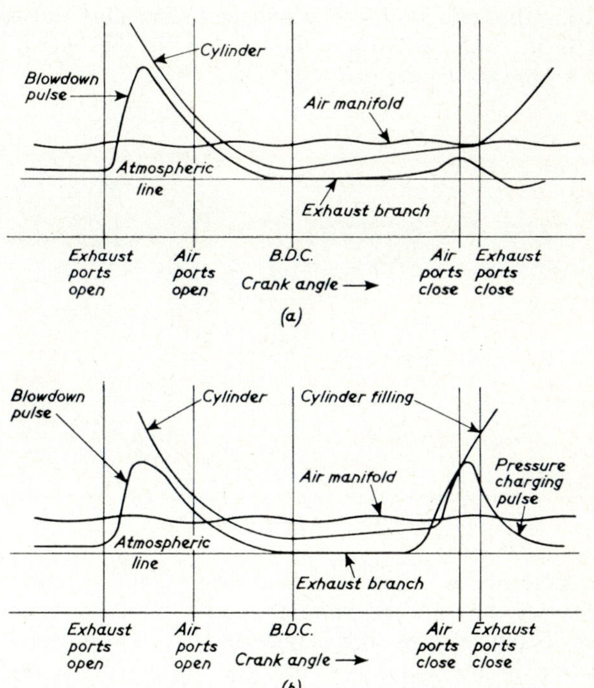

FIG. 3–8a—*Gas exchange period two-stroke cycle engines not turbocharged.*

FIG. 3–8b—*Gas exchange period two-stroke cycle engine with exhaust pulse pressure charging, not turbocharged.*

through the exhaust ports during the corresponding and inevitable exhaust lag at port closing. It works by reversing the pressure gradient between the cylinder and exhaust ports after the air ports have closed not merely retaining the trapped charge at air manifold pressure but augmenting it by trapping the exhaust pulse. The back flow of gases from the exhaust manifold branch into the cylinder which this implies is not detrimental as these gases at the end of the scavenge period contain a very high proportion of unburnt air. Exhaust pulse pressure charging is capable of application to other types of port controlled engine and is based on arranging the cylinders together in groups of three, four and sometimes five having equal firing intervals. This system is equally applicable to turbocharged two stroke cycle engines (see Ref. 5) as is illustrated in Figure 3–9(b).

3.8. *Turbocharger energy balance*

The cycle on which the turbocharger operates was outlined in Chapter 2 and Figure 2–6 refers to the conditions at the salient points. The turbocharger is a free running unit operating on a gas turbine cycle in which the output appears as a pressure drop from the compressor delivery to the turbines inlet. The engine is the combustion chamber and also at the same time, the load.

In the case of the four stroke cycle engine, the pressure in the cylinder during the induction stroke will be higher than during the exhaust stroke, and in this way, positive work is added to the output of the engine from the turbocharger via the piston.

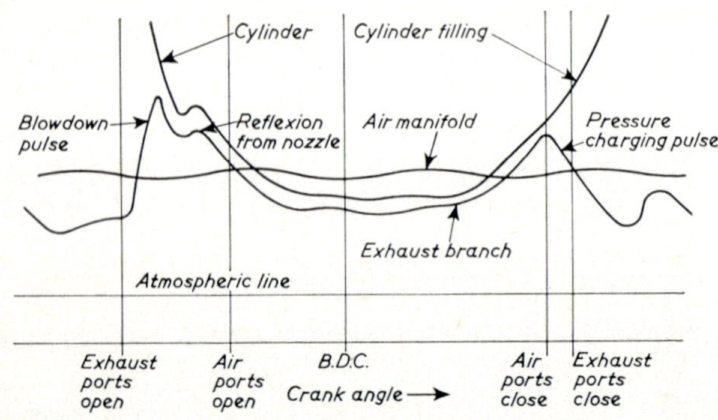

FIG. 3–9—*Gas exchange period, turbocharged two-stroke cycle engine with exhaust pulse pressure charging.*

When operating at other than the design point, *e.g.* at reduced loads, the turbocharger will not be running under matched conditions and in consequence its efficiency is likely to be less. As a result of this, there will be a reduction in the output of the gas turbine cycle and under some circumstances, there can be a negative output and the engine cylinder will have to provide work to the turbocharger. In other words, the engine will revert to pumping losses as for a naturally aspirated engine. This will often be the case at very low loads and during starting.

In the two stroke cycle engine, the energy in the exhaust is relatively low as the necessity to use high scavenge ratios in order to obtain the pure charge results in a large proportion of unburnt air in the exhaust which is therefore at a relatively low temperature.

Turbocharger efficiency is therefore of great importance if the output from the turbocharger cycle is not to be too low. This leads to a need with all two stroke cycle engines for the lowest pressure drop across the cylinders compatible with the pattern of scavenging necessary to produce a pure charge.

The loop scavenge engine is at something of a disadvantage because of this requirement and operation away from the design point usually requires assistance to the turbocharger which in this case can be provided only by crankshaft power, driving a positive displacement blower in series or parallel with the turbocharger, or by providing an auxiliary fan driven from a separate power source, usually in series with the turbocharger compressor. The positive displacement blowers may take the form of reciprocating scavenge pumps or rotary blowers, or assistance from the underside of the pistons in cross-head engines.

When a turbocharged engine (two or four stroke cycle) is required to carry an increased load, the response to the controls is first to introduce extra fuel into the cylinder. The burning of this fuel results in greater heat in the exhaust gases, which in turn pass on extra energy to the turbine, causing the turbocharger to run at a higher speed and provide an increased air supply to the cylinder. The air supply is therefore bound to lag behind the fuel supply and in consequence an increase of load in steps above a certain magnitude will result in exhaust smoke and can even lead to stalling of the engine. In an engine-alternator set used for power generation, it is necessary to limit the steps of load increase not only to avoid stalling or smoke but also to avoid unacceptable temporary speed variations.

In engines used for propulsion, acceptance of an increase in load is associated with increasing the speed of the propeller. Here, the controls are usually arranged to operate so that the fuel supply to the cylinders is limited to that which can be completely burned without undue smoke under the air supply obtaining at the time. To do this, the controls are linked to either air manifold pressure or to turbocharger rev/min. Some high speed two stroke cycle engines designed primarily for automotive applications are fitted with crankshaft driven positive displacement blowers in series with the turbocharger to provide rapid acceleration.

3.9. *Turbochargers*

Most turbochargers take the form of a centrifugal single stage compressor, driven by a single stage turbine, forming a self-contained free running unit. The turbine is usually axial flow in the larger sizes and radial flow in the smaller.

3.10. *Centrifugal compressors*

Figure 3–10 shows in diagrammatic form a section of a centrifugal

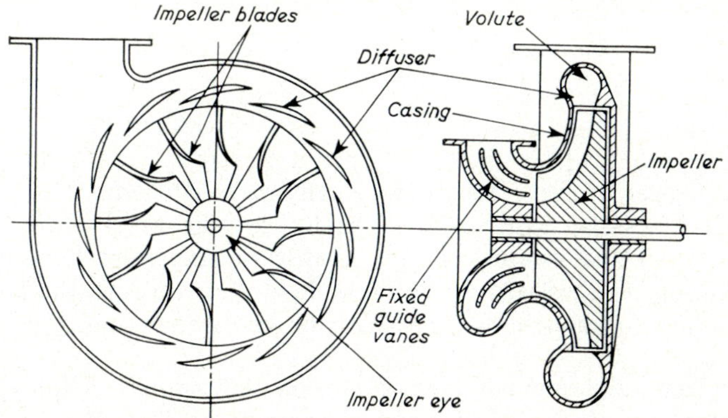

FIG. 3–10—*Diagram of centrifugal compressor.*

compressor with the nomenclature used. Air is led, sometimes by fixed guide vanes, to the eye of the impeller which consists of a rotating disk having radially disposed vanes which transmit the driving torque to the air. The air then flows to a diffuser which converts the kinetic energy of the air to static pressure, this process being continued in the scroll or volute which also gathers the air and delivers it. The compressor is a steady flow machine continuously imparting energy to the air flowing through it, this energy being converted into static pressure. Figure 3–11 shows the velocity

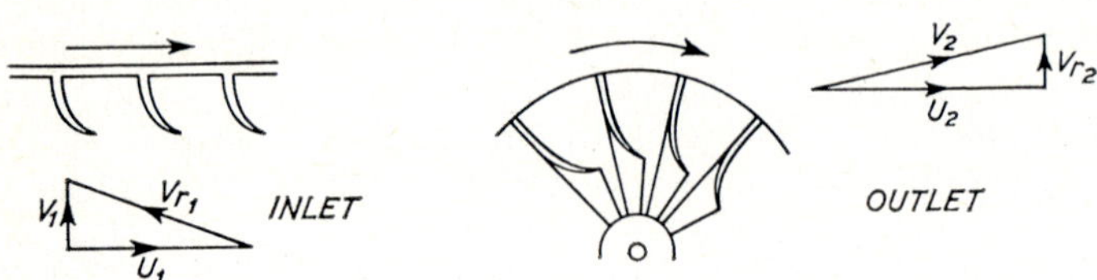

FIG. 3–11—*Velocity triangles.*

triangles of the air at the entry to the impeller eye and the exit from the impeller tip. U_1 and U_2 represent the impeller speeds at the appropriate radius and V_1 and V_2 represent the velocity of air entering and leaving the impeller. V_{r1} is the relative velocity of the air at entry and V_{r2} is the relative velocity at exit.

Most turbocharger impellers have radial vanes and these are assumed for the two diagrams shown, whilst Figure 3–12 illustrates forward curved

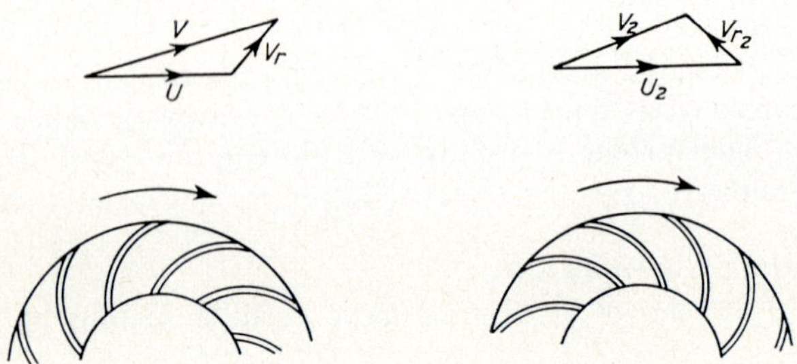

FIG. 3–12—*Velocity Triangle, curved vanes.*

and backward curved vanes. Impellers can be designed as shrouded, as shown in Figure 3–13, or open. The open design is by far the more popular as it is far easier to make and is marginally more efficient.

In all cases the exit velocity absolute of the air V_2 will have two components, a radial component and a whirl or tangential component. The radial component is related to the mass flow of air through the compressor and the whirl component is related to the energy put into the air by the impeller.

From the fundamental laws of motion the torque on the impeller is equal to the rate of change of angular momentum of the air. For air enter-

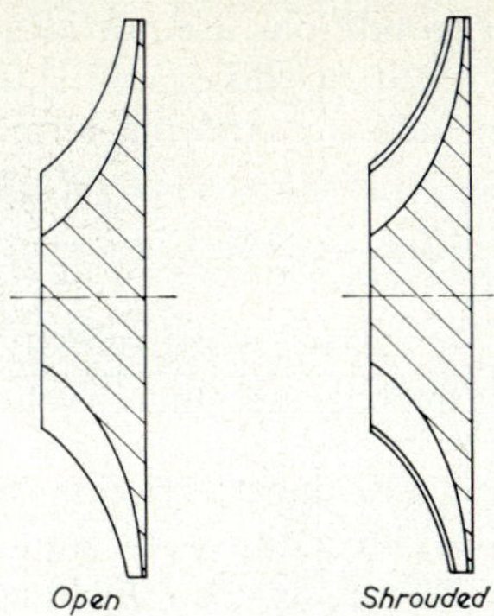

FIG. 3–13—*Open and shrouded impellers.*

ing with no angular velocity, that is entry velocity axial, and taking unit mass flow, then torque

$$T = V_w \frac{R}{g} \tag{3.2}$$

where V_w equals the tangential or whirl velocity of the air leaving the impeller tip and R equals the radius of the impeller. The angular velocity of the impeller ω equals the linear velocity U at the tip, divided by the tip radius R and hence the rate of change of energy or the work done per unit mass flow of air is

$$W = V_w \frac{R\omega}{g} = V_w \frac{U}{g} \tag{3.3}$$

For the ideal case with purely radial vanes the air will leave the tip with a whirl velocity equal to the impeller peripheral velocity that is $V_w = U$ and hence

$$W = \frac{U^2}{g} \tag{3.4}$$

This whirl velocity leaving the impeller represents the energy which is converted to static pressure in the diffuser and the volute so that the whole of the work done by the impeller can be equated to the work of adiabatic compression for this ideal compressor. The work of adiabatic compression is given by $JC_p\,(T_2 - T_1)$, hence

$$JC_p\,(T_2 - T_1) = \frac{U^2}{g} \tag{3.5}$$

It should be noted that this is true only for compressors having radial vaned impellers. Forward or backward curved blades result in a value for V_w which is different from U and which in fact varies with the throughput or mass flow of air through the impeller. Radial vanes give very suitable characteristics for the turbocharger compressor and are easy to manufacture. Their work capacity is insensitive to change of mass flow and they are nearly always used in turbochargers.

In an actual impeller, conditions are not ideal and the velocity of whirl will be less than U as a result of eddies and skin friction. In practice it is therefore customary to include a modifying term in the equation

$$JC_p\,(T_2 - T_1) = \sigma\,\frac{U^2}{g}$$

Where $$\sigma = \frac{\text{actual whirl velocity}}{\text{impeller peripheral velocity}}$$

There are also other losses due to friction, heat loss and leakage past the vanes and as these also affect the work input another modifying term, termed the power input factor, is also included so that equation (3.5) becomes

$$JC_p\,(T_2 - T_1) = \phi\,\sigma\,\frac{U^2}{g}$$

Both ϕ and σ are determined from observed experience and experiment.

The torque and work done must be matched by those of the turbine, see Sections 3.12 and 3.16.

The centrifugal force imparted to the air by the rotating impeller causes the particles of air to move from the small radius of the eye to the large radius at the impeller and at the same time it increases the pressure of the air. Roughly speaking, about half the pressure rise takes place in the impeller in this way and the remaining half takes place in the diffuser where the outward velocity is converted into static pressure.

The pressure ratio required of a turbocharger compressor may vary from about 1·2 to 1 up to 3 to 1, according to the duty it has to perform in connexion with the engine for which it is intended. The higher the pressure ratio required the higher the tip speed of the impeller and this may reach 1,500 feet per second. If this is the case and a slip factor of 0·9 is assumed, the tangential velocity of the air leaving the impeller tip is 1,350 feet per second and this must be reduced in order to recover the energy as static pressure.

The air leaving the impeller takes up a free vortex motion combined with a radial component velocity which decreases with increased radius, and as a result a natural diffusion takes place. The space into which a turbocharger must fit, however, does not permit a construction which would depend wholly on this natural diffusion, and in consequence this air is guided into channels in a diffuser, these channels or passages being of diverging section formed by a number of vanes which have thin profile leading edges, the tips being two or three centimetres away from the impeller tip. This clearance space allows some diffusion by the free vortex principle and also avoids buffeting impulses which would be communicated to the impeller if the tips of the diffuser approach it too closely. The channels in the diffusers are arranged as divergent passages to slow the velocity of the air down in as short a distance as possible without resulting in a breakdown of the flow and by so doing increase its static pressure, Figure 3–14. The exit velocity from the diffusers will be down to about 120 metres per second and diffusion is carried out further in the volute which collects the air that is delivered from each of the diffuser passages.

Diffusion is obtained in a shorter length by a larger number of small passages, but the throat (which is the smallest part of the diffuser passage) is a critical part of the design and diffusers with small throats have steep

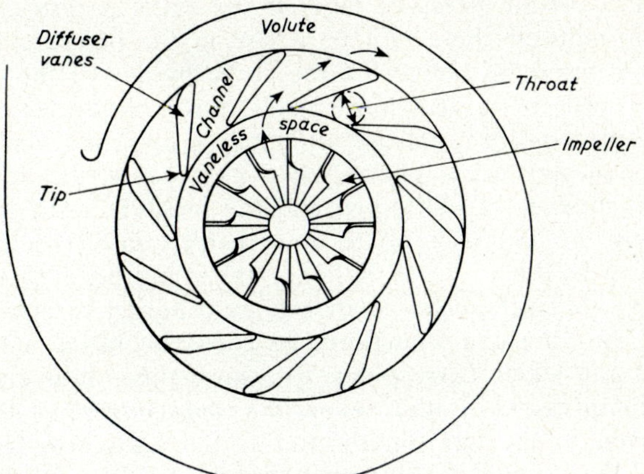

FIG. 3–14—*Diagram of diffuser.*

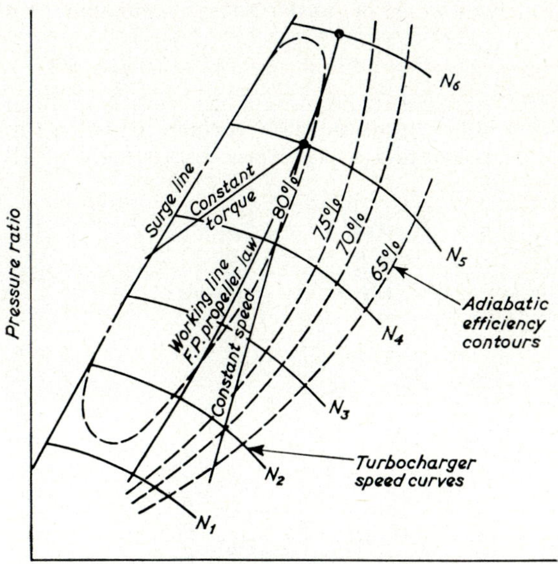

FIG. 3–15—*Compressor characteristics.*

characteristics which are not easily matched to engines. Diffusers which have larger throats and also diffusers which have a large vaneless space between the impeller and the diffuser have flatter characteristics which are more easily matched, although their optimum efficiencies may not be so high as the more critical type of diffuser. Figure 3–15 shows a typical set of compressor characteristics, the ordinates are pressure ratio across the compressor and the abscissae are mass flow. The figure shows the lines of constant speed and lines of constant efficiency.

This figure also shows the surge line. When a compressor is run at a given set speed, according to the delivery conditions it will give a certain mass flow at a certain pressure ratio. If the outlet is restricted so that the mass flow is reduced there will be a rise in the pressure ratio, and this process will continue as shown on the curve by any of the constant speed lines. If the process is continued mass flow will decrease without any marked rise in pressure ratio and ultimately a point is reached where what is known as surging takes place. When this occurs, the flow through the diffuser and impeller breaks down and the pressure ratio across the compressor drives the air back in the reverse direction. This can result in a sufficient drop in the pressure ratio for the compressor to recover and to start to deliver again but immediately to be followed by a fresh breakdown of the flow. Surging can sometimes be cured by fitting a diffuser with a small throat and which corresponds to a lower mass flow characteristic for the whole compressor. In other words the characteristic as shown in Figure 3–15 is moved to the left slightly and as the mass flow for the required condition is the same it follows that this mass flow is then further away from the surge line. However, a penalty will be paid in decreased efficiency. This feature is of importance in matching turbochargers to engines.

3.11. *Axial flow turbines*

As its name implies an axial flow turbine is one in which the general direction of flow is parallel to the rotor axis. Figure 3–16 shows a dia-

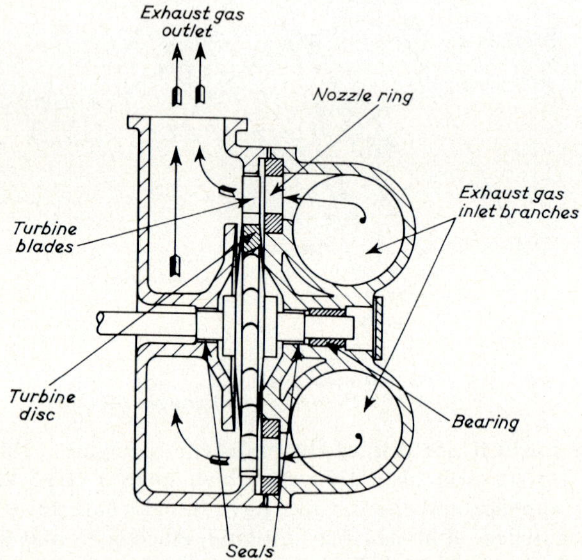

FIG. 3–16—*Diagram of Axial flow turbine.*

grammatic sketch of this type of turbine. The gas enters at the turbine entry casing which is divided into chambers which keep apart the gas streams from the exhaust systems of the different cylinder groups and lead them to the separate segments of the nozzle ring. The gas flows through the nozzle blades losing pressure and gaining velocity. It impinges on the rotor blades and in passing through them drops further in pressure, increasing in relative velocity and transferring energy to the rotor. As it leaves the rotor it is gathered in the outlet-casing and conducted to the exhaust outlet.

Blading is often classified as impulse or reaction, an impulse design being one in which the whole of the pressure drop takes place in the nozzle whilst with the reaction design some of the pressure drop takes place in the rotor. Strictly a wholly reaction turbine would be one in which the gas would be conveyed to the rotor at low velocity with negligible pressure drop. All the pressure drop would then take place in the rotor and the gas issuing at high velocity would drive the rotor by reaction. No practical gas turbine operates in this way with 100% reaction; most designs use some pressure drop in the nozzles and some in the rotor.

3.12. *Thermodynamic analysis*

The velocity at exit from the nozzle is determined by the drop in total heat through the nozzle blades. Nozzles are invariably of convergent passage only design. Increase of relative velocity as the gas passes through the moving blades is determined by the drop in enthalpy (total heat) through the rotor. This may be followed in Figure 3–17 which indicates the

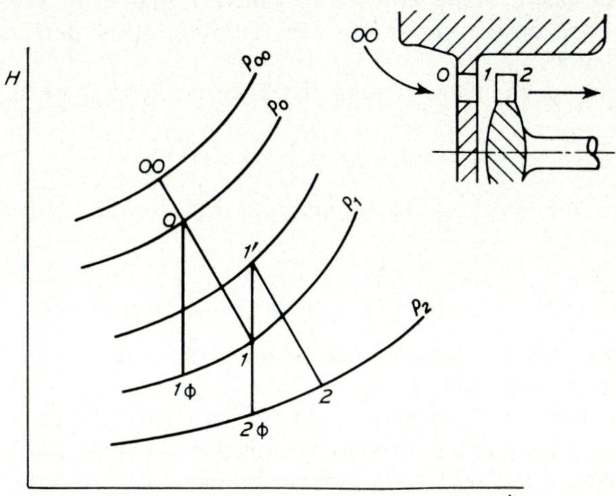

FIG. 3–17—*Enthalpy-entropy diagram for turbine.*

processes on an enthalpy–entropy chart. P_{oo} to P_o represents the small loss of pressure as the gas flows up to the nozzle ring. The isentropic heat drop 0–1ϕ is available for conversion to energy but owing to the nozzles being less than 100% efficient there is an increase of entropy so that the vertical distance from 0–1 represents the actual fall in enthalpy. Considering conditions relative to the moving blades the vertical distance 1^1–1 can be taken as representing the enthalpy equivalent of V_{r1}, the relative velocity at entry. 1^1–2ϕ will then represent the enthalpy available and 1^1–2 the enthalpy equivalent of V_{r2} the relative velocity at exit.

In order to maintain radial equilibrium of pressure in the gas as it flows through the space between the nozzles and the rotor, blading designs frequently attempt to follow a free vortex pattern. This, together with the increase in blade velocity as the radius increases, results in a varying amount of reaction which increases from root to tip. Figure 3–18 shows

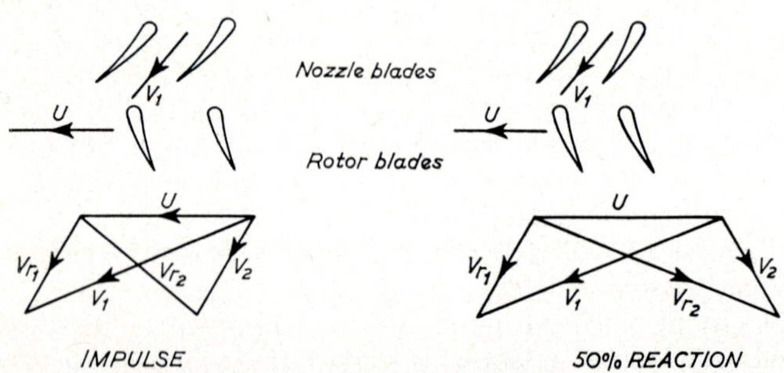

FIG. 3–18—*Turbine velocity triangles.*

velocity triangles at inlet and outlet of the blades in a typical case. Also indicated is the blade shape which is designed to match the relative velocity of the gas under design conditions yet retaining good performance when running off match.

The torque on the rotor is obtained from the change of momentum

$$\text{torque} = r\,\frac{(V_{w1} - V_{w2})}{g} \tag{3.7}$$

and similarly, the work done is equal to the torque multiplied by the angular velocity, U_r

$$W = \frac{U}{g}\,(V_{w1} - V_{w2}) \tag{3.8}$$

The torque and work done must be matched to that of the compressor, equations (3.2), (3.3) and (3.6).

In some cases turbochargers used for two stroke cycle engines carry on their shaft a gearwheel driven by speed increasing gearing from the engine crankshaft in order that any deficiency of torque may be made good. The gearing may embody a freewheel arrangement as mechanical assistance is often required only for starting and low loads.

3.13. *Aerodynamic design*

The losses referred to in the thermodynamic analysis are inevitable in an actual turbine and fall into four classes.

Profile losses, which are associated with the friction of the gas on the blade surfaces as they deflect it. The design of the blading can influence this type of loss, together with the number of blades and pitching and also the aerodynamic loading. The designer has to balance the claims of blade shape and geometry required for the aerodynamic considerations against those for the friction loss.

The second class of loss is associated with the end effects. Nozzle blades are almost invariably shrouded whilst rotor blades are not. Both shrouded and unshrouded blades have end effects and the boundary conditions of passages and blades upstream can be carried over into the next section producing effects which result in some loss of energy.

The third category of losses are the casing losses at both inlet and outlet which may result in something of the order of a 10% loss in total head. A large part of this is in the outlet casing where there is a failure to recover energy from the gas leaving the blades at fairly high velocity. Casings could undoubtedly be designed which would not have this loss so high, but would take up far more space and it is usually considered that the space saved is worth the loss of efficiency.

The remaining sources of loss are the blade clearances. These vary according to the design and size of the turbocharger and are the result of leakage paths around the blades. The loss is less in large turbochargers and increases as size is diminished and is usually quite unsatisfactory in axial flow turbines for small flows.

3.14. *Turbine characteristics*

Typical characteristics of axial flow turbines are shown in Figures 3–19 and 3–20. The flow is affected by the speed at which the turbine is run but

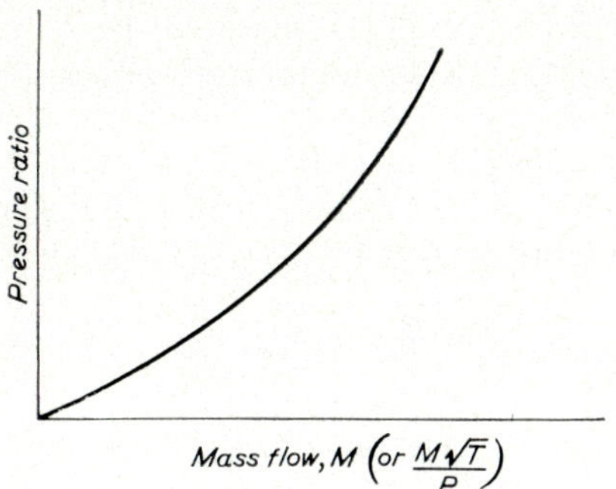

FIG. 3–19—*Mass flow characteristic.*

only slightly so that usually the one curve shown in Figure 3–19 representing mass flow against pressure drop is used for all speeds of the turbine. The other characteristic is the efficiency which is plotted against U/C where U is the blade speed and C is the velocity equivalent of the adiabatic heat drop. Again there are small variations with different pressure ratios and often one characteristic curve is quoted. The variations are greatest at the position of optimum efficiency as Figure 3–20 shows.

The turbine has to be matched to the compressor which it drives, such that the actual torque of the turbine is equal to the actual torque of the compressor and that the work done for appropriate mass flows is equal in both cases. The design also has to cater for off-match conditions and in the case of a pulse system of charging, these can be very wide, so much so that a designer may choose different blading for different applications. For instance, in the case where the turbine is connected to a three cylinder group, the pressure variations are comparatively small and approximate

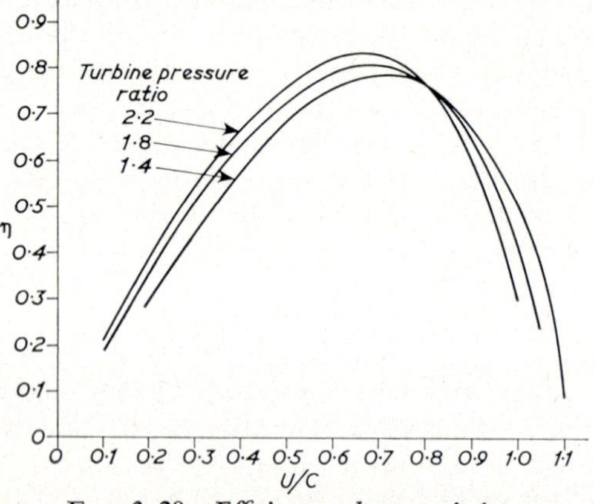

FIG. 3–20—*Efficiency characteristics.*

to constant pressure so that a high reaction design is advantageous. In other cases where part of the nozzle is not running full all the time, a high reaction design will have windage losses and an impulse design is better. Quite often, nozzle outlet angles are compromised in order to permit a relatively high mass flow through a small casing or in order to enable a given frame size to cover a wide number of applications.

3.15. *Radial flow turbines*

Radial flow turbines are so called because the exhaust enters the rotor in a radial direction. For low mass flows, they have the advantage that

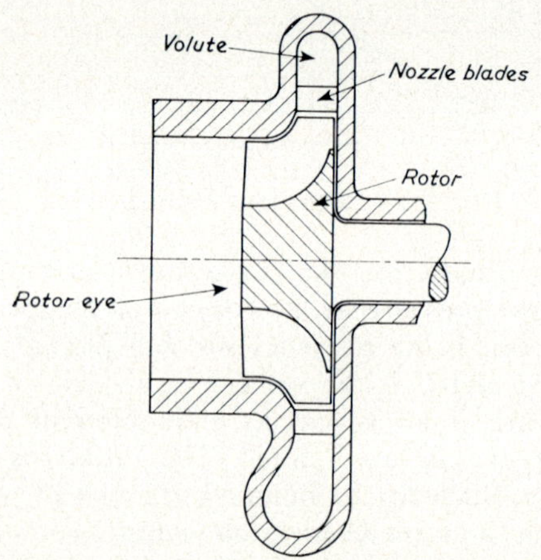

FIG. 3–21—*Diagram of radial flow turbine.*

clearances are relatively small and manufacture is comparatively easy. They are difficult to adapt to high mass flows because the rotor becomes massive when it has to cope with large quantities of gas. The geometry lends itself to low casing losses.

The shape of the rotor is similar to that of a centrifugal compressor but with wider passages to cope with the greater specific volume of the exhaust gas at the higher temperatures. Figure 3–21 shows the arrangement diagrammatically and the velocity of the gas entering the rotor can be defined by radial and whirl components. The gas leaving the rotor similarly can be defined by axial and whirl components. The transfer of energy from the gas to the rotor follows the fundamental laws of motion so that:

$$\text{torque} = r_1 V_{w1} - r_2 V_{w2} \tag{3.9}$$

and the work done for unit mass flow is given by

$$W = \frac{U}{g} \quad (V_{w1} - V_{w2}) \tag{3.10}$$

The mechanical design of turbochargers and examples of both axial and radial flow types are described in Chapter 19.

REFERENCES

1. WALLACE, W. B., *High Output Medium Speed Diesel Engine Air and Exhaust System Flow Losses,* 1967 Proc. I. Mech. E.
2. WHITEHOUSE, N. D., STOTTER, A., GOUDIE, G. O., PRENTICE, B. W., *Method of Predicting Some Aspects of Performance of a Diesel Engine Using a Digital Computer,* 1962 Proc. I. Mech. E., Vol. 176 No. 9.
3. SCHWEITZER, PAUL N., *Scavenging of 2-stroke Cycle Diesel Engines,* Macmillan and Co., Ltd.
4. CARTER, H. D., *Loop Scavenge Diesel Engine,* Proc. I. Mech. E. 1946, Vol. 154.
5. HENSHALL, S. H., *Turbocharged Loop Scavenge Diesel Engine,* Proc. I. Mech. E., 1963, Vol. 177 No. 6.

BIBLIOGRAPHY

FERGUSON, T. B., *The Centrifugal Compressor Stage,* London, Butterworth, 1963.

SHEPHERD, D. G., *Principles of Turbomachinery,* New York, Macmillan, 1956.

AINLEY, D. G., *The Performance of Axial Flow Turbines,* Proc. I. Mech. E., Vol. 159, 1948.

CRAIG, H. and JANOTA, M., *The Potential of Turbochargers as Applied to Highly Rated 2-stroke and 4-stroke Engines,* CIMAC Paper B14, London, 1965.

HORLOCK, J. H. and BENSON, R. S., *The Matching of 2-stroke Engines and Turbochargers,* CIMAC Paper A10, Copenhagen, 1962.

BROWN BOVERI, *Turbochargers and Gas Turbines,* CIMAC Congress, London, 1965.

CHAPTER FOUR

Combustion and Fuel Injection

4.1. *Nature of fuels*

Fuels for diesel engines are obtained from crude petroleum which, when distilled, yields gasoline, kerosene, gas oil, diesel oil, fuel oil, lubricating oil stocks and waxes. All these oils are composed of mixtures of hydrocarbons, each of which has a formula of the general form Cx Hy where both x and y can range from unity to several hundreds. The lighter fractions in which x and y are small are gases, for example Methane, the chief constituent of natural gas, has the formula CH_4. The range extends to heavy fractions where x and y are large and the molecular structure is complex. Hydrocarbons are classified into groups having similar molecular structure and members of the same group exhibit largely similar properties and patterns of behaviour during combustion.

The paraffins are one important group. They have the general formula C_nH_{2n+2} and some of the simpler members are shown below

CH_4 methane
C_2H_6 ethane
C_3H_8 propane
C_4H_{10} butane

The paraffins occur naturally in petroleum but some of the other groups are produced by refinery processes breaking down complex molecules into simpler ones. Such a process is "cracking" in which the hydrocarbon is subjected to heat or a catalyst and which results in a greater yield of fuels that are light or have special properties than would result from simple distillation. Most gasoline is produced by cracking. Certain hydrocarbon groups occur solely or mainly in breakdown products, a typical series being the Olefines having the formula C_nH_{2n} with $n = 2$ as the first number of the series.

C_2H_4 ethylene
C_3H_6 propylene
C_4H_8 butylene

The various fuels derived by distillation and other means consist of mixtures of hydrocarbons and are described as lighter or heavier in relation to their neighbours without any well defined boundaries. In fact there is considerable overlap between the divisions. A light fuel is more volatile than a heavy one and has a lower viscosity and a lower specific gravity. The heaviest fuels are the residuals which remain after the other grades have been removed by distillation. A heavy viscous residual fuel may be thinned by mixing with it a small quantity of light fuel, the result being a blended fuel of intermediate viscosity. The combustion properties of such blended fuels are usually the same as those of the residual fuels from which they originate, the chief reason for the blending being to reduce the viscosity to make them easier to handle. Crude petroleum is found in many parts of the world and the proportions of the constituent hydrocarbons vary according to the source.

Although composed predominantly of carbon and hydrogen, liquid fuels may contain sulphur: up to about 1% in light diesel oils and up to 3% or 4% in heavy residual fuels. In combustion sulphur oxides are formed which combine with moisture to form corrosive acids detrimental to the engine. The energy from the combustion of sulphur is negligible and it is preferable for a fuel to contain as little sulphur as possible.

4.2. *Chemistry of combustion*

In complete combustion a fixed amount of fuel combines with a fixed amount of oxygen and liberates a definite amount of energy as heat in the process. The overall process can be illustrated using the basic chemistry of combustion of the elementary fuels, carbon and hydrogen, according to their proportions in the fuel and regardless of their combination into hydrocarbons. Using the usual chemical symbols and equations

$$C + O_2 \rightarrow CO_2 + 393{\cdot}8 \text{ MJ/kmol}$$

$$2H_2 + O_2 \rightarrow 2H_2O + 261{\cdot}06 \text{ MJ/kmol} \qquad \text{(liquid at 25°C)}$$

The combined proportions by weight are obtained by considering the molecular weights

$$H_2 = 2; \; C = 12; \; O_2 = 32$$

thus 12 parts by weight of carbon combine with 32 parts oxygen to form 44 parts of CO_2 with a release of 393·8 MJ/Kmol.

$$C + O_2 \rightarrow CO_2 + 393{\cdot}8 \tag{4.1}$$

$$12 + 32 \rightarrow 44 \tag{4.1a}$$

and similarly

$$2H_2 + O_2 \rightarrow 2H_2O + 261{\cdot}06 \tag{4.2}$$

$$4 + 32 \rightarrow 36 \tag{4.2a}$$

Suppose a fuel to consist of 87% carbon and 13% hydrogen by weight. Then of 1 kilogramme of this fuel 0·87 kilogramme would be carbon, the combustion of which would liberate

$$0{\cdot}87 \times \frac{393{\cdot}8}{12} = 28{\cdot}55 \text{ MJ.}$$

The remaining 0·13 kilogrammes would be hydrogen and its combustion would liberate

$$0{\cdot}13 \times \frac{261{\cdot}06}{2} = 16{\cdot}96 \text{ MJ.}$$

The gross calorific value of the fuel would thus be

$$28{\cdot}55 + 16{\cdot}96 = 45{\cdot}51 \text{ MJ/kg.}$$

The latent heat of steam at 25°C is 583 cal/g. The mass of steam formed by the combustion of 0·13 kg of hydrogen $= 0{\cdot}13 \times 36/4 = 1{\cdot}17$ kg and the latent heat of this steam $= 1{\cdot}17 \times 2{\cdot}441 = 2{\cdot}85$.
The net calorific value of this fuel would thus be

$$45{\cdot}51 - 2{\cdot}85 = 42{\cdot}65 \text{ MJ/kg (18 240 BTU/lb).}$$

The mass of oxygen required to burn 1 kilogramme of this fuel can be found as follows:

$$C + O_2 \rightarrow CO_2$$

$$12 + 32 \rightarrow 44$$

$$0{\cdot}87 + 0{\cdot}87 \times \frac{32}{12} \rightarrow 0{\cdot}87 \times \frac{44}{12}$$

$$0{\cdot}87 + 2{\cdot}32 \rightarrow 3{\cdot}19 \qquad (4.1b)$$

$$2H_2 + O_2 \rightarrow 2H_2O$$

$$4 + 32 \rightarrow 36$$

$$0{\cdot}13 + 0{\cdot}13 \times \frac{32}{4} \rightarrow 0{\cdot}13 \times \frac{36}{4}$$

$$0{\cdot}13 + 0{\cdot}975 \rightarrow 1{\cdot}17 \qquad (4.2b)$$

Oxygen per kg of fuel $= 2{\cdot}32 + 0{\cdot}975 = 3{\cdot}295$ kg. (4.3)

This oxygen is provided as part of the air. Air is a mixture of oxygen and nitrogen with a small proportion of carbon dioxide and rare gases which are usually considered as part of the nitrogen content. The proportions by volume are 20·9% oxygen and 79·1% nitrogen and by weight 23·2% oxygen and 76·8% nitrogen.

The air required per kg of fuel is therefore

$$\frac{3{\cdot}295}{0{\cdot}232} = 14{\cdot}15 \text{ kg.}$$

4.3. *Air/fuel ratio*

The exact proportion of air to fuel for complete combustion is known as the theoretical or Stoichiometrical mixture and is expressed by weight for liquid fuels. For most diesel fuels it lies between 14 and 14·5 to 1. Mixtures with less air than the theoretical are known as "rich" and those with excess air are known as "lean" or "weak".

The diesel combustion process requires considerable excess air, air/fuel ratios of 30 to 1 and more are not uncommon. The carbon in the fuel burns completely to form carbon dioxide, CO_2. There is no partial combustion of carbon to form carbon monoxide, CO, as in the gasoline engine. Analysis of the exhaust gases may be carried out, the results being based on volumes at room temperature as is usual with gas analysis. The H_2O in the exhaust gas will have condensed so that its volume is negligible and the constituents will be CO_2, O_2 and N_2, the first two being measured by absorption and the last one usually being obtained by difference.

The proportions of these gases that are to be expected may be derived from a consideration of the chemical equations together with Avagadro's hypothesis which states that equal volumes of all gases at the same temperature and pressure contain the same number of molecules. Thus

$$C + O_2 \rightarrow CO_2$$

$$1 \text{ volume} + 1 \text{ volume} \rightarrow 1 \text{ volume} \tag{4.1c}$$

$$2H_2 + O_2 \rightarrow 2H_2O$$

$$2 \text{ volumes} + 1 \text{ volume} \rightarrow 2 \text{ volumes (steam)} \tag{4.2c}$$

Taking as an example the fuel already used, of one hundred volumes of air used in combustion of a stoichiometric mixture 20·9 volumes are oxygen. From equation (4.3)

$$\frac{2{\cdot}32}{3{\cdot}295} \times 20{\cdot}9 = 14{\cdot}8$$

volumes of this oxygen combine with the carbon in the fuel.

Of the 100 volumes of air 79·1 volumes are nitrogen. As the theoretically correct mixture strength has been assumed there will be no free oxygen and as already explained the H_2O will have condensed to a negligible volume of water, therefore the exhaust gas will occupy

$$14{\cdot}8 + 79{\cdot}1 = 93{\cdot}9 \text{ volumes}$$

The CO_2 is therefore $14{\cdot}8/93{\cdot}9 = 15{\cdot}7\%$ of the exhaust gas and the N_2 is $79{\cdot}1/93{\cdot}9 = 84{\cdot}3\%$ of the exhaust gas.

Now consider $x\%$ excess air to be supplied, that is the air to fuel ratio

$$= \left(1 + \frac{x}{100}\right) \times \text{theoretical.}$$

The air supplied $= (100 + x)$ volumes and in the exhaust gas there will be 14·8 volumes of CO_2, 79·1 volumes of N_2 associated with the burnt oxygen, $(0{\cdot}791\ x)$ volumes of N_2 associated with the excess air and $(0{\cdot}209\ x)$ volumes of O_2. The exhaust gas will occupy a total of $(93{\cdot}9 + x)$ volumes and the fraction of CO_2 will be given by

$$y = \frac{14{\cdot}8}{93{\cdot}9 + x}$$

thus if y, the percentage of CO_2 in the exhaust gas is obtained by analysis the excess air for this particular fuel may be calculated from the following equation

$$x = \frac{14{\cdot}8}{y} - 93{\cdot}9$$

Table 4.1 shows the constituent proportions and the theoretical air required for three grades of fuel.

TABLE 4.1

		Light Diesel Oil	*Marine Diesel Oil*	*Heavy Fuel Oil*
Specific gravity 15·5°C		0·870	0·940	0·960
Analysis:	% C	87·1	86·1	86·1
	% H	12·7	12·2	11·9
	% S	0·2	1·4	1·7
	% Ash	—	—	0·3
Viscosity:	Centistokes at 50°C ...	2·8	15	200
	Seconds Red. No. 1 at 100°F	32	90	1800
Theoretical air/fuel ratio ...		14·41	14·16	14·1
Calorific value	gross MJ/kg	45·22	43·96	43·54
	net MJ/kg	42·32	41·45	41·03

4.4. *Range of inflammability*

For every hydrocarbon and for each fuel, being a mixture of hydrocarbons, there are limiting air fuel ratios in both rich and lean regions, beyond which combustion is not self propagating. The range between these limits is known as the range of inflammability. The limits vary very widely for different hydrocarbons and are affected by temperature and pressure.

When fuel is injected into a diesel engine cylinder during the combustion process it mingles with the air to form local mixtures that vary in strength, temperature and pressure throughout the combustion chamber space and throughout the combustion process time. An exact knowledge of the limits of inflammability is of little practical significance but their existence helps to account for the completeness or otherwise of combustion under various circumstances.

4.5. *Exhaust smoke*

Local regions of over rich mixture result in incomplete combustion giving rise to the presence of carbon particles in the exhaust. These particles colour the exhaust grey or brown and in extreme cases dense black smoke can be emitted. Such conditions are the result of incorrect distribution of the fuel resulting from maladjustment of fuel injection equipment or shortage of charge air as a consequence of badly functioning turbochargers or inlet or exhaust arrangements. Any attempt to overload a diesel engine by increasing fuel will result in shading of the exhaust.

White smoke is occasionally produced, sometimes quite densely. This is usually the result of faulty fuel injection which permits fuel to impinge on the comparatively cool surfaces in the combustion chamber or even in the exhaust ducts, from which it is vaporized but not burnt. The subsequent condensation into droplets as it leaves the exhaust stack gives the white appearance. Blue colouring in the exhaust is generally taken as a sign that excessive quantities of lubricating oil are entering the cylinder, most probably as a result of worn piston rings and liners or worn valve stems and guides. Lubricating oil entering the combustion chamber can also contribute to white and dark smoke.

4.6. *The diesel combustion process*

The combustion process in the diesel engine cylinder may be regarded as taking place in three phases. Phase 1 is a delay period from the start of fuel injection until the cylinder pressure commences to rise above the compression pressure curve. The first minute droplets entering the combustion chamber are heated by the air and vapour forms from their surfaces. This vapour mixes with the air to form an ignitable mixture and combustion begins. During this time other fuel droplets are entering and being distributed throughout the combustion space and are similarly being prepared for ignition by vaporization and mixing with oxygen.

Once combustion is initiated it spreads rapidly through the mixture of vaporized fuel and air resulting in a steep rate of pressure rise; this part is Phase 2.

After the flame has spread throughout the chamber the pressure and temperature are so high that combustion is accelerated to the extent that fuel burns as it enters at the nozzle. The rate of pressure rise is much lower and dependent on the rate at which fuel is injected. Phase 3 consists of this controlled burning period.

Figure 1 shows a diagrammatic pθ indicator diagram, illustrating the

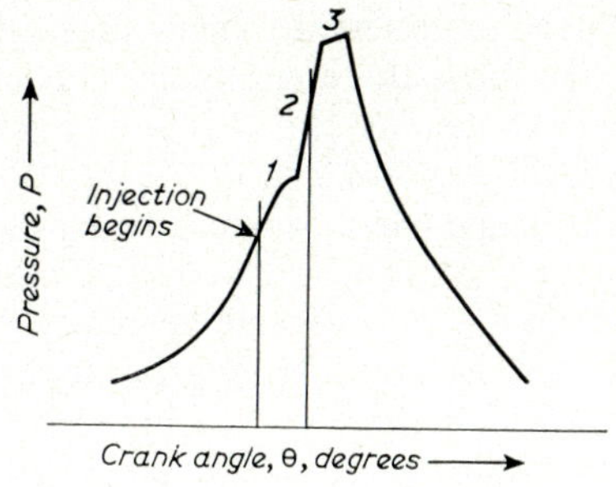

FIG. 4–1—*Three phases of combustion.*

three phases as quite distinct. This diagram may be compared with an actual diagram such as that shown in Figure 2, in which the phases merge gradually as with most natural processes.

4.7. *Ignition delay*

A combustible mixture ignites spontaneously when a certain limiting temperature is exceeded. This temperature is known as the ignition temperature and it varies with mixture strength and pressure level. The delay in igniting fuel in the diesel cylinder is due partly to the time required to form the mixture and raise it to the ignition temperature, and partly by the chain reactions which various groups of hydrocarbons follow in setting off on the processes of combustion.

The longer the delay period the more fuel enters the cylinder during it with consequent steep pressure rise as burning rapidly spreads with uncontrolled violence through the combustion chamber during Phase 2. The rapid increase in pressure makes itself apparent as the noise known as "diesel knock". Also isolated pockets of lean mixture strength may be formed in the combustion chamber which reach ignition conditions during the rapid rise of pressure and temperature and may detonate. This adds to the noise but is not necessarily harmful.

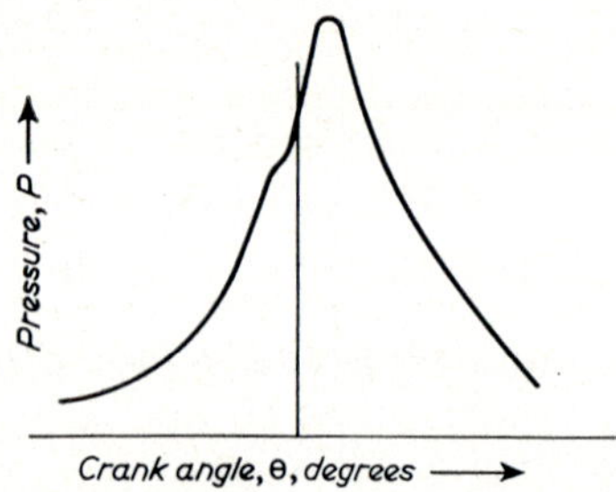

FIG. 4–2—*Combustion from time base indicator diagram.*

From the point of view of the delay period it is desirable to use as high compression ratio as the cycle considerations of Chapter 2 will permit. The higher temperature of the compressed charge which results from an increased compression ratio will initiate combustion more rapidly.

Fuels vary in their readiness to ignite according to the groups of hydrocarbons of which they are composed. The measure by which this property is classed is known as the "Cetane number". The fuel being considered is compared in performance with a mixture of cetane and a pure fuel, the percentage of cetane in the mixture being the cetane number rating given to the fuel under trial. The method is similar in some respects to octane rating for gasolines but it should be realized that gasoline engines require fuels of high volatility and high ignition temperature, whereas diesel engines use fuels of low volatility but require low ignition temperatures. Furthermore, the use of fuels of reduced octane values in a gasoline engine will affect its efficiency and its power output, but whilst the use of a fuel of reduced cetane number in a diesel engine may result in rougher running it will not affect the efficiency or the power output appreciably over a very wide range.

The injection of fuel as a well dispersed spray of fine droplets promotes the reduction of a delay period as it presents a large surface area of fuel in relation to its quantity. This permits its temperature to be raised quickly as well as leading to thorough mixing of fuel and air. In order to make maximum use of the air in the combustion chamber the spray must also have adequate penetration. There are problems in reconciling these somewhat conflicting requirements and practical difficulties in producing injection equipment to give the desired results.

Air movement plays a large part in the third phase of combustion but also assists in the mixing of the air during the delay period.

Designers and development engineers working in the diesel combustion field mostly find their efforts have to be directed towards shortening the delay period. However, it should be realized that it is possible to have too short a delay period as a result of which, particularly if coupled with low penetration, the second phase of combustion will be so localized that the third phase cannot be made to utilize all the air available, with consequent very low output albeit with a smooth running engine.

4.8. *Air movement*

The rate of pressure rise during the second phase of combustion is influenced chiefly by the length of the delay period, nevertheless air movement can help in avoiding extreme rates of pressure rise. Without air movement stagnant pockets of over rich or over lean mixtures may form and detonate as pressure and temperature rise. Such stagnant pockets can be avoided by using air movement to ensure thorough and even mixing of fuel and air.

Air movement is of greatest importance in the third phase of combustion for here the conditions of pressure and temperature in the cylinder are such that the fuel will burn as it issues from the nozzle provided that oxygen is available. Air movement at least ensures that combustion products are swept away and replaced by fresh air; additionally it disperses partially burnt fuel and combustion products throughout the cylinder to find the oxygen necessary to complete their combustion. This air movement may be relatively orderly swirl or it may be highly turbulent. It can be generated during the induction process or during compression and during combustion itself.

The motion set up when movement is formed during the induction process is generally in the form of smooth rotary swirl about the cylinder axis. The inlet valves may be provided with shrouds to cause the incoming air to take a tangential path into the cylinder, or a similar effect may be obtained by the location and disposition of the inlet valves and the shape of the induction passages through the cylinder head. In uniflow two stroke engines the inlet ports are shaped to give tangential flow to the entering air; loop scavenge two stroke engines are usually arranged with air and exhaust ports symmetrical about a vertical plane in order to preserve the all important scavenge path, even at the sacrifice of air swirl.

The incoming air follows a helical path into the cylinder and during the ensuing compression this becomes of diminishing pitch as the piston rises and the motion tends to form a vortex in the combustion chamber at the end of the compression stroke. It is usual to intensify this effect by designing the combustion chamber as shown in Figure 4–3. The rim of the piston closely approaches the cylinder head confining the swirling air to a chamber of smaller dimensions than the cylinder bore, causing the speed of rotation to rise correspondingly. This effect is sometimes termed "squish".

Air movement generated during compression and combustion is

associated with pre-combustion chambers. As the air is compressed from the cylinder into the combustion chamber it passes through a throat or series of holes disposed to give either a swirl motion or turbulence in the small pre-combustion chamber. The swirl rotation so obtained is greater than that resulting from inlet induced swirl. As the fuel is injected the piston commences to descend, the burning fuel and air mixture rushes out through the throat into the increasing volume in the cylinder and becomes thoroughly mixed in its turbulent passage. The fuel is thus enabled to find the oxygen required for complete combustion.

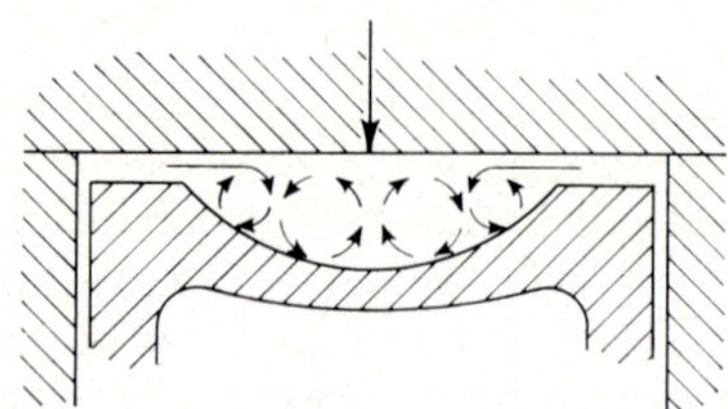

FIG. 4–3—*Combustion chamber with squish.*

Frequently the throat to the pre-combustion chamber is formed in a member made from heat resisting steel and mounted in a manner that insulates it from the cooling spaces. This member reaches a high temperature when in operation which may be 800°C or more at full load but is certainly always above 300°C even at low loads. It acts as a regenerator storing some heat from the combustion process and surrendering it to the charge during compression. This ensures an adequately high temperature for ignition and complete combustion under all running conditions. Because of the rapid air movement in pre-combustion chambers the fuel injection need be at only a low pressure relatively and a coarse spray from either a single hole or a pintle. The spray may be directed at the hot member and intentionally impinge upon it to cause the fuel to vaporize and ignite readily. By these methods the delay period may be controlled to take only a short time. The Ricardo Comet designs of combustion chamber are of this sort and an example is illustrated in Figure 4–4.

Some pre-combustion chambers have a series of smaller holes in place

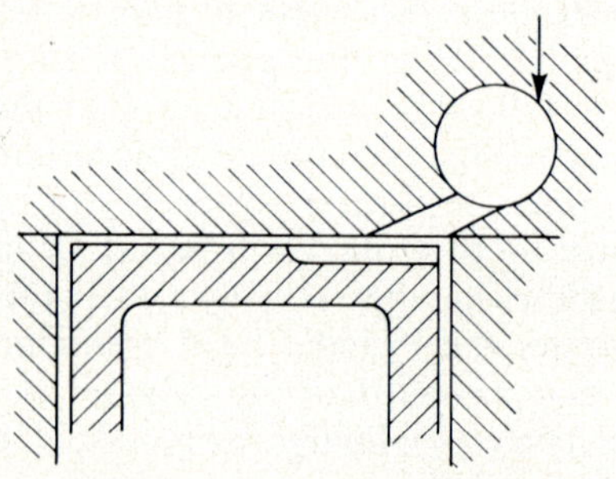

FIG. 4–4—*Ricardo "Comet" combustion chamber.*

of one throat, distributing the burning fuel and air in all directions into the cylinder. Variations on the design retain some air in a separate compression space outside the pre-combustion chamber where it awaits the rush of partially burnt products and so completes their combustion in the cylinder itself.

4.9. *Energy losses during combustion*

By whatever means the swirl or air movement is generated it will cause a certain loss of energy. The inlet induced swirl is the lowest in speed and demands only a small pressure drop across the inlet valve passages. This pressure loss can become considerable at increased speeds and shrouded valves are generally avoided in the high speed engines of today. The motion of the air also raises the heat transfer coefficients increasing energy losses to the coolants but again inlet induced swirl is the least objectionable from this aspect.

Movement set up during compression and combustion involves pressure losses during the passage into and out of the chamber through the throat or "pepper pot" holes. These losses are usually dissipated as heat to the coolant. The rather violent motion of the air results in high heat transfer coefficients and the construction of a chamber in the cylinder head often results in a fairly large surface area by which heat is transmitted directly to the coolant.

Pre-combustion chamber hot member engines are characterized by their smooth running, complete combustion with clean exhaust and a tendency to high fuel consumption, together with difficulties in starting in very cold conditions.

The drive for improved thermal efficiency has led to attention being focused increasingly on direct injection combustion chambers. These have tended to develop from the simple bowl shape of Figure 4–5 to the toroidal

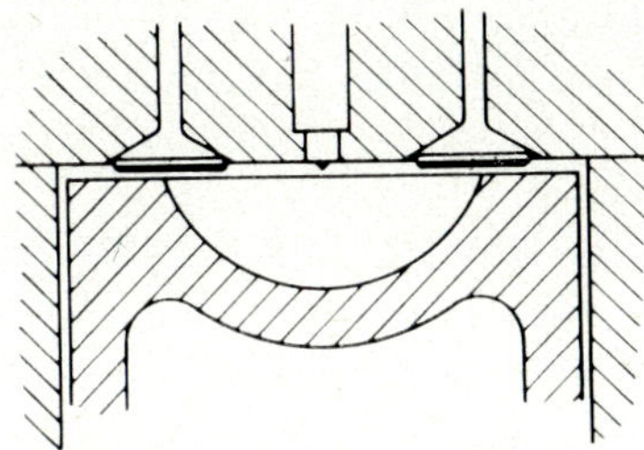

FIG. 4–5—*Bowl shaped combustion chamber.*

form shown in Figure 4–6. The mechanical considerations of fuel injection favour the production of fuel sprays of identical form, shape, fineness and penetration from the identical holes in the nozzle. The toroidal shape confines the swirling air to regions where such identical jets can reach it easily, whereas the simple bowl results in a central pocket of low air movement and requires a rather different spray form to fill it with droplets of fuel.

Because of the losses resulting from swirl the highest thermal efficiencies are obtained in larger cylinders and lower speed engines where quiescent conditions can still provide adequate combustion. There is obviously an incentive to extend quiescent combustion to high speed engines but in such applications it is far more demanding on the fuel injection equipment. Finely dispersed droplets from many holes yet with sufficient penetration are required to yield a well dispersed mixture of fuel and air throughout

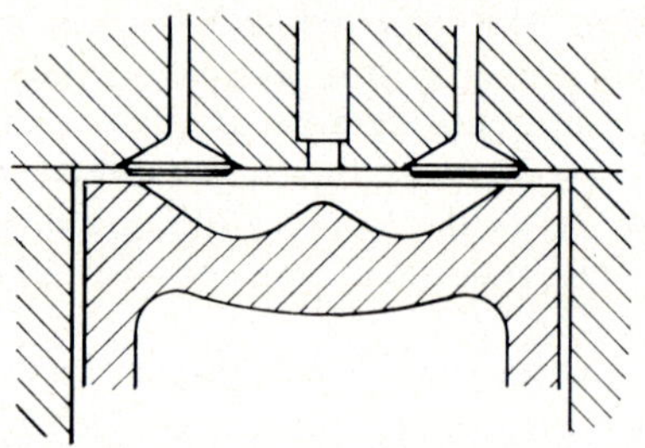

FIG. 4–6—*Toroidal combustion chamber.*

the compression space and this has to be achieved in a very short time. Modern developments of fuel injection equipment are directed towards making this possible, but at the time of writing the majority of engines both medium and particularly high speed use air swirl to achieve complete combustion. The higher the engine speed the greater the need for swirl to mix fuel and air. If the process is to be performed by fuel injection alone the pressure in the fuel lines and hence the mechanical difficulties become very great.

4.10. *Rate of fuel injection and rate of heat release*

Figure 4–7(a) shows a diagram of rate of fuel injection with crank

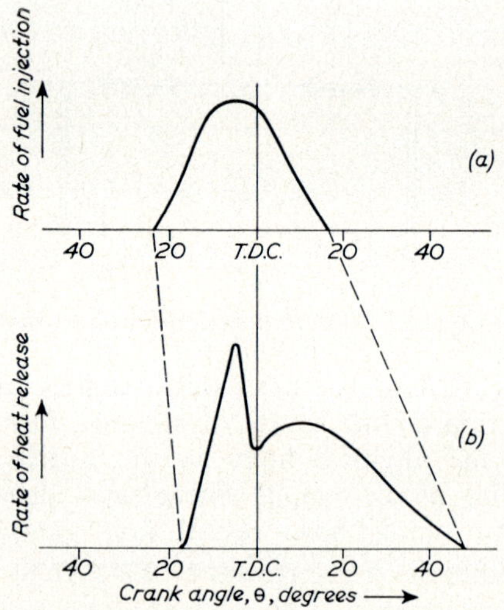

FIG. 4–7—*Rate of fuel injection and rate of heat release.*

angle that is typical of most fuel injection systems. The rate increases rapidly at first then tends to flatten out and finally falls to zero a little less quickly than it rose. The rate of heat release depends on how this fuel burns and Figure 4–7(b) indicates the heat release in relation to the rate of fuel injection as shown at (a). The delay period is responsible for the lag between the start of fuel injection and the start of heat release. When combustion starts the fuel which has been distributed throughout the cylinder burns rapidly causing a rapid rise in heat release and a sharp peak as it all becomes consumed. The heat release rate then falls to a lower value consistent with the rate of fuel injection and then rises again as this rate also increases until a second but lower peak is reached. Notice that when fuel injection has ceased unburnt or partially burnt fuel is still in the combustion chamber and this continues to find oxygen for combustion so that heat release goes on for a longer period than injection and dies away at a much lower rate.

4.11. *Combustion temperature, dissociation and variable specific heats*

It may be supposed that if the heat supplied by the fuel is divided by the specific heat of the air, as is generally understood and which is the value at normal temperatures, the temperature reached during combustion will be obtained. Actually such a calculation yields far too high a temperature being almost double that which is attained in reality. The discrepancy is due to three factors:

(1) the specific heats of gases increase with increasing temperature,

(2) at high temperatures the molecular products of combustion, CO_2 and H_2O dissociate with absorption of heat into carbon monoxide and oxygen, and hydrogen and oxygen respectively. The molecules divide according to the formulae

$$2CO_2 \rightleftharpoons 2CO + O_2$$

$$2H_2O \rightleftharpoons 2H_2 + O_2$$

(3) heat is lost to the walls of the combustion chamber.

Dissociation is dependent upon temperature, pressure and the relative proportions of the different gases in the mixture. These properties together with those of variable specific heats have been fully documented by Keenan and Kay (Ref. 1) and in combination with the work of Eichelberg, Annand and Lynn (Refs. 2, 3 and 4) may be used for the purposes of cycle analysis and the prediction of temperatures during combustion as outlined in Chapter 2. The actual temperatures reached may be deduced from the pressures obtained from an indicator diagram and Figure 4–8 shows the pattern of heat release, pressure and temperature attained in a typical high output cycle.

4.12. *Fuel injection*

Fuel injection systems are designed to raise the pressure of the fuel to a sufficiently high value to ensure adequate dispersion into the combustion chamber, to admit it to the chamber at the instant it is required and then to cut off the supply sharply according to the output required from the engine.

There are two basic systems in use, the common rail in which fuel is maintained at high pressure and admitted to the cylinders by timed valves and the jerk pump system in which the fuel is raised to the injection pressure only during the injection period. Medium and high speed engines almost without exception use the jerk pump system and a description of the components forming this type of system follows.

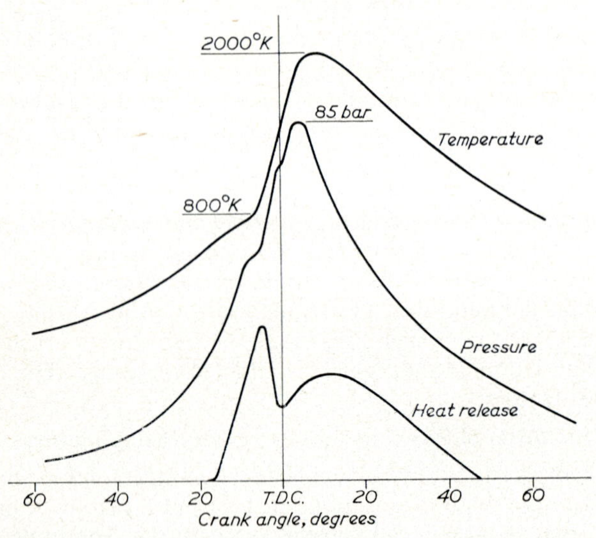

FIG. 4–8—*Relation between heat release, temperature and pressure during combustion.*

4.13. *Fuel pump*

The operation of a fuel pump is shown in Figure 4–9. It is actuated by a cam and roller tappet follower. When the follower is on the base circle of the cam the pump plunger is at the bottom of its stroke and the inlet port is open allowing fuel in the gallery to flow into and fill that portion of the barrel above the plunger. The plunger is a close fit in the barrel. As the cam rotates the plunger rises and seals off the inlet port and at this point of the stroke the pumping action starts. By means of adjustments provided in the tappet follower this event can be timed in a precise relationship to the crank angle. Further upward movement of the plunger causes the fuel to be raised in pressure and expelled through the delivery valve to the injector.

A helical groove extends from the top of the plunger part way down its cylindrical surface. Angular rotation of the plunger about its axis so positions this groove in relation to the inlet port to give a greater or lesser stroke between the sealing of the port by the top of the plunger and its communication with the helical groove. In some designs a separate spill port is provided but as this also is made to communicate with the inlet gallery the action is precisely the same. When the port is opened to the groove the high pressure in the fuel above the plunger is released to the low pressure in the gallery and pumping ceases although the plunger continues to move upwards. The amount of fuel delivered will vary in accordance with the effective length of stroke as determined by the angular position of the plunger. This angular position is controlled by a rack and pinion, the latter forming a sleeve having an axial slot or slots to receive a projection on the plunger. The rack position, therefore, determines the quantity of fuel supplied. It will be seen that the period over which fuel is injected bears a definite relationship to the quantity.

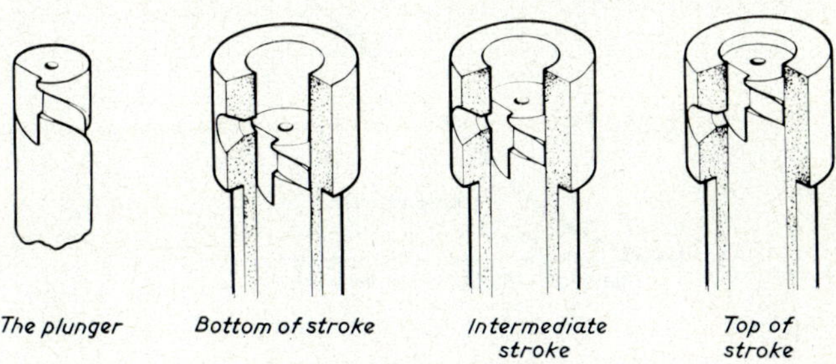

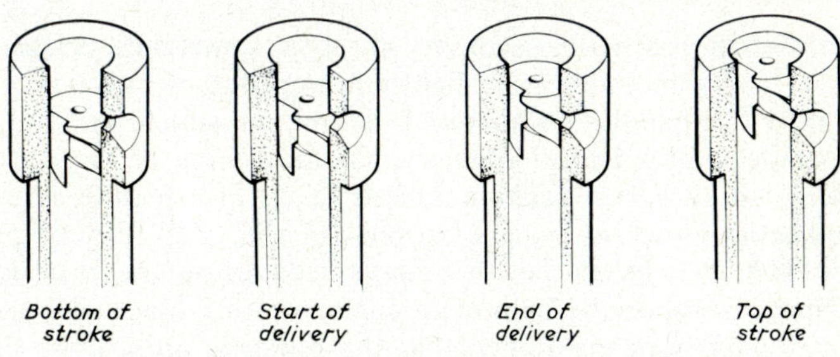

FIG. 4–9—*Fuel pump action*

The operating cam is usually designed to give displacement, velocity and acceleration characteristics of the form displayed in Figure 4–10. The commencement of the stroke must be a period of smooth acceleration to avoid heavy mechanical loads on the operating gear. By the time the inlet port is closed the phase of uniform velocity has started and the whole of the injection at the highest load on the engine is arranged to occur during this portion of the lift. After the constant velocity period deceleration takes place to bring the plunger to rest at the top of its stroke, the design of the cam and the plunger spring being arranged to maintain contact between the follower and the cam. If the engine is direct reversing the fuel cam is made symmetrical, the two flanks being mirror images of one another. Unidirectional engines may have the trailing flank designed for minimum operating loads. As the plunger returns the delivery valve reseats and fuel is drawn in through the spill port and helical groove until this is sealed, the filling process being completed through the inlet port at the bottom of the stroke ready for the next injection.

These characteristics are the most usual form of fuel pump plunger motion but variations are sometimes found; for instance, the injection period need not be one of constant velocity attempting a constant rate of

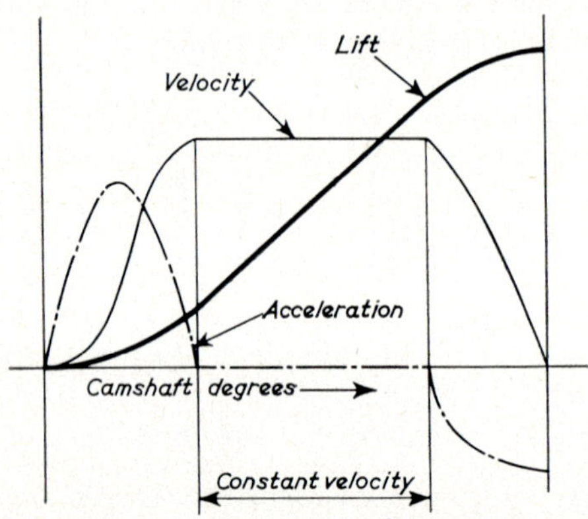

FIG. 4–10—*Typical fuel cam characteristics*

injection, although it usually is made so. It is sometimes designed as a slightly rising and sometimes as a slightly falling rate of injection. Another variation more frequently encountered is one in which the top of the plunger is made with a helical edge so that the timing of the start of the injection may be varied in relation to the load. This feature is useful in certain applications such as marine propulsion with a fixed pitch propeller where the load and speed vary together. Advancing the injection with increasing speed is especially helpful to engines with a long delay period.

It has already been mentioned that the plungers are a close fit in the barrels to ensure sealing at the high pressures but some leakage of fuel between the plunger and barrel is inevitable and, in fact, is necessary to provide lubrication for the relative movement. Pumps for high speed engines are usually grouped into a single unit as in Figure 4–11 and are complete with camshaft, cams and followers within the one casing. The lubrication of all these parts is carried out by the fuel oil. Larger medium speed engines usually have separate fuel pumps one to each cylinder operated by cams mounted on the same camshaft that carries the inlet and exhaust cams. This camshaft, the cams on it and their followers are lubricated by lubricating oil as part of the engine lubricating oil system. It is essential to avoid a fuel leakage from plunger lubrication finding its way into the camshaft section as dilution of the engine lubricating oil would result. Figure 4–12. shows a typical flange mounted fuel pump and Figure 4–13 an arrangement whereby the fuel is prevented from mixing with the lubricating oil and is drained away separately. Flange mounted pumps are

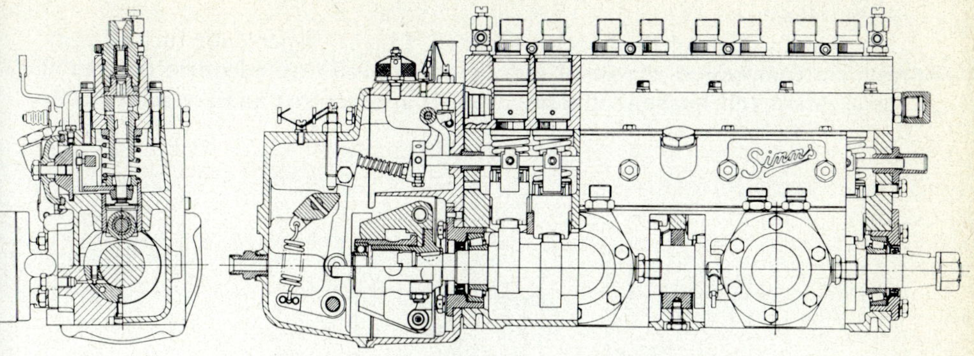

FIG. 4–11—*Monobloc type fuel pump.*

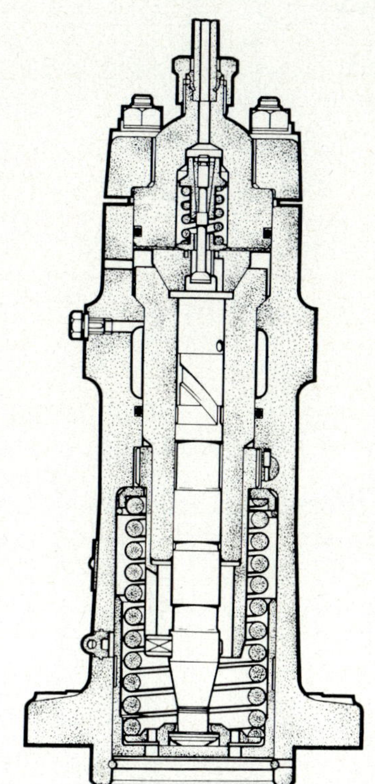

FIG. 4–12—*Flange mounted pump.*

sometimes made with roller and tappet as an integral part of the pump as in Figure 4–14. Here fuel contamination of the lubricating oil is prevented by forced lubrication of the plunger, lubricating oil being introduced to a groove in the bore of the barrel level with the control rack. Another groove a little higher up the bore collects any leakage of fuel going down the plunger together with the small leakage of lubricating oil up the plunger and conveys it to the low pressure fuel inlet gallery.

4.14. *Injector*

The function of the injector Figure 4–15 is to disperse the fuel throughout the compressed charge of air. To accomplish this, adequate degrees of penetration and fineness of spray droplets are required and are achieved by

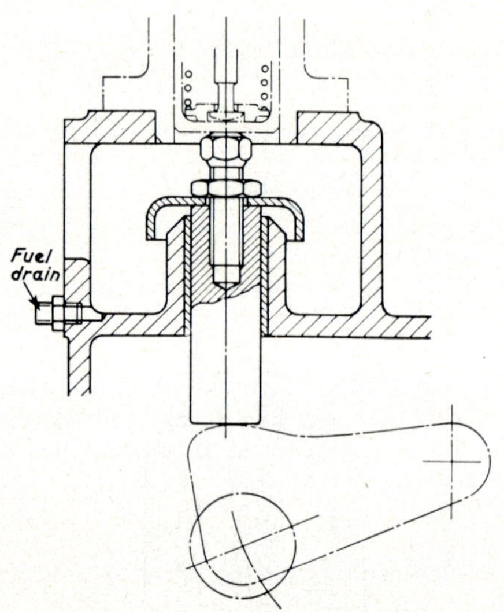

FIG. 4–13—*Fuel pump tappet.*

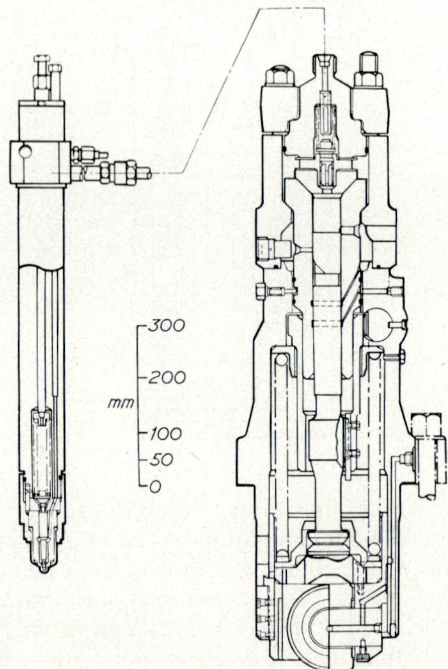

FIG. 4–14—*Flange mounted pump with integral roller and tappet*

passing the fuel at high velocity through small bore holes in the injector nozzle. The high pressure of the fuel necessary to do this must be created sharply at the commencement of injection and must be just as sharply dropped when injection ceases in order to avoid deterioration of the spray formation into a dribble or jets. The usual way of meeting these requirements is by a spring loaded differential valve the operation of which is

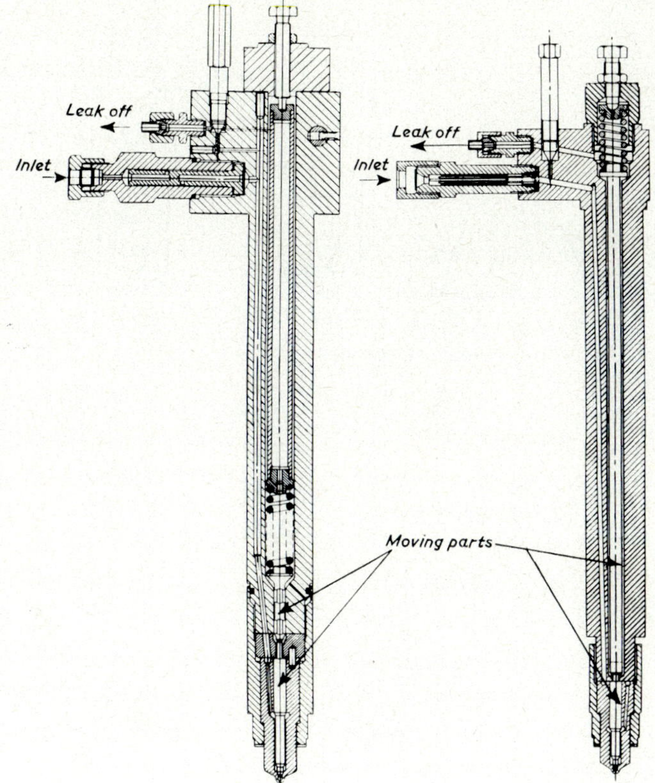

FIG. 4–15—*Fuel injectors.*

shown in principle in Figure 4–16(a) and will be understood from a study of Figure 4–16(b). The pressure generated by the fuel pump acts upon the annular part of the needle valve, A, and when the upward force on the needle due to this pressure exceeds the force, F, exerted by the compressed spring the needle lifts off its seat. The fuel pressure is then able to act over the whole cross sectional area of the needle which immediately snaps to the fully open position. The lift of the needle is quite small, usually a fraction of a millimetre. When the pressure in the fuel pump is released by the spill action the needle valve is immediately re-seated by the spring. For many years it was customary to locate the spring at the end of the injector remote from the nozzle, when there was more space to accommodate it. This design, shown on the right of Figure 4–15, incorporated a long spindle of relatively high inertia. Modern developments have permitted the design shown on the left of Figure 4–15 in which a compact spring is located inside the injector near to the nozzle enabling the spindle to be of

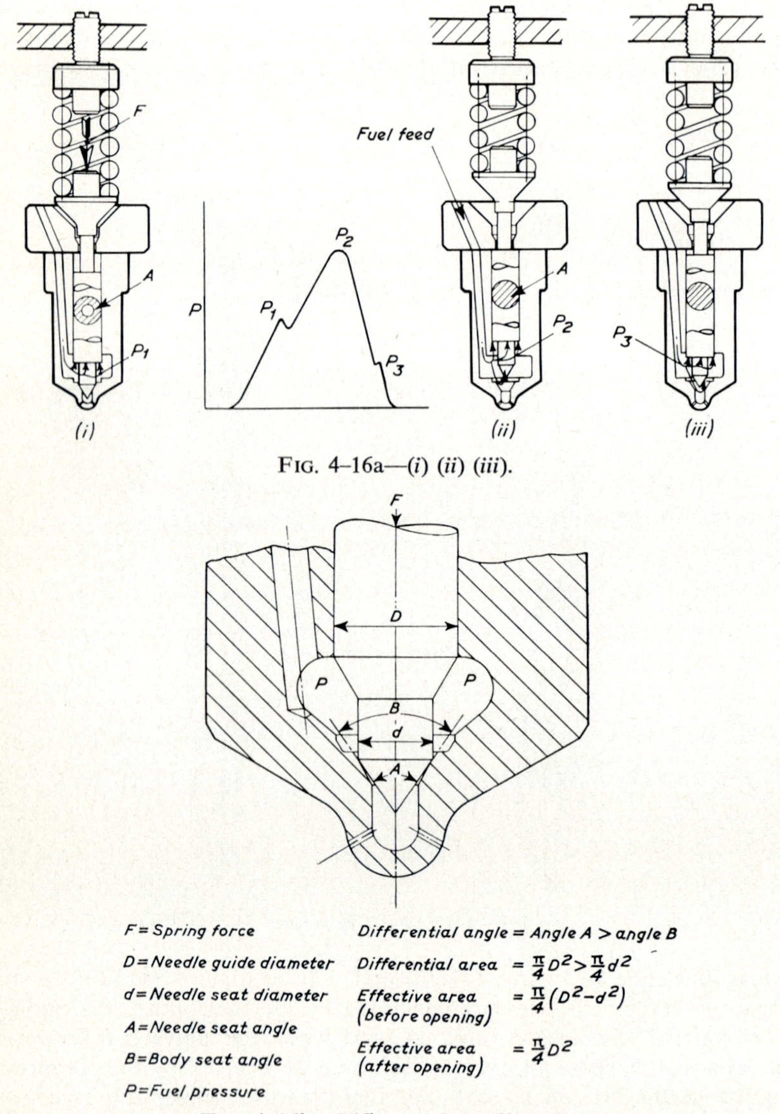

FIG. 4–16a—(*i*) (*ii*) (*iii*).

FIG. 4–16b—*Differential needle valve.*

extremely short length and low inertia. Thereby, the closing action of the needle can be accomplished rapidly yet with minimum impact loading on the seat.

The opening pressure P_1 at which injectors are set may be about one hundred bars for precombustion chamber designs and about two hundred bars for direct injection engines. The needle closing pressures P_3 are lower than the opening pressures according to the design of the differential portion of the valve. The actual opening pressure is determined by the engine manufacturer in collaboration with the fuel injection equipment maker and for the correct setting the manufacturer's instruction book should always be consulted. The maximum pressure P_2 reached during the injection process is far higher than the needle opening pressure.

The relationship between the fuel pressure in the nozzle and the gas pressure in the cylinder can also be seen in Figure 4–17. During the

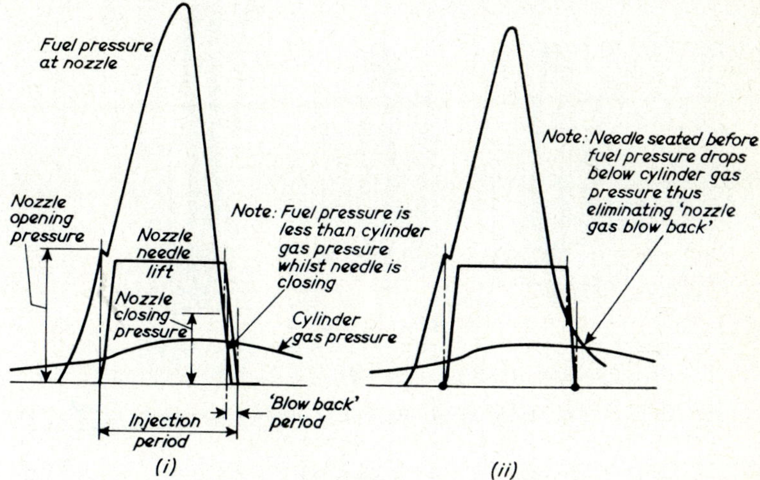

FIG. 4–17—*Fuel pressure cylinder gas pressure and needle lift diagrams.*

period that the needle is open the fuel pressure must exceed the gas pressure if blow back and consequent needle sticking are to be avoided. Although a clean end of injection is always essential too sharp a drop in fuel pressure may lead to the situation in Figure 4–17 case (i) in which the hot gas can flow back into the nozzle. Case (ii) is the result of a more satisfactory design of injection system the needle being completely seated before the fuel pressure falls below the cylinder gas pressure.

Nozzles for direct injection engines are usually of the multiple orifice type as shown in Figure 4–18, the disposition of the holes and their number being arranged to distribute the fuel into the combustion chamber. The dimensions of the sac below the needle valve seat and the diameter and length of the holes in the nozzle all play vital roles in achieving the correct spray formation. Typically the number of holes varies from about four to ten depending on the engine bore and the amount of air swirl. The holes can be drilled as small as 0·1 millimetre or as large as 1 millimetre.

In Figure 4.19 is shown a nozzle in which the end of the needle valve has a profiled extension that protrudes into the valve body orifice to form an annular spray hole. This is known as a pintle type nozzle and the shape of the pintle can be controlled to vary the annular orifice area as the valve lifts. Depending on the pintle profile the spray can be made to issue in the form of a hollow cone with an included angle as high as 60°. This type of nozzle is usually used with pre-combustion chamber engines.

For reliable service and satisfactory operation the nozzle tip must not be allowed to get too hot. Allowable temperatures range from 150°C to 250°C depending on the type of steel used for the nozzle. To control the temperature at this level the designer must pay attention to the circulation of the cooling water around the injector when designing the water passages in the cylinder head.

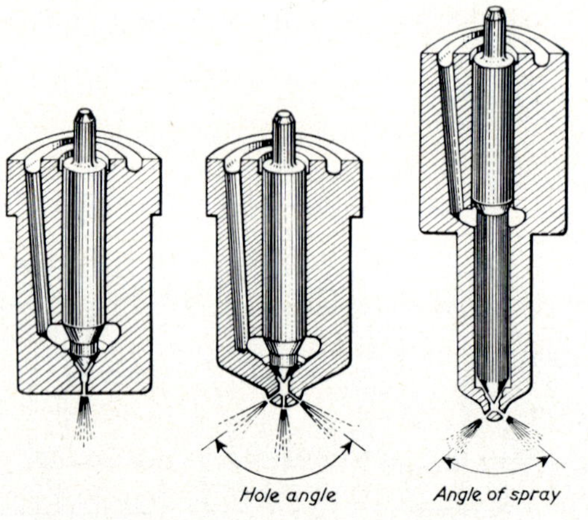

FIG. 4–18—*Multiple orifice injector nozzles.*

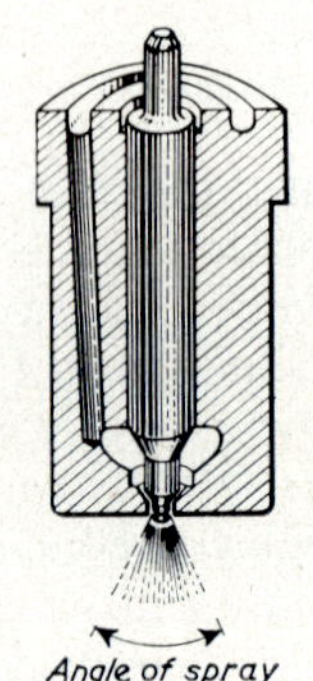

FIG. 4–19—*Pintle nozzle.*

For highly rated engines and particularly when heavy fuel is used it is important to keep the nozzle tip even lower in temperature. Some heavy fuels in conjunction with excessive nozzle temperatures lead to the formation of trumpets or cones round the orifices which interfere with combustion, Figure 4–20. Intensively cooled injectors incorporate internal cooling passages as can be seen in Figure 4–21, the coolant used is fuel oil or distilled water the latter being more effective. The coolant is led down one passage, circulated around the nozzle tip and returned via the other.

The cost of such injectors is considerably higher than that of uncooled types.

On the other hand over-cooling of nozzles by running with low cooling water temperatures at low loads can lead to trouble of a different kind: corrosion of the nozzle. The cause is that the temperature of the nozzle has fallen to a point at which the sulphur trioxide and moisture in the combustion gases can condense in the form of sulphuric acid resulting in highly corrosive conditions. Examples appear in Figures 4–22 and 4–23.

FIG. 4–20—*Formation of carbon trumpets.*

The nozzle needle valve is lapped into a close fit with its guide in order to secure its prompt operation by the fuel line pressure. Lubrication of the needle and guide is essential however and is provided by the slight leakage that takes place between the two surfaces. This leakage of fuel is piped away from the injector by a connexion at the top which can be seen in the various illustrations and is conducted usually to the inlet fuel line of the fuel pumps.

The satisfactory operation of the engine depends a great deal upon the condition of the injectors. They should be cleaned and reset at regular intervals determined by the operating conditions and the fuel used. It is essential for servicing to be carried out in scrupulously clean conditions, and for dismantling to be performed with the correct tools. A nozzle held by carbon deposits in a nozzle nut should be extracted using a copper or brass tubular drift as in Figure 4–24 and certainly not by striking the tip in the region of the precisely drilled holes. Correct sizes of probes and wires are necessary for cleaning out the nozzle holes and passages adequately yet without damage.

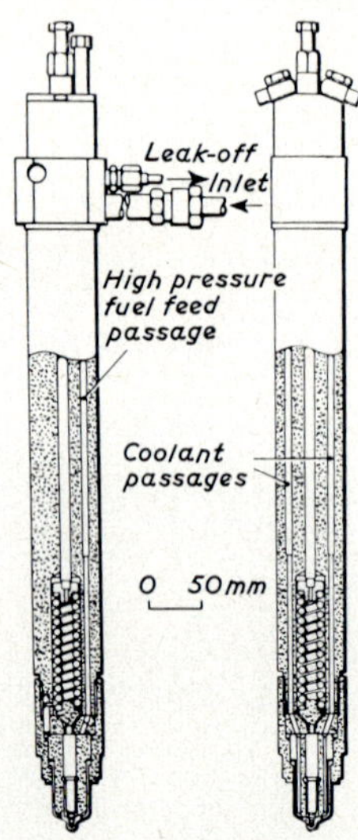

FIG. 4–21—*Integral cooled injector.*

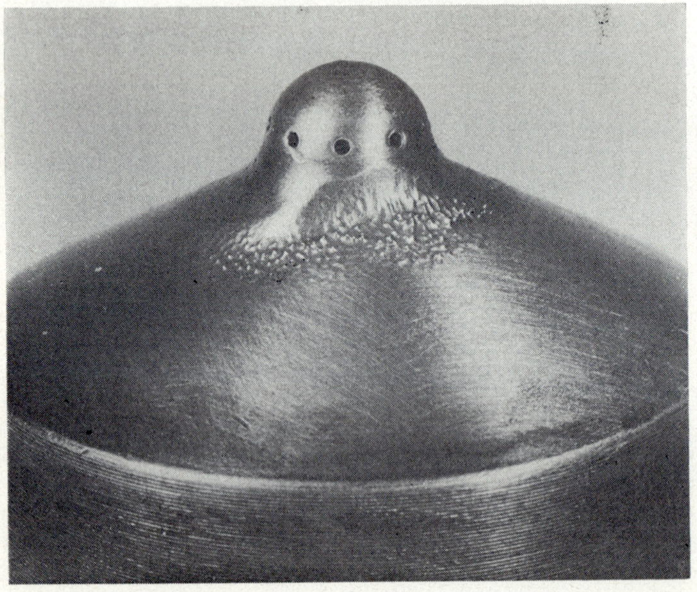

FIG. 4–22—*Cold corrosion of nozzle tip.*

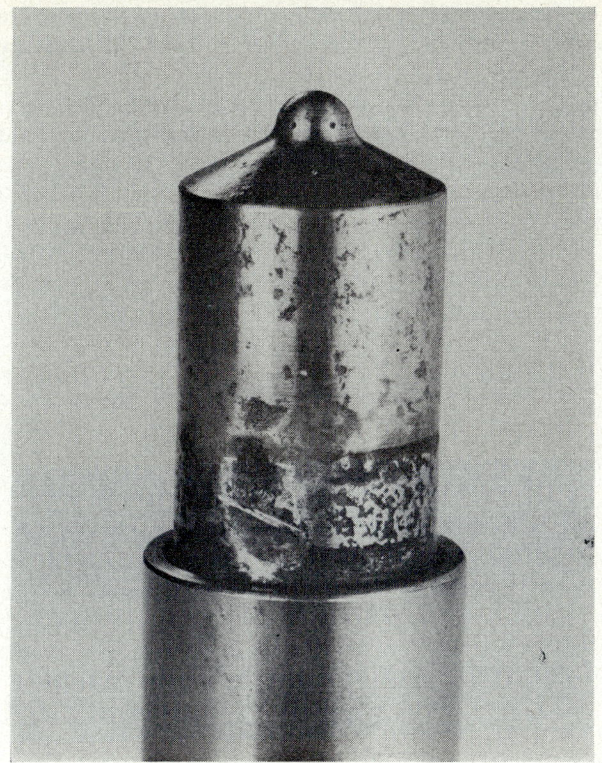

FIG. 4–23—*Cold corrosion of nozzle body.*

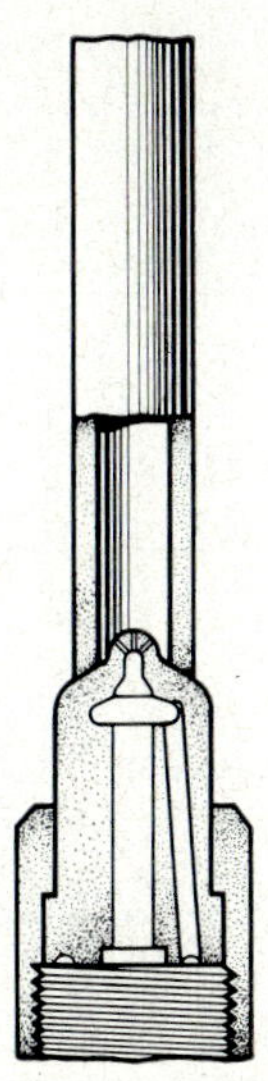

FIG. 4–24—*Extracting a nozzle from a nozzle nut.*

4.15. *Delivery valve*

When the helical edge of the pump plunger uncovers the spill port in the pump barrel at the end of the delivery stroke the pressure of fuel is immediately reduced and the delivery valve promptly drops on to its seating. If this valve were a simple non-return valve the fuel left in the high pressure pipe between the valve and the injector would gradually flow through the injector until its pressure fell to the reseating pressure. Such gradual falling pressure would almost certainly lead to dribbling at the injector nozzle. The delivery valve is therefore designed to perform a

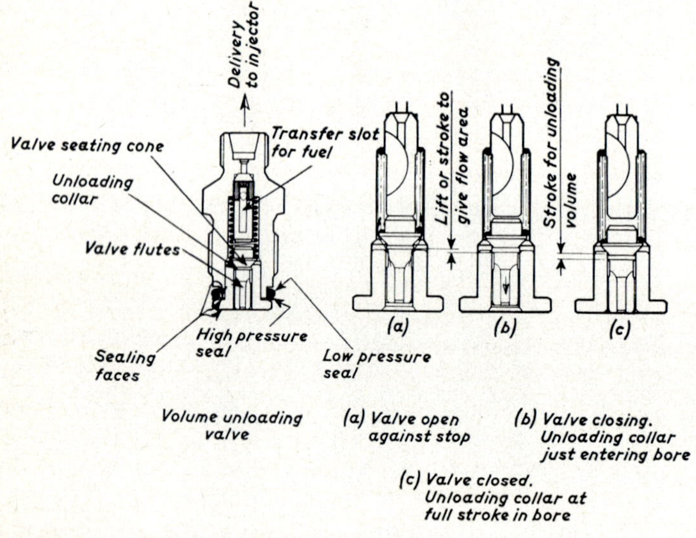

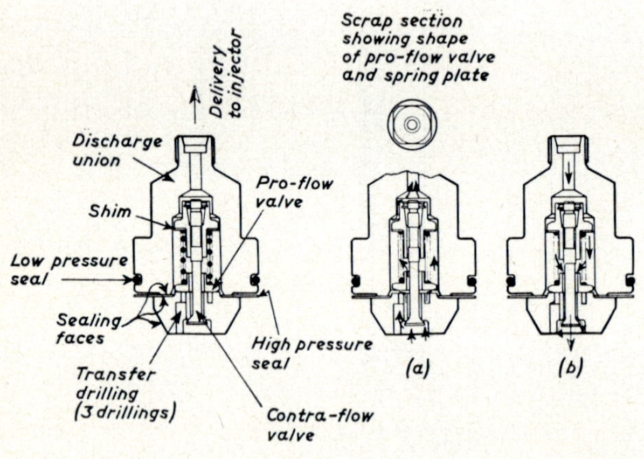

FIG. 4–25—*Action of delivery valve.*

function of unloading the pressure pipe as well as acting as a non-return valve. The usual form is illustarted in Figure 4–25 where it can be seen that the delivery valve is constructed as a mitre valve with a small piston below the conical seat and just above the wings. This piston is a close fit in the delivery valve guide. When the pump is delivering fuel the pressure of the fuel raises the delivery valve until the fuel can escape through the longitudinal flutes and over the face of the piston and the valve seat and along to the injector nozzle. When the pump plunger releases the pressure in the barrel the delivery valve snaps back on to its seat and in doing so the small piston part of the valve enters the guide and increases the volume above the delivery valve by an amount equal to the travel of this piston before the valve seats itself. This action has the effect of suddenly reducing the pressure in the pipe and so causes the needle valve in the nozzle to snap back on to its seat quickly without dribble.

In article 4.14 attention was drawn to the importance of the rate of decay of fuel pressure in the nozzle relative to the gas pressure in the cylinder as illustrated by Figure 4–17. The proportions of the delivery valve influence the characteristics of the fuel injection system in this respect. Furthermore, it will be appreciated from Figure 4–17 that the fuel pressure during the injection process is in the form of a pulse. The pulse is generated at the fuel pump and takes a finite time to travel along the high pressure pipe to the injector, which it does at the speed of sound. As with any pulse within a closed pipe, reflections of changes of pressure occur at the ends of the pipe according to whether these are closed or open. When the pressure falls in the fuel pump and the delivery valve closes and unloads the pipe the pulse in the high pressure pipe may continue to oscillate between the delivery valve and the injector. The magnitude of the first reflection may be sufficiently high for it to reopen the injector needle valve when it reaches it so causing secondary injection. Secondary injection will cause fuel to enter the cylinder when the piston is descending on the firing stroke and will result in late and poor combustion. In addition the injector needle and seat in their open condition will be subjected to high temperature from the cylinder and partial oxidation of the fuel inside the nozzle can take place resulting in choking and damage to the seats. Figure 4–26 shows a

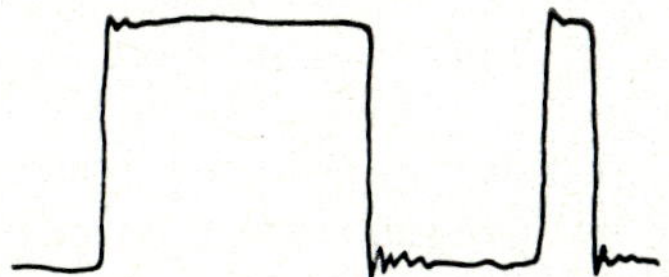

FIG. 4–26—*Needle lift record showing secondary injection.*

needle lift diagram with secondary opening. The unloading of the high pressure fuel line by the delivery valve must be arranged so that the residual line pressure is low enough for the reflections of this type to be at a level which will not cause secondary injections. On the other hand excessive unloading can lead to cavitation when the engine is running at low loads. Cavitation causes damage to all parts of the injection system but particularly to needle valve seats, pump plungers and barrels. Its nature, effects and measures taken to combat it are covered in detail by Clifton, (Ref. 5).

The problems of avoiding both secondary injection and cavitation whilst designing for a satisfactory rate of pressure decay at all loads and speeds become severe on highly rated engines. Modified forms of delivery valve are sometimes used in which the piston portion appears above the seat or both above and below the seat, designed to give unloading and also dashpot actions to control the rate of closing. Another approach has been to design the delivery valve to unload the pipe down to a predetermined pressure, instead of by a fixed volume, as described by Cockburn and Gröschel (Ref. 6). Clifton (Ref. 5) also describes this type of valve which is shown in the lower part of Figure 4–25 and deals with many facets of the detail design of modern fuel injection equipment.

4.16. *High pressure pipes*

The pipe connecting the pump and injector is subject to stress from high pulsating pressure and vibration. Its design is vital to the correct operation of the injection system; its bore, outside diameter and length are carefully established during the development stage of the engine and it is usually manufactured in special material to withstand the arduous conditions.

Various forms of end fitting are used to ensure leak proof connexions at the pump and injector; some of these are shown in Figure 4–27.

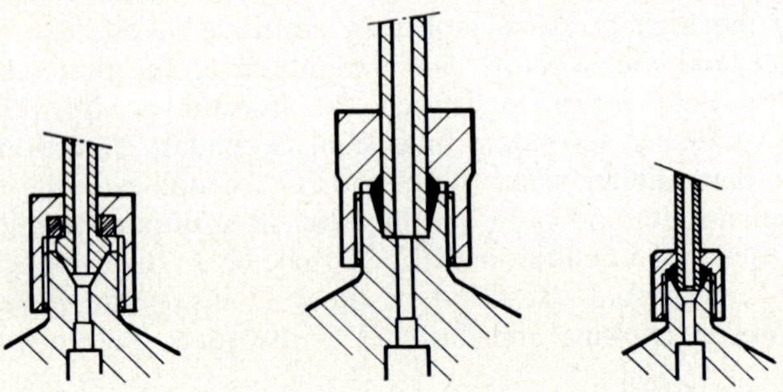

FIG. 4–27—*High pressure pipe connexions.*

As leakage of high pressure fuel can be dangerous, particularly in unmanned engine rooms, the high pressure pipe is sometimes enclosed in a sheath forming a double walled pipe. The passage between the inner and outer walls conveys any leakage to a position which is safe but where it will be noticed. To do this it is necessary to seal the outer pipe also at the connexions, but only against a low pressure, in a way which permits a leaking high pressure connexion, as well as a defective pipe to be detected. One solution is shown in Figure 4–28.

4.17. *Inspection*

Cleaning of the fuel itself is essential and equipment for carrying out this process is described under fuel systems in Chapter 10. Careful inspection of the fuel injection equipment during regular maintenance is often of great help in diagnosing trouble at an early stage before real damage is done. As well as the obvious harm resulting from abrasive dirt the presence of water in the fuel, especially sea water, can cause deterioration by corrosion. An early indication of water contamination is discoloration of the pump plungers and barrel bore, the surfaces becoming blackened in appearance as illustrated by Figure 4–29. If elements are observed in this condition, immediate remedial action should be taken to remove the source of water ingress.

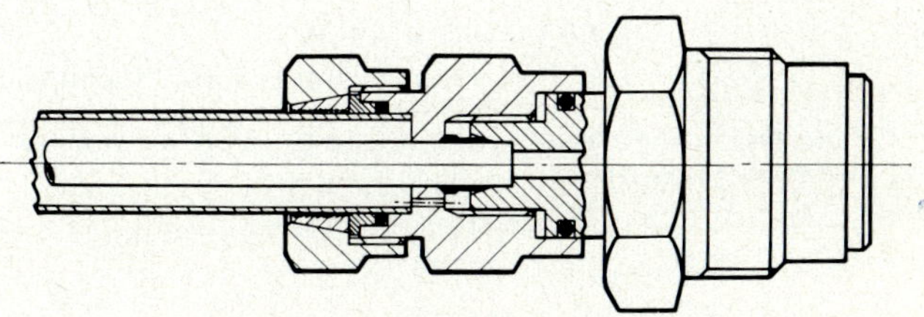

FIG. 4–28—*Connexion for double-walled pipe.*

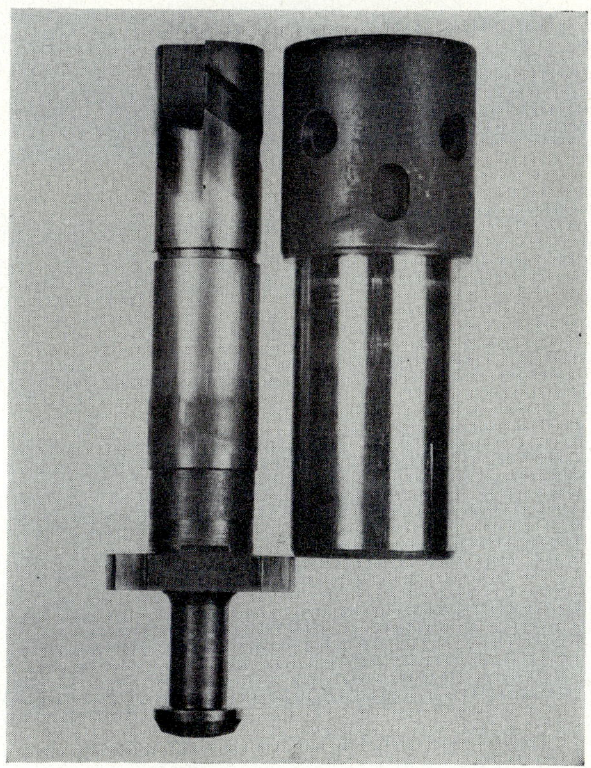

FIG. 4–29—*Water staining of pump element.*

REFERENCES

1. KEENAN and KAY, *Gas Tables,* J. Wiley and Sons, 1950.
2. EICHELBERG, G., *Some New Investigations on Old Combustion Engine Problems,* Engineering, London, 1939.
3. ANNAND, W. J. D., *Heat Transfer in the Cylinder of Reciprocating Internal Combustion Engines,* Proc. I. Mech. E., 1963, Vol. 177 No. 36.
4. LYN, W. T., *Relation Between Fuel Injection and Heat Release in a Direct Injection Engine and the Nature of the Combustion Process,* Inst. of Mech. Eng. Proc. of the Automobile Divn. No. 1, 1960/61.
5. CLIFTON, W. L., *Current Development in Fuel Injection Equipment,* DEUA Publication 315, July, 1967.
6. COCKBURN, K. W. M. and GRÖSCHEL, E. R., *The Design and Development of a Highly Rated Fuel Injection System,* CIMAC, 1968, paper A.15.

BIBLIOGRAPHY

GOODGER, E. M., *Petroleum and Performance,* Butterworth Scientific Publications, 1953.

BURMAN, P. G. and DELUCA, F., *Fuel Injection and Controls for Internal Combustion Engines,* Technical Press, 1962.

CHAPTER FIVE

Thermal Loads

5.1.

The high temperatures of the combustion process raise the temperature of the surfaces forming the combustion chamber. These surfaces would obviously become very hot indeed unless measures were taken to keep them cool. Similar conditions arise from the continuous passage of hot exhaust gas through the exhaust system and in particular at the turbine of turbocharged engines. From the point of view of thermal efficiency, cooling of these parts is undesirable as it represents a loss of energy, some of which could have been turned into useful work, but from a practical aspect it is very necessary. Inadequate cooling or excessive combustion temperatures may lead to trouble in a number of different ways. Components may deform to such an extent that they cease to function correctly, extreme examples leading to such difficulties as gasket leakage or piston seizure. Some components may be subjected to high stresses which may arise from two interconnecting components having different coefficients of expansion and undergoing a temperature change, or from the thermal expansion of a hot part of a component being wholly or partially restrained by the cool remainder of the component, or from a combination of materials, shapes and temperature distribution which precludes free expansion.

This situation may be aggravated by the deterioration of the physical properties of the materials at high temperatures so that they can no longer withstand the stresses which they were able to do when cool.

There is also the phenomenon of creep. A load which is well below that required to cause fracture or even yield can, nevertheless, result in continuous permanent deformation if applied to a material maintained at high temperature.

In practice failure from thermal stressing almost always takes the form of initiation and propagation of cracks as a result of repeated heating and cooling in combination with mechanical loads.

Distortion and resultant stressing are not the only thermal effects which lead to failure. The materials themselves may deteriorate by oxidization or by attack from the hot combustion gases. Lubricating oil films may be destroyed leading to heavy abrasive wear or even seizure whilst the products of combustion of the lubricants may well be of an undesirable nature and a source of trouble both within the cylinder and elsewhere in the engine.

5.2. *Stress due to thermal effects*

Any estimation of thermal stress must begin from a knowledge of the temperature distribution throughout the component concerned. Precise determination of such a temperature distribution pattern and the subsequent derivation of stresses in a component as intricate as a cylinder head is so involved as to have defied calculation up to the present time. Modern computing techniques are only just beginning to bring within reach analytical methods capable of solving problems of this complexity and extent. In consequence published work goes little further than an estimation of the order of stresses obtained by reducing the components to idealized forms made up from simple geometric shapes such as thin hollow cylinders, rings and flat circular plates and by assuming regular patterns of temperature distribution. (Ref. 1).

In this chapter the design of thermally loaded parts is discussed in relation to qualitative assessments of the probable nature of stresses arising from observed temperature distribution patterns.

5.3. *Determination of metal temperatures*

Although it is usual to measure and continually record temperatures of fluids (*e.g.* lubricating oil, water, or exhaust gas) on engines in service, it is rarely that metal temperatures are monitored. This is partly because there is no necessity to do so, as a proved design operated correctly according to the fluid temperatures is unlikely to give trouble, and partly because the points of vital interest are extremely difficult to reach by ordinary temperature measuring equipment.

Thermal loading is of such great importance in the development of engines that various techniques of metal temperature measurement have been devised for use under research laboratory or experimental test shop conditions.

For static components such as cylinder heads or cylinder liners small thermal couples are probably the most useful tool and by means of an arrangement for traversing through the metal thickness (Ref. 2) these can yield valuable information not only on temperatures but on temperature gradients as well.

Simple thermo-couples can also be used in moving components, notably pistons, provided that the signals can be transmitted successfully, which may be done by flexible leads, intermittent contacts or telemetry. There is a limit to the number of points which can be monitored by thermocouples in the relatively small pistons of medium speed engines and other devices are used to overcome this difficulty. The conventional method was the use of fusible sentinels. These were small plugs of metals and alloys, the melting point of which was known. By introducing them into the piston in groups covering a range of temperatures at certain strategic points, running a number of tests and observing which plugs melted a picture of piston temperatures could be built up. A quicker and more reliable method has been devised by the Shell Laboratories. (Ref. 3). Small threaded plugs of hardened steel which are known as "Templugs" are inserted in a corresponding flat bottomed hole in the surface to be tested. After the engine has been run up to load and kept there for a period the templugs are removed and a check made of their hardness. By means of the known relation between the hardness relaxation and temperature for the selected plug material the local temperature which existed during the run can be ascertained. If the temperature at a sufficient number of points on the component can be discovered then isothermal lines (that is lines of constant temperature) can be drawn on a sectional drawing. This process can be assisted by analogue techniques. The heat which is conducted through the metal may be assumed to flow along a path having a direction normal to the isotherms. The rate of heat flow will be proportional to the temperature gradient. The conduction of heat through the metal can be described by an electrical analogy, in which temperature corresponds to voltage and heat flow to current. This fact permits the use of analogues in the form of water tanks made to the shape of a section of the component being considered or the use of sensitized paper which is electrically conducting. By their assistance the isotherm and heat flow pattern may be built up from relatively few known temperatures at various points on the component. Methods of this kind have enabled pictures to be built up of the temperature and heat flow pattern in a number of components.

5.4. *Materials for thermally stressed parts*

Cast iron is the most popular material for the parts bounding the combustion chamber, particularly the cylinder heads, fairly high strength grades being used. Sometimes nodular cast iron and sometimes alloyed cast steel is used for high output engines. Many engine designs use cast iron also for pistons although light alloys are frequently employed for this component in high speed engines because of reduced inertia effects. Steel is becoming increasingly popular in the form of crowns for the two-piece piston designs of high output engines.

A high thermal stress is not necessarily successfully countered by increasing the strength of the material used. Thermal loading almost invariably leads to the imposition of certain strains whatever material is chosen and the permissible safe strain is given by:

$$\frac{\text{permissible safe stress}}{\text{modulus of elasticity}}.$$

A material having a high tensile strength may be no better than one with a low tensile strength if their elastic moduli are in the same ratio.

Desirable factors are high thermal conductivity to avoid large temperature gradients and low coefficients of expansion to avoid high strains. Thus materials resistant to thermal loads are those in which the factor

$$\frac{\text{ultimate tensile strength} \times \text{thermal conductivity}}{\text{elastic modulus} \times \text{coefficient of expansion}}$$

is relatively high.

5.5. *Piston cooling*

The design of pistons pivots a good deal on thermal effects. Trunk pistons are so buried in the engine that they must inevitably dispose of the heat they receive by means of lubricating oil which makes contact with them. With every piston design some heat is transmitted through the oil film separating it from the cylinder liner thence through the liner and into the jacket cooling water or the air if the engine is an air cooled engine. In the case of the simplest designs this path is the chief means of disposal of heat which the piston receives but it is also supplemented by heat that is carried away by lubricating oil splashed on the inside walls and the underside of the crown of the piston by the running gear. Pistons designed for engines of higher output require more positive methods of heat disposal. A considerable increase in the heat extracted from the piston by lubricating oil splash can be made by arranging for a jet of oil issuing from the top of the connecting rod to be sprayed on the underside of the piston crown.

Other methods of cooling are in the form of chambers and passages in the crown through which oil is conducted. In some designs such oil ducts are intended to run full and the passages may be purposely of small cross sectional area in order to obtain a high velocity of oil flow as this results in better heat transfer to it. Other designs use chambers of larger dimensions with wide apertures through which both oil can escape and air enter; these chambers are intended to operate only partially full and to promote violent splashing as a result of the piston motion. This type of piston cooling is often termed the cocktail shaker design, the very high local velocities giving a high rate of heat transfer. Illustrations of the different designs are given in Figure 5–1.

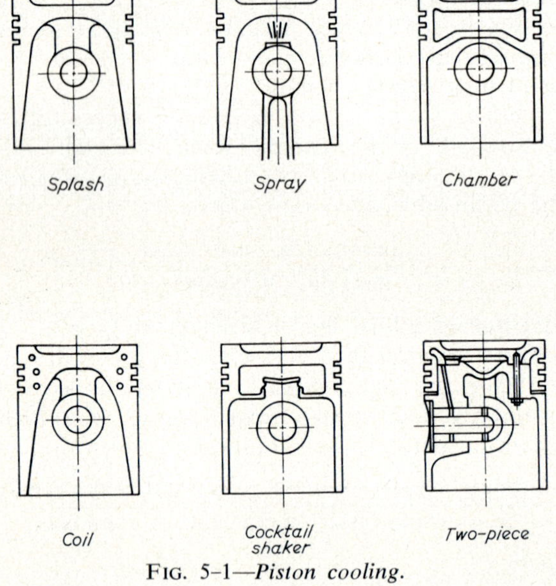

FIG. 5–1—*Piston cooling.*

5.6. *Thermal stresses in piston crowns*

The temperature distribution of a simple splash cooled piston is shown in Figure 5–2. Most of the heat flow is radially outwards into the water cooled liner and thermal stresses are due mainly to a radial temperature gradient in the crown. A simple form of oil cooled piston is shown in Figure 5–3. Here most of the heat is removed by transfer to the oil in the chamber below the crown resulting in a high axial temperature gradient. The conditions in most combustion chambers are such that there is a greater intensity of heat release nearer to the edge of the bowl forming the

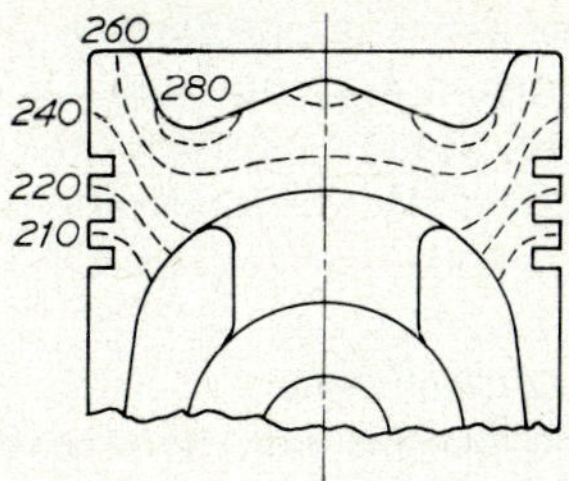

FIG. 5–2—*Temperature distribution splash cooled piston.*

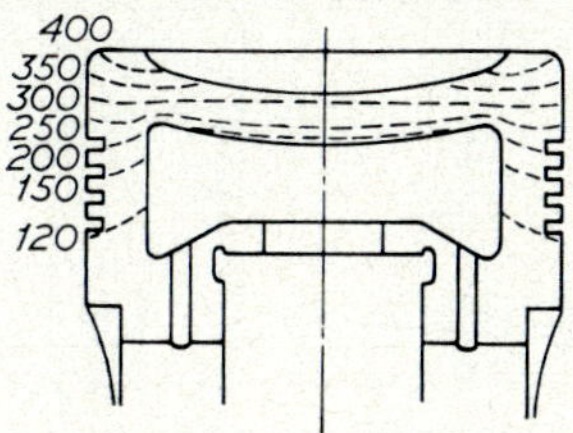

FIG. 5–3—*Temperature distribution chamber cooled piston.*

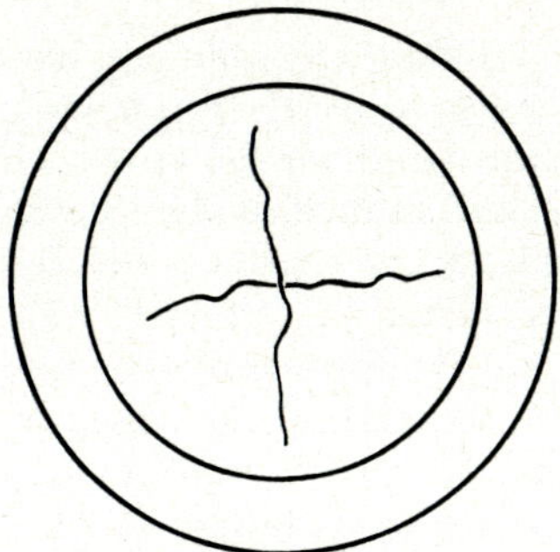

FIG. 5–4—*Star cracking of piston crown.*

combustion chamber with the result that the isothermal pattern is complicated and although most of the isothermals tend to be parallel to the shape of the crown of oil cooled pistons there is also a radial temperature gradient as well.

A common method of failure when a piston is subjected to excessive thermal load is that of star cracking on the crown, see Figure 5–4. The material of the crown at the gas side tends to expand and is restrained from doing so by the cooler material below it and around it. The hot surface material is therefore in a state of compression which will be augmented by the mechanical stresses resulting from the intermittent pressures of the engine operating cycle. If the temperature becomes high enough for excessive creep to occur this compressive stress is relieved and subsequently, when the engine is shut down or the load reduced so that the piston cools, residual tensile stress is set up. This residual tensile stress may be high enough to initiate a crack and continued operation with repeated heating and cooling can cause this crack to propagate. The material immediately on the hot surface of the piston may fluctuate cyclically a little in temperature due to the cyclic fluctuations of temperature within the cylinder as the engine operates but this effect usually does not penetrate very deeply into the material.

5.7. *Thermal stresses in piston walls*

Another way in which pistons fail under excessive thermal loads is a failure through the wall by cracking. Figure 5–5 illustrates the deformed

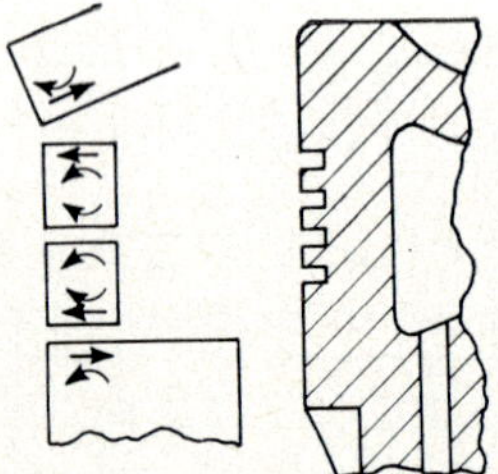

FIG. 5–5—*Nature of thermal stresses in chamber cooled piston wall.*

shape which the piston attempts to take up under the influence of the thermal loading. It applies particularly to chamber cooled pistons. Under a constant axial temperature gradient the crown, if free, would take up an expanded shape with the hot surface tending to become convex. Although the axial gradient through the crown usually varies with the radius so that the thermal expansion is partially restrained with accompanying internal stresses, for simplification an approximation to the thermally distorted shape is that obtained by considering the shape of the crown under mean axial temperature gradient. Down the wall of the pistons through the ring belt the temperature is continually falling. If this fall is linear then the cylindrical wall will tend to become conical and if it were not restrained by the crown it would remain unstressed. In most cases, however, the temperature drop is non-linear and thermal distortions lead to internal moments being set up stressing the material as indicated in Figure 5–5. As can be seen there will be a tensile stress on the inside of the wall below the crown. At the same time the gas pressure on the crown will tend to deflect it and the wall as shown in Figure 5–6, and this will occur every cycle of the engine. The combination is a fluctuating tensile stress on the inside of the wall which, at excessive thermal loads, can lead to a crack forming and propagating from the inside of the piston outwards, usually into a ring groove.

5.8. *Piston ring groove temperatures*

If piston rings are to seal properly and to operate smoothly without scuffing they must be lubricated. Lubricating oils generally available will oxidize heavily if they are heated too much. To achieve satisfactory operation of rings the grooves must be situated where the temperature does not exceed 220°C or 230°C. In fact it is preferable that the temperature of the top ring groove should be less than 200°C.

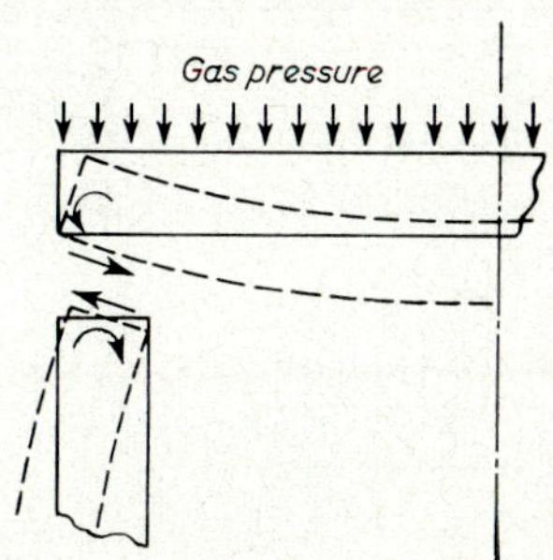

FIG. 5–6—*Nature of stresses due to gas pressure on piston crown.*

5.9. *Light alloy piston design*

In a splash cooled piston which disposes of most of its heat to the cylinder walls it is desirable to provide a broad path for the heat flow if the rings are to be kept cool. In Figure 5–2 can be recognized the shape which has proved successful for this type of piston where the material behind the rings is full, narrowing only below the ring belt. This is typical of light alloy pistons in the smaller high speed engines. This same shape is also used for spray cooled pistons in which a proportion of the heat is dissipated by means of a lubricating oil jet.

A means of directing the heat dissipation away from the ring belt which is widely employed in highly loaded large aluminium-silicon alloy pistons is that of casting in a cooling coil as shown in Figure 5–7. This coil carries the lubricating oil used for cooling and forms a heat barrier between the crown and the ring belt as can be deduced from the isothermal pattern as shown in the figure. This design protects the ring belt without taking heat from the combustion chamber unnecessarily.

5.10. *Two-piece piston design*

The pistons of high output engines may also require cooling of the crown as well as protection of the ring belt and a highly developed design of two piece piston appears in Figure 5–8. The two piece construction separates the functions of the crown, converting gas pressure into mechanical thrust and coping with the thermal loads; from the functions of the skirt, transmitting the thrust to the connecting rod and side thrust to the liner wall whilst permitting relative motion under well lubricated conditions.

The crown is designed as a thin and, therefore, relatively lowly strained, heat barrier supported by a sturdy annulus carrying the gas load to the trunk. The ring belt is suspended from the crown and, as it does not have to transmit the gas load, the connecting wall to the crown is also made thin to form a heat dam between the crown and the rings. Being a component of relatively small mass and of simple form it is easily machined resulting in close control of its dimensions. It is made of high strength heat resisting alloy steel.

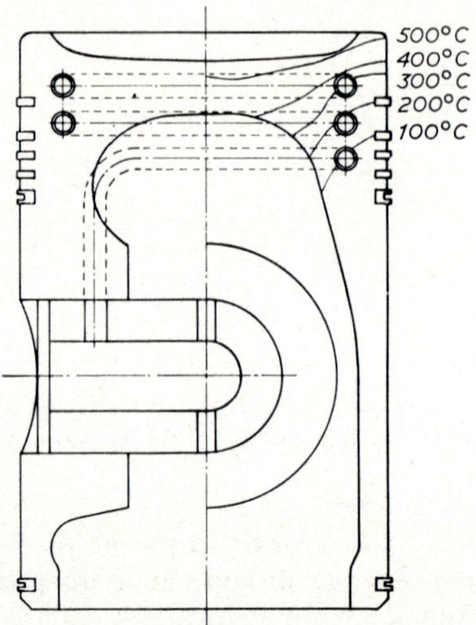

FIG. 5-7—*Light alloy piston with cooling coils.*

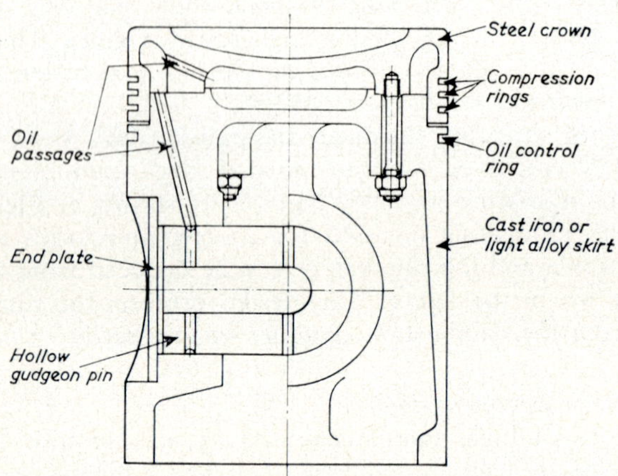

FIG. 5-8—*Two-piece piston.*

The lower or trunk part of the piston is made of less expensive material such as cast iron or hard wearing silicon aluminium alloy if low inertia is desirable. Its more complex form is produced by casting. It is comparatively cool and so provides good conditions for lubrication of the sliding motion between the piston and liner and the oscillating motion between the pin and connecting rod bush.

The crown and skirt fit together to form passages through which the cooling oil is circulated at high velocity with consequent good heat transfer behind the ring belt. Some cooling of the crown is accomplished in the chamber formed behind it before the oil finally leaves the piston.

5.11. *Cylinder heads*

The majority of cylinder heads are of water cooled design although a few high speed engines use air cooled heads. In all cases the flame plate is subject to intense thermal loading.

All cylinder head flame plates are perforated by holes to accommodate various valves and fittings. (The loop scavenge two stroke engine does not have inlet and exhaust valves in the cylinder head but nevertheless it does have openings for the injector and other fittings, and although these are relatively small they give rise to the same difficulties as large holes though perhaps not to the same extent). All these openings are surrounded by walls of the form shown in Figure 5–9 in which is also indicated the

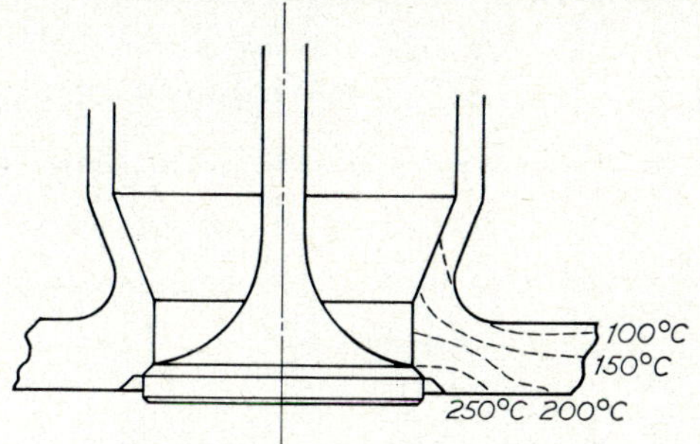

FIG. 5–9—*Temperature distribution at valve pocket.*

trends taken by isotherms in the vicinity of these apertures. It is in the nature of things that the corners get hot as their geometry dictates that they are the farthest removed from the coolant. The designer attempts to keep the temperature within bounds by rounding off the material on the gas side and arranging to get the water as close as possible into the corner on the inside.

Some cylinder head fittings become a great deal hotter in operation than the cylinder head itself and thereby suffer a much greater expansion. An extreme example is the pre-combustion chamber an illustration of which is shown in Figure 5–10 and which shows the importance of providing sufficient clearance for fittings of this nature.

When cylinder heads fail it is often by cracking on the flame plate starting from the corner of an opening. Typically the crack extends between the exhaust valve and the injector pockets in two-valve heads or between the two exhaust valve pockets in four-valve heads as illustrated in Figure 5–11. The mechanism of failure is the same as that already described for piston crowns suffering star cracking.

A highly rated engine demands intensive cooling of the flame plate which is usually attempted by incorporating a partition deck above and parallel to the flame plate and as close to it as manufacturing techniques permit so that the cooling water is made to pass rapidly across the back of the flame plate. Many designs constrain the flow to small passages between the exhaust valve seats in order to obtain good heat transfer in this locality.

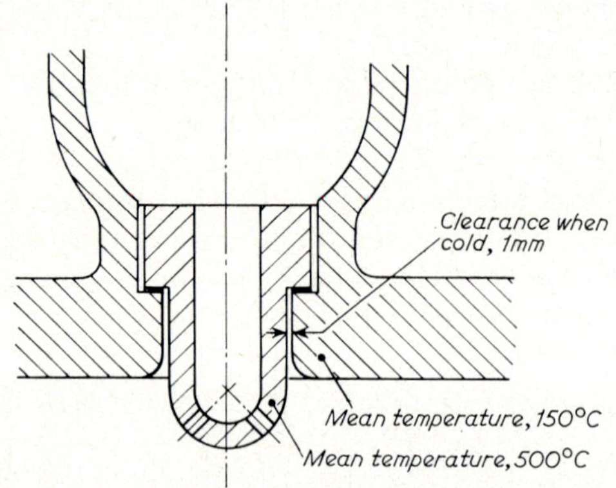

FIG. 5–10—*Pre-combustion chamber nozzle temperatures.*

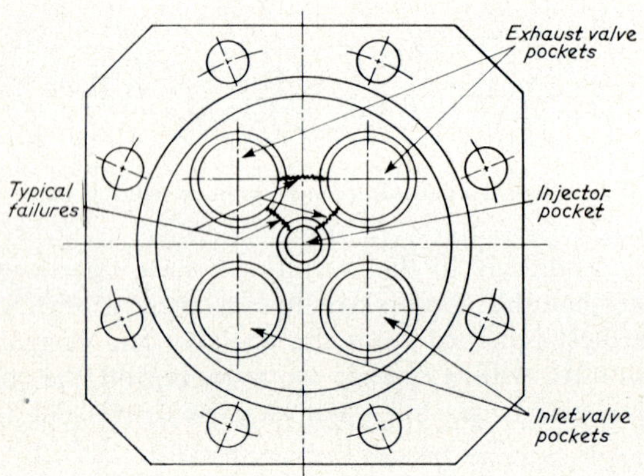

FIG. 5–11—*Typical cracking of cylinder head due to thermal stress.*

It is certain that at many hot spots on the cylinder head the high rate of heat transfer can only be achieved by local boiling of the water and the rapidly moving stream helps to achieve better results by quickly carrying away the bubbles of vapour so that they can condense and mix with the remainder of the water in cooler regions.

If designs of this degree of sophistication are to work well in practice it is necessary by adequate maintenance to ensure that the water side of the passages are kept free from scale and that all vent holes are maintained open so that steam pockets can be relieved instantly if they should form.

5.12. *Valves*

The surfaces of the valve heads form an appreciable proportion of the combustion chamber surface area and so receive a good deal of heat. Inlet valves are kept comparatively cool by the air stream flowing over them when open and rarely present problems of thermal loading. Exhaust valves operating in a stream of hot gas can only dispose of the heat they receive through their seats and stems. Heat disposal through the stem is dependent on heat transfer across the lubricating oil film between the stem and the guide. Clearance here should be kept as low as possible consistent with operating requirements. Valves seating directly in the cylinder head material transfer heat to the cooling water reasonably well but it is fairly common practice to use seat inserts which are either pressed or screwed into position. These inserts can become hot, see Figure 5–12 and being

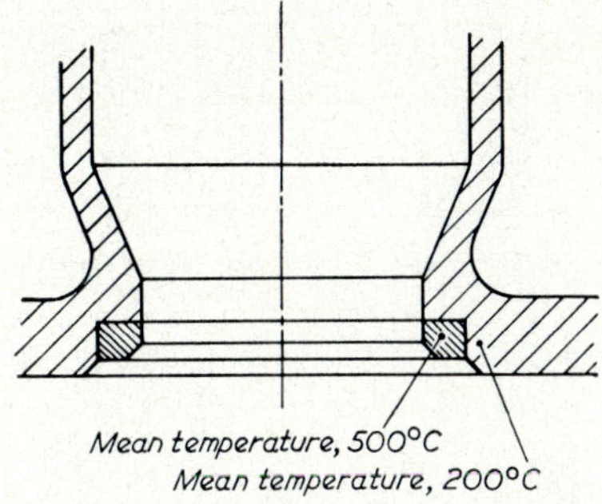

FIG. 5–12—*Valve seat insert in cylinder head.*

restrained by the cooler cylinder head present thermal strain problems of their own as well as leading to hotter valves. The justification for their use is that these disadvantages are more than offset by making them of a harder wearing material than that of the cylinder head itself and arranging for them to be replaceable at comparatively low cost.

Failure of exhaust valves generally takes the form of deterioration of the valve seat as a result of the hot gas passing over it. Particles of ash or carbonaceous deposits may be trapped between the valve and seat leading to pitting. In severe cases or if neglected this can result in tracking across the seat of the high pressure high temperature combustion gases ending in guttering of the valve. When engines are run on heavy fuels which contain sodium and vanadium the probability of the formation of deposits is increased and valve failure in the manner just described becomes much more likely. The melting range of ash containing sodium and vanadium salts commences just below 600°C, and if the valve seat temperature can be kept below about 550°C, the build up of this type of deposit can be prevented with consequent improvement in valve life. The achievement of such low valve seat temperatures is not easy however.

Medium speed engines burning heavy fuel are generally those of fairly large bore and often the exhaust valves are arranged to be in cages which can be removed without disturbing the cylinder head. This enables maintenance work on exhaust valves to be carried out more speedily. Separate valve cages add to the thermal loading problems as it becomes more difficult to get cooling water close to the valve seat. This, however, is now being done successfully on some modern engines intended for high output on heavy fuels. Figure 5–13 illustrates cooled and uncooled cages and Figure 5–14 compares the temperature distribution in the corresponding valves.

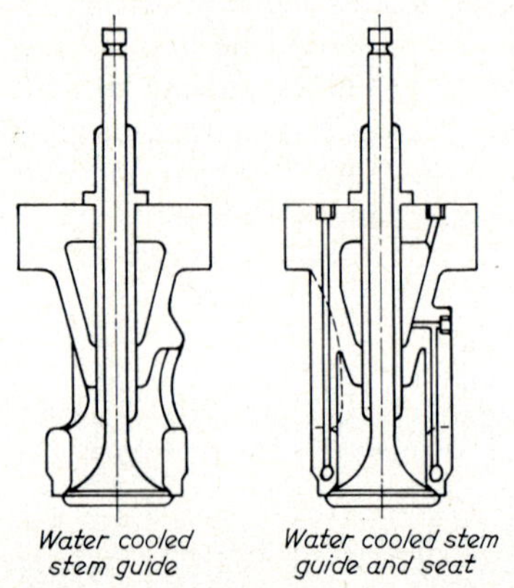

FIG. 5–13—*Cooling of exhaust valve cages.*

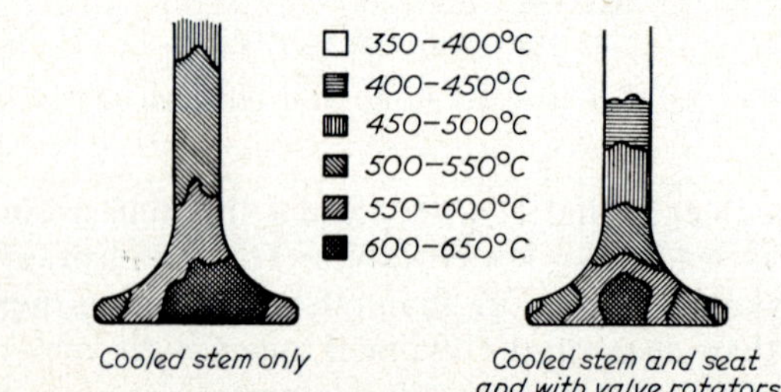

FIG. 5–14—*Temperature distribution in exhaust valves in cages having cooled and uncooled seats.*

Another approach to the problem is to cool the valve itself and this is illustrated in Figure 5–15. The valve stem and head is made hollow and a tube down the centre of the stem forms passages for the cooling fluid, usually water, to reach the valve head and to return. Flexible connexions are used at the top of the valve to convey the water to and from it.

Both the water cooled valve and the water cooled cage rely on modern manufacturing techniques to produce the necessary fine passages carrying the cooling water to the seat in either the valve or the cage. It is essential to use closed circuits containing only clean inhibited water supplying the cooling to these parts and to maintain them in good condition as any blockage can have disastrous results.

A third method of cooling exhaust valves which avoids the problems of cooling water and small passages uses a small quantity of air bled from the air manifold and directed down the stem of the exhaust valve and on to its head. The cooling effect achieved at the seat is understandably smaller than that given by other methods.

In some designs the cooling of the exhaust valve is assisted by drilling down the stem and partially filling the chamber with sodium or a sodium salt and then sealing at the top. The filling, which is liquid at operating

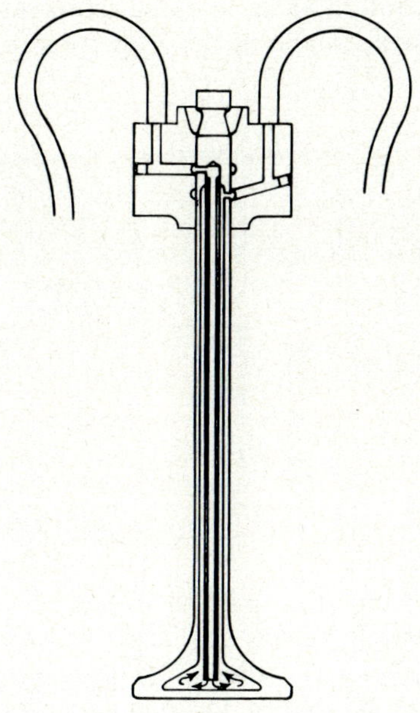

FIG. 5–15—*Water cooled exhaust valve.*

temperatures is splashed about by the motion of the valve and repeatedly transfers heat from the head to the stem. Sodium cooled valves have been used in several high speed high output gasoline engines but are not used so often in diesel engines. The dissipation of heat through the stem requires close contact with the guide over a substantial area and in most medium speed engine designs the need to provide adequate lubrication and to avoid difficulties with distortion result in clearances too large for sodium cooling to be applied with advantage. In addition, this form of cooling is concentrated at the centre of the head and has only a small effect at the valve seat face of a large valve.

In order to avoid local hot spots on the valve head seat and uneven build up of deposits some designs incorporate valve rotators: a ratchet type mechanism which slowly turns the valve, indexing it a fraction of a revolution each time it is opened by the valve gear. The use of this type of mechanism results in the temperature distribution pattern being axisymmetric and has proved its worth on many engines operating on heavy fuel.

5.13. *Cylinder liner*

The upper end of the cylinder liner encloses part of the combustion chamber but it rarely presents problems of thermal loading. On most designs this portion of the liner registers in the cylinder block and is located by it at this point. The construction is usually such that the liner itself is well supported by the block in a strong back fashion, leaving the liner in compression during operation at its working temperature.

Figure 5–16 shows the shape of the isotherms in the upper portion of

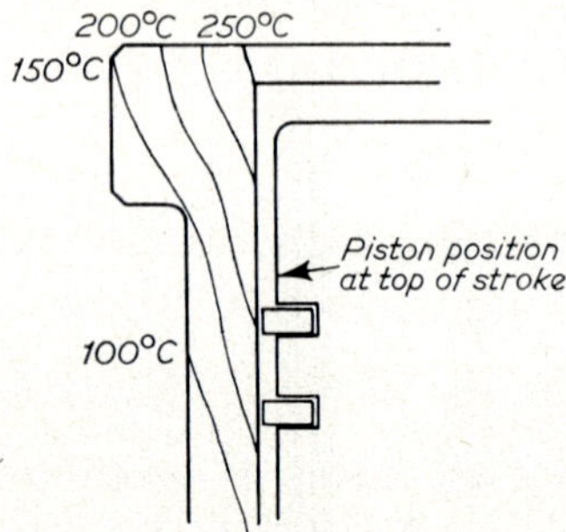

FIG. 5–16—*Isotherms in cylinder liner.*

the cylinder liner and from their pattern it can be seen that the height to which the cooling water is taken by the jacket has an important bearing on the degree to which the liner is cooled. The designer's problem is to ensure that the liner is sufficiently cool at the point at which the top piston ring reaches the end of its travel so that lubrication is satisfactorily maintained.

5.14. *Exhaust gas turbine*

The turbine blades of exhaust gas turbochargers operate continuously in a high temperature environment and, being difficult to cool because of their situation, attain the temperature of the exhaust gas which is constantly flowing across them. This results in metal temperatures quite as high as those of parts subjected to the combustion temperature intermittently.

The blades are exposed to bending as a result of the gas impinging upon them and also to centrifugal force from their rotation. Gas bending can be to some extent offset by designing-in centrifugal bending in the opposite direction which is a known technique in gas turbines. With turbochargers, however, there is an added difficulty in applying this technique because of the pulsating nature of the exhaust gas flow. Thermal loading is met by a combination of design techniques and choice of materials which retain their strength at high temperatures. Failure from thermal effects is rare.

REFERENCES

1. Dennis, R. A. and Radford, J. M., *Piston Stresses—Theoretical and Experimental Developments,* Symp. Thermal loading of diesel engine. Proc. I. Mech. E., 1964–65 Vol. 179 part 3C.
2. Ricardo and Co., *Heat Flow Measurement,* Ministry of Supply ARC Technical Report R & M No. 2171. November 1946.
3. *Screws Tell Temperature,* Engineering News, 6 December, 1962, 1.

BIBLIOGRAPHY

Fitzgeorge, D. and Pope, J. A., *An Investigation of the Factors Contributing to the Failure of Diesel Engine Pistons and Cylinder Covers,* Trans. N.E. Coast Institute of Engineers and Shipbuilders, 1955, 71, 163.

Dearden, A., *Residual Thermal Stresses in Compression Ignition Engines,* Trans. Inst. Marine Engrs., June 1962, 74 (No. 6).

Enderby, L. R. and White, D. J., *Axisymmetrical Method for Determining Thermal Stresses in Pistons,* J. Strain Analysis 1968, 3, 146.

Scholes, A. and Strover, E. M., *A Method for the Stress Analysis of Pistons,* Symposium. Computers in Internal Combustion Engine Design, Proc. Inst. Mech. Engrs., 1967–8, 182 (Pt 3L), 169.

Fiskaa, G., Iversen, P. and Sarsten, A., *Computer Calculation of Stresses in Axisymmetric Thermally Loaded Components,* Symposium. Computers in Internal Combustion Engine Design, Proc. Inst. Mech. Engrs., 1967–8, 182 (Pt 3L), 152.

CHAPTER SIX

Mechanical Loads, Inertia, Balancing and Bearings

6.1.

The mechanical loads which the component parts of a diesel engine are called upon to withstand are comprised of static loads arising from bolted assembly and dynamic loads. The dynamic loads are due to forces arising from two sources: the fluctuating gas pressure in the cylinder and the inertia forces. In some cases, such as the tension in the cylinder head studs, the fluctuating load is due entirely to the gas pressure, whilst in others, such as the bending of the frame to resist internal out of balance forces it is due entirely to inertia forces, but generally a component experiences dynamic loads which are compounded of forces from both these sources. The range of loading from the gas pressure in the cylinder is easy to determine but the range of loading resulting from a combination of this with inertia forces is not so simple. The maximum loads may well result from high gas pressures at low speed in the case of some components, whilst other components experience their highest load when the engine is run at its maximum rev/min.

6.2. *Inertia forces—rotating masses*

When a mass rotates about a centre other than its centre of gravity it has a centrifugal force,

$$F = M\omega^2 r$$

where M is the mass, ω is the angular velocity and r is the radius of its centre of gravity from the centre of rotation. If this mass is supported on a shaft in bearings, as shown in Figure 6–1, then the bearings will experience a load equal to the centrifugal force. A mass such as this can be balanced by introducing another mass with its centre of gravity diametrically opposed to the first mass and at a position so that its moment is equal. In other words since ω^2 is the same for the two masses then Mr must be equal but opposite. An example of this is shown in the single crank illustrated in Figure 6–2. The crankpin and large end of the con-

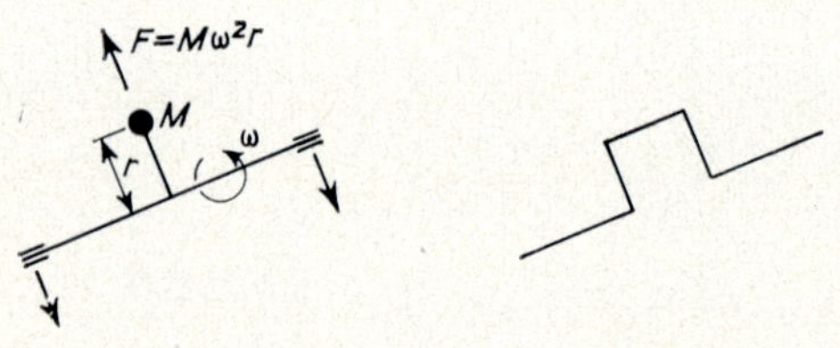

FIG. 6–1—*Single unbalanced revolving mass.*

necting rod give rise to an unbalanced centrifugal force which can be neutralized by fitting two balance weights on the webs which have a combined mass moment and centre of gravity which is diametrically opposite that of the crank itself.

When a shaft carries a number of eccentric masses, if the sum of the moments of all the masses about the axis of rotation is equal to zero for all angular positions the assembly is said to be in static balance. In this condition the polygon of forces for the centrifugal forces is closed, see Figure 6–3.

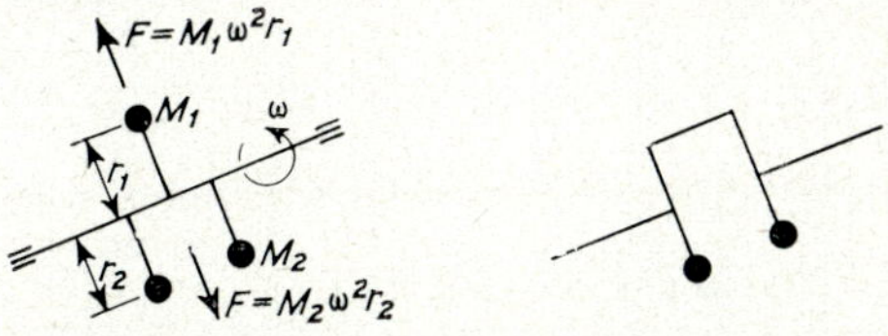

FIG. 6–2—*Balanced revolving mass.*

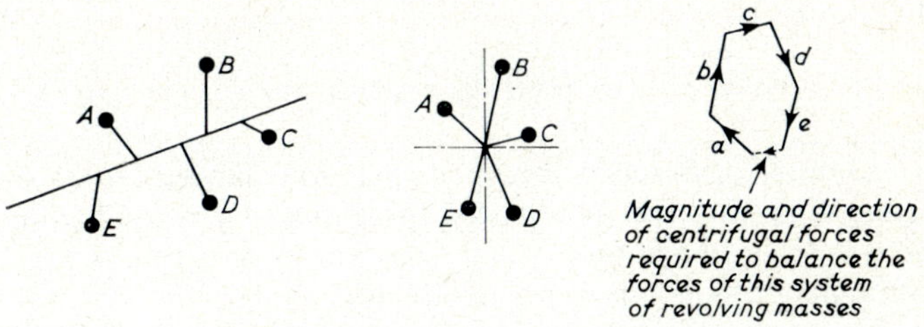

FIG. 6–3—*Polygon of centrifugal forces.*

The crankthrow elements of multi-cylinder diesel engines are almost invariably identical with each other and in the case of engines having more than two cylinders these cranks are usually spaced evenly so that the rotating forces are in static balance.

However, a shaft and mass assembly can be in static balance but not in dynamic balance. This is illustrated in Figure 6–4 where masses, although diametrically opposite each other, are not in the same plane. When a shaft with masses in a position such as this is revolved, the centrifugal force of the two masses will create a couple which has to be resisted by the bearings, and again the bearings are loaded. To achieve dynamic balance all the masses must be arranged so that there are no out of balance couples. This may be done by adding balance weights or it may be possible by suitably arranging the angular position of the cranks. When dynamic balance is achieved the polygon of couples in any reference plane is a closed figure.

If there is an unbalanced couple in the system it will result in the same residual vector being required to close the polygon whatever reference plane is chosen. If there is an unbalanced force it will result in a residual closing vector varying with the reference plane and becoming zero when the reference plane coincides with the plane of the force.

The magnitude of out of balance forces and couples is proportional to the square of the angular velocity, and, therefore, proportional to the square of the speed of the engine.

6.3. *Reciprocating masses*

The out of balance force of a rotating mass is a rotating centrifugal force but the out of balance force of a reciprocating mass is a reciprocating cyclic force along the line of operation.

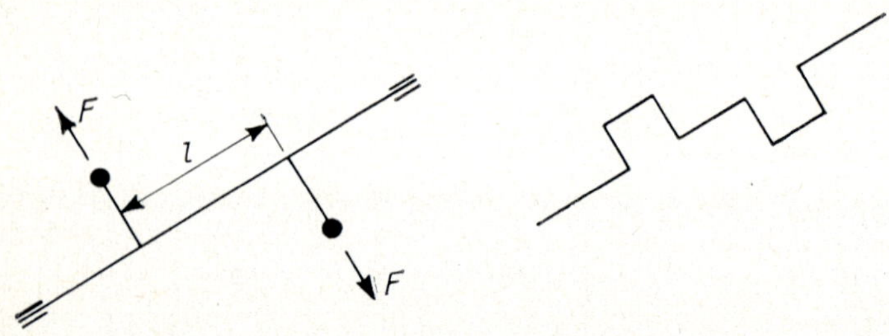

FIG. 6-4—*Two revolving masses giving a couple.*

The essential difference between the unbalanced force due to a reciprocating mass and the unbalanced force due to a revolving mass is that the former varies in magnitude but is constant in direction, while the latter is constant in magnitude, but varies in direction.

If the connecting rod were infinitely long the motion of the piston would be the component of the rotation of the crankpin in line with the piston axis. Because the connecting rod has a finite length the motion departs from simple harmonic and is affected by the length of the connecting rod. When the piston is pulled down from top dead centre by a connecting rod which is short compared to the crankshaft throw, see Figure 6–5 it is displaced further than if the connecting rod were long, because the sideways movement of the bottom end of the connecting rod increases the vertical displacement. Similarly, when the piston is moved upwards from a position near to bottom dead centre by a short connecting rod the swing of the rod is to some extent neutralized by the swing of the crank, and the vertical displacement of the piston is less than occurs near to top dead centre. Thus the acceleration downwards is not quite the same as the acceleration upwards. Figure 6–6.

Text books on Theory of Machines show that the acceleration of the piston can be expressed with sufficient accuracy for most practical purposes by the equation,

$$f = \omega^2 r \left(\cos\theta + \frac{\cos 2\theta}{n}\right)$$

where f is the acceleration of the piston
ω is the angular velocity of the crank
r is the radius of the crank
θ is the inclination of the crank to the i.d.c.
n is the ratio of connecting rod length to crank length

If M_R is the mass of the reciprocating parts, then the inertia force is given by:

$$F = M_R f = M_R \omega^2 r \left(\cos\theta + \frac{\cos 2\theta}{n}\right)$$

$$= M_R \omega^2 r \cos\theta + M_R \omega^2 \frac{r \cos 2\theta}{n}$$

$$= Fp + Fs$$

where $Fp = M_R \omega^2 r \cos\theta$ is termed the primary disturbing force, and represents the inertia force of a mass having simple harmonic motion, and

$$Fs = M_R \omega^2 r \frac{\cos 2\theta}{n}$$

is termed the secondary disturbing force and represents the correction required in order to allow for the obliquity of the connecting rod.

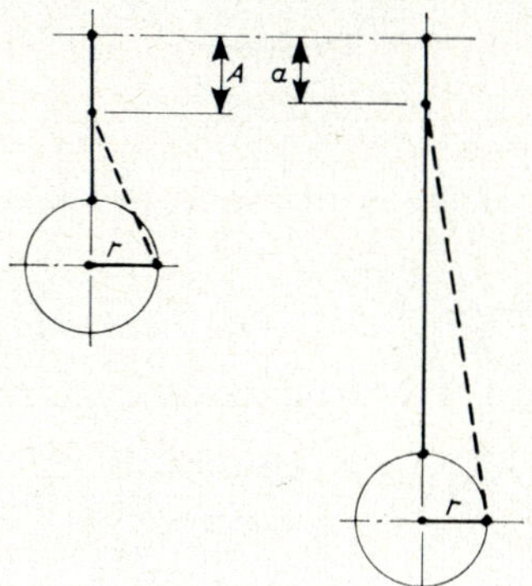

FIG. 6–5—*Effect of connecting rod length.*

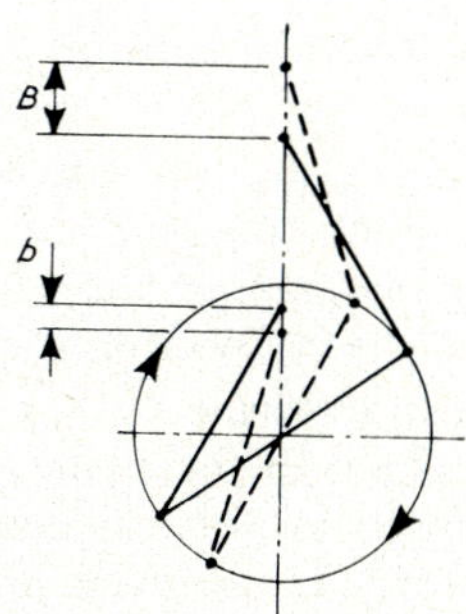

FIG. 6–6—*Effect of connecting rod on piston motion.*

Figure 6–7 shows the primary and secondary forces plotted over one complete rotation of the crankshaft and how they add together to give the overall result. Top dead centre is taken as the zero position. The secondary force is much smaller in magnitude than the primary, its relative size is proportional to the crankthrow connecting rod length ratio $(1/n)$ and it oscillates at twice the frequency.

6.4. *Balancing of engines, single cylinder*

Both revolving and reciprocating masses have to be considered. The revolving mass consists of the unbalanced portion of the crank plus that

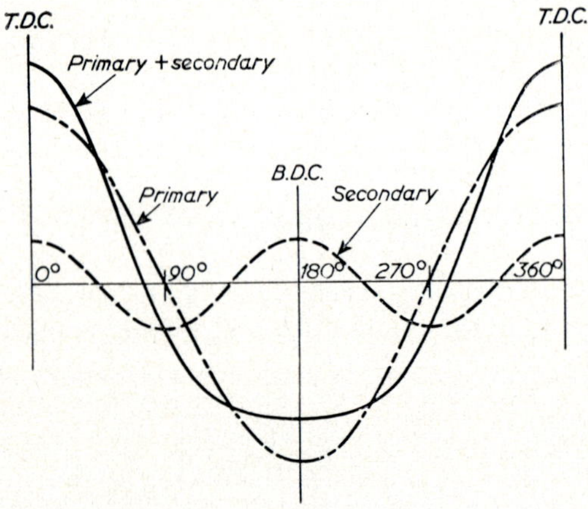

FIG. 6–7—*Primary and secondary inertia forces for crank/connecting rod ratio* $= \frac{1}{4}$

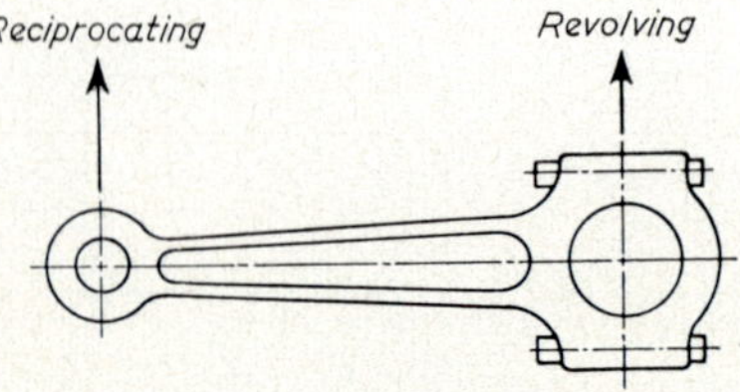

FIG. 6–8—*Connecting rod, reciprocating and revolving portions.*

part of the connecting rod which can be regarded as revolving. The reciprocating mass consists of the piston, gudgeon pin and cross head if any, plus that part of the connecting rod which can be regarded as reciprocating. The division of the connecting rod into reciprocating and revolving parts is usually taken as the reactions at the centre of the large end and the centre of the small end when the connecting rod is supported horizontally at these two positions only, as in Figure 6–8.

The reciprocating forces act in the plane of the cylinder and for convenience the engine is regarded as vertical engine in the text that follows. The primary reciprocating force is, as given in paragraph 6.3,

$$Fp = M_R \omega^2 r \cos \theta$$

and this can be regarded as the vertical component of a centrifugal force, caused by a mass, M_R, rotating at radius r, the radius of the crankpin and coinciding with it.

The secondary force, again as shown in paragraph 6.3, is given by,

$$Fs = M_R \frac{\omega^2 r \cos 2\theta}{n}$$

which can be written as:

$$Fs = M_R \frac{(2\omega)^2}{4n} r \cos 2\theta$$

This is equivalent to the vertical component of a centrifugal force, caused by a mass, $M_R/4n$ at the same radius r, the radius of the crankpin, but rotating at twice the speed. The secondary force can, therefore, be imagined as the vertical component of such a mass on a "secondary" crank which rotates at twice the speed of the actual crankshaft, in the same direction and which coincides with the actual crankshaft at inner dead centre. When the actual or primary crankshaft is at 90° to i.d.c. the secondary crank will be at 180° to i.d.c. and when the actual or primary crank is at 180° to i.d.c. the secondary crank will be at i.d.c. This is illustrated in Figure 6–9.

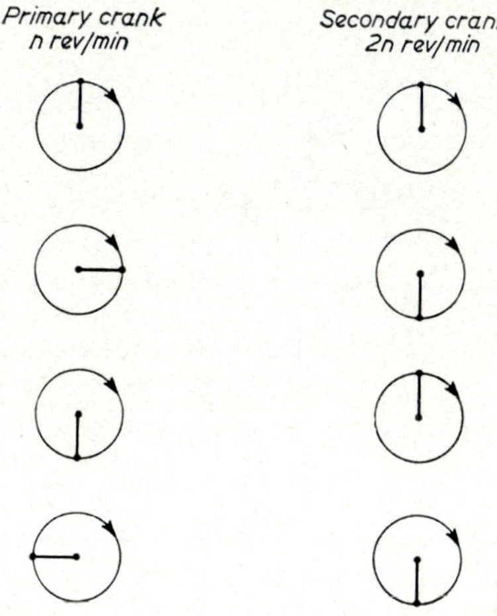

FIG. 6–9—*Primary and secondary cranks.*

The revolving masses can be balanced by counter weights fitted on the crankshaft, arranged so that their combined centre of gravity is in the same plane as the crank and their centrifugal force is equivalent but opposite to that arising from the crank.

The primary reciprocating force cannot be balanced by a single rotating counter weight although it is possible to balance it by two such weights rotating in opposite senses as shown in Figure 6–10. The two weights are phased so that their centrifugal forces add together in the vertical plane in opposition to the primary reciprocating force but cancel out each other in the horizontal plane. In the same way, secondary reciprocating forces can only be precisely balanced by two counter weights rotating in opposite senses at twice the crankshaft speed. This arrangement is well known as the "Lanchester Balancer", and is shown in Figure 6–11.

It is common practice to fit to the crankshaft counter weights larger than those required to balance only the revolving forces in order to achieve what is known as partial balance of the primary reciprocating

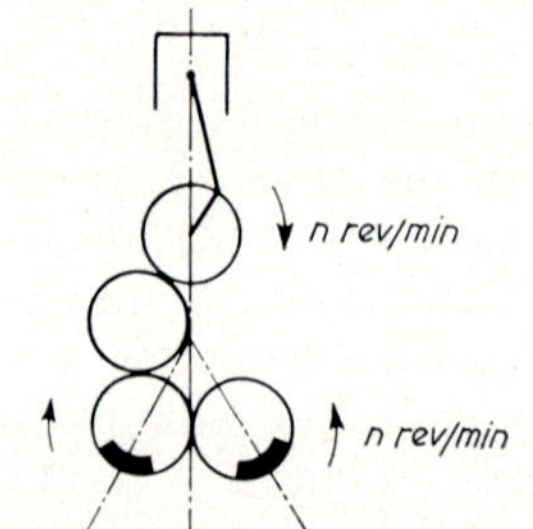

FIG. 6–10—*Primary force balancer.*

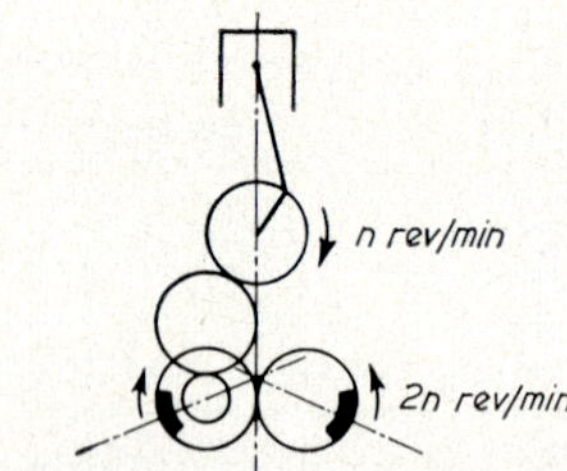

FIG. 6–11—*"Lanchester" secondary force balancer.*

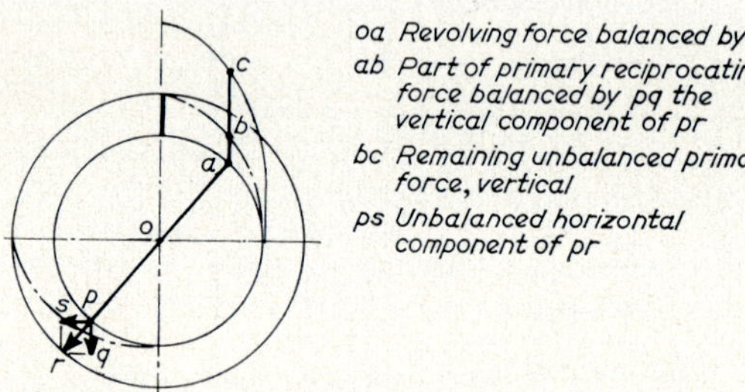

FIG. 6–12—*Partial balance of primary reciprocating force by revolving counterweight.*

forces. The revolving counter weights have a vertical component which is set against the primary reciprocating disturbing force but it also has a horizontal component and this will remain unbalanced. The mass of the counter weight and its radius can be chosen so that the vertical and horizontal out of balance forces remaining are smaller in magnitude than the original vertical force. This effect is illustrated in Figure 6–12.

The most common degree of partial balance used is that in which 50% of the primary reciprocating force is balanced so that vertical and horizontal out of balance forces are equal in magnitude.

6.5. *Balancing multi-cylinder in-line engines*

It is unusual for multi-cylinder engines to consist of anything other than a number of identical cylinder assemblies which, for balancing purposes, may be represented by a number of identical reciprocating masses. Each of these will give rise to a reciprocating cyclic force and by suitably arranging the angular intervals between the cranks cancellation of these forces is possible to a greater or lesser degree.

Engines with two and odd numbers of crank throws will have residual out of balance forces or couples or both. It is common to add rotating balance weights to the crankshaft of such engines in order to partially balance these effects in the vertical plane at the expense of inducing complementary but smaller effects in the horizontal plane.

For convenience the plane of the cylinders will be regarded as vertical which is nearly always the case with in-line marine engines. The primary force arising from each reciprocating mass can then be regarded as the vertical component of a rotating mass, M_R, at the crankpin radius, r, just as in the case of the single cylinder engine.

For there to be no unbalanced reciprocating forces the algebraic sum of the vertical components must be zero at all angular positions of the crankshaft. In Figure 6–13 the polygon of centrifugal forces for a system

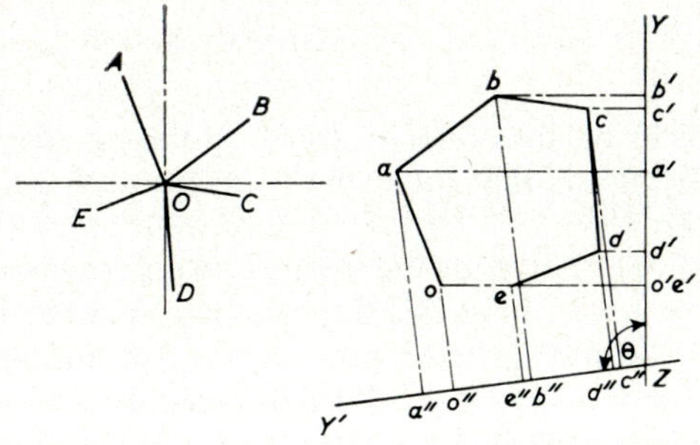

FIG. 6–13—*Vector polygon of centrifugal forces.*

of primary cranks is drawn and the vertical components are projected on to the plane ZY and, as can be seen, their algebraic sum is zero. If this crank is rotated clockwise through an angle θ then it is equivalent to rotating the vertical plane anti-clockwise by the same angle to position ZY^1. In this position there will be a residual vector, $o\ e$, and the vertical component of $o''\ e''$, is a measure of the unbalanced primary force. For no such force to arise the vector polygon representing the centrifugal forces of the primary cranks must close. This condition is just the same as that for the rotating forces. In the same way the primary reciprocating couples correspond to the rotating couples.

In systems where there are residual vectors these represent unbalanced forces or couples in the vertical plane only, as distinct from the rotating forces or couples arising from rotating out of balance. The vertical forces or couples are cyclic alternating and have a maximum value equal to the value of the vector.

In the same way the secondary forces arising at each reciprocating mass can be regarded as the vertical component of the centrifugal force arising from a mass $M_R/4n$ at r rotating at twice the speed of the crankshaft and coinciding with it at inner dead centre. The balance of secondary forces and secondary couples can be assessed by considering the vector summations of secondary cranks.

To determine completely the state of balance of an engine, the rotating forces and couples, the primary reciprocating forces and couples and the secondary reciprocating forces and couples must all be investigated.

Engines with three or more crank throws usually have them arranged symmetrically so that revolving and primary reciprocating forces are balanced.

6.6. *Balancing four cylinder in-line engines*

An example is seen in Figure 6–14 of a four throw crankshaft. The

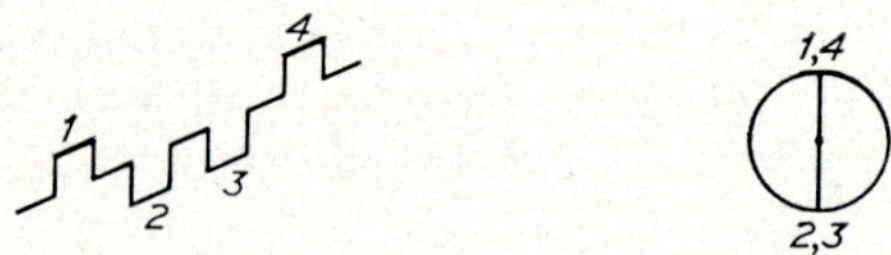

FIG. 6–14—*Four throw crankshaft* 180 *degree firing interval.*

rotating forces are in balance and so are the rotating couples as is obvious from inspection. It follows that the primary reciprocating forces and couples are also in balance.

If the shaft is considered in the position where number one throw and number four throw are at inner dead centre then number two throw will be at 180° from this position. The secondary crank for number two throw will, therefore, be 360° from i.d.c. that is, it will coincide with the i.d.c. position. The same is the case with number three secondary crank. The secondary cranks for numbers one and four will be with the primary cranks at i.d.c. so that the secondary cranks are all in phase, and the total maximum secondary force will equal:

$$4F_s = 4\frac{M_R}{n}\omega^2 r$$

The secondary couple will, of course, be zero.

This high secondary out of balance force is notorious on four cylinder four stroke engines having this form of crankshaft. It is, however, a most popular form as it results in equal firing intervals of 180°. The secondary disturbing force can be balanced by using Lanchester harmonic balancers, one at each end of the engine, at the expense of length, cost and complication, but very few designs go to this trouble.

Another form of four cylinder crankshaft is shown in Figure 6–15; this is most commonly used for four cylinder two stroke cycle engines where the firing intervals are 90°. Sometimes it is used for marine auxiliary

FIG. 6–15—*Four throw crankshaft* 90 *degree firing interval.*

naturally aspirated four cylinder four stroke cycle engines, the uneven firing and out of balance couples being more tolerable in this application than the unbalanced secondary forces of the previous example.

With this arrangement revolving, primary and secondary forces are balanced, but there are unbalanced revolving, primary and secondary couples. The revolving couple can be balanced by counter weights as indicated in the Figure 6–15, each pair having an equivalent effect equal to 2/3 *M.r.*

6.7. *Balancing of six cylinder in-line engines*

The firing intervals of a six cylinder four stroke cycle engine when evenly spaced occur at 120° and the crankshaft shown in Figure 6–16 in which the cranks are spaced at 120° in pairs is the most common arrangement. This shaft results in complete inherent balance of all external forces and couples.

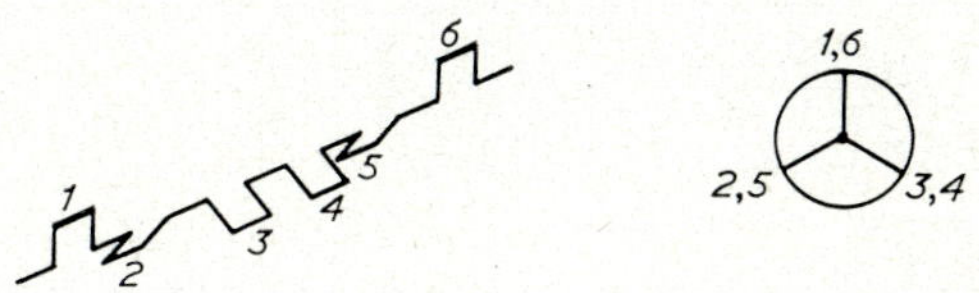

FIG. 6–16—*Six throw crankshaft* 120 *degree firing interval.*

The six cylinder two stroke cycle engine has 60° firing intervals and requires a crankshaft with 60° spacing of the cranks. The most commonly used form is shown in Figure 6–17 and this crankshaft has complete external balance except for a small secondary couple.

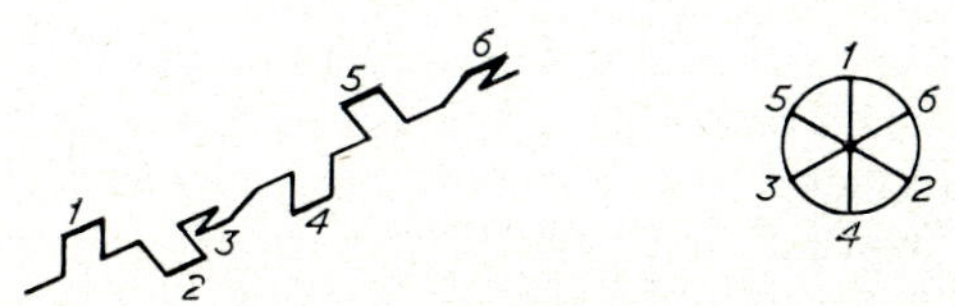

FIG. 6–17—*Six throw crankshaft* 60 *degree firing interval.*

6.8. *Balancing of eight cylinder in-line engines*

Crankshafts with eight throws have many possible arrangements and, indeed, it is obvious that the larger the number of cylinders the more arrangements are available. The considerations to be met are those of even torque from equal firing intervals, grouping of the cylinders for turbo-charging, where this applies, and the external balance of both forces and couples. Some examples are given in Table 6.1 and others may be found in the references and Bibliography at the end of the chapter.

TABLE 6.1

8 Cylinder Crank Arrangements
All Forces Balance

Crank Arrangement	Firing Order	Primary Couples	Secondary Couples
1,8 (top); 3,6 (left); 4,5 (right); 2,7 (bottom)	16258374 13258674	Nil	Nil
1,8 (top); 2,7 (left); 3,6 (right); 4,5 (bottom)	17438256 12468753	Nil	Nil
1,8 (top); 2,7 (left); 4,5 (right); 3,6 (bottom)	12358764 17348265	Nil	Nil
1,6 (top); 3,8 (left); 4,7 (right); 2,5 (bottom)	18276354	Nil	$\frac{8M_R r L\omega^2}{n}$
1,4 (top); 5,8 (left); 6,7 (right); 2,3 (bottom)	15264837 18264537	Nil	$\frac{16M_R r L\omega^2}{n}$

L = pitch of cylinder centres.

6.9. *Internal couples*

If Figure 6–16 is examined it will be appreciated that this six cylinder crankshaft which is inherently in complete balance is in the form of two three-cylinder crankshafts, one of which is a mirror image of the other as illustrated in Figure 6–18. Each three cylinder portion has complete balance of forces, but has unbalanced revolving, primary and secondary couples. These unbalanced couples of each three cylinder portion balance and neutralize each other in the complete six cylinder crankshaft but they tend to cause the shaft to whirl as shown in Figure 6–19. This movement has to be restrained by the bearings, particularly the centre bearings and by the frame supporting the bearings. In some engine designs, the pitch of the middle two cylinders is increased to accommodate a larger centre bearing.

The six cylinder crankshaft has been used here only as an example; internal couples of the same nature arise in the great majority of engines

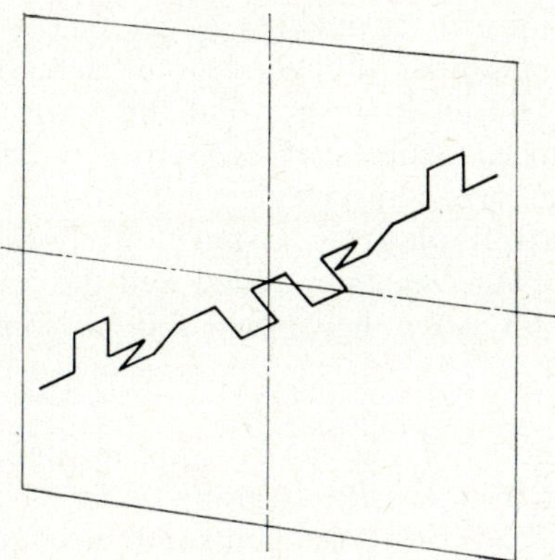

FIG. 6–18—*Six throw crankshaft as three throw plus mirror image.*

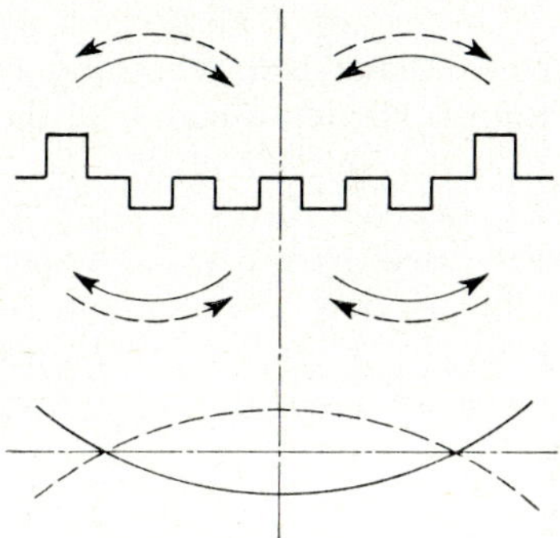

FIG. 6–19—*Internal Couples.*

and impose mechanical loads on the bearings and the engine frame. Crank web balance weights are sometimes used to counteract these effects even when they are not required to achieve external balance.

6.10. *Vee engines*

Crank throw arrangements of Vee engines usually correspond with those of the equivalent in-line engines as similar considerations obviously apply. Considering the engine overall, the firing intervals depend on vee angle, for example, a 60° vee 12 cylinder engine has even firing intervals but a 60° vee 16 cylinder engine has not. With large numbers of cylinders slightly uneven intervals are of no consequence and the vee angle is maintained at the same figure throughout a production range of engines for the sake of ease of manufacture.

The two large end bearings on one crankpin may be arranged as side by side, fork and blade or articulated. The motions of the pistons in the first two cases are just the same as those of in-line engines but in the third case the pistons connected to the articulated rods experience a slightly modified motion. This modification is small and usually ignored in approximate calculations for balancing.

Unless the total number of cylinders is eight or less the crankshaft arrangement can be chosen to ensure that there are no unbalanced forces. The residual unbalanced couples of each bank (and forces if they exist) may be resolved into two components, vertical and horizontal where, for convenience, the vertical plane is assumed to bisect the vee angle. The components in each plane are then added and the results expressed as unbalanced couples (or forces) in the vertical and horizontal planes.

6.11. *Gas forces*

The cyclic variations of the gas pressure in the cylinder can be traced from the indicator or from the Pθ diagram.

To evaluate the loads on components in the running gear it is necessary to examine the gas loads transferred through the geometry of the engine; the piston, the crank and the connecting rod, at frequent intervals throughout a complete cycle. This will have to be done for, at least, the maximum and minimum speeds of the engine. The process would be very tedious if it were not that computers have helped greatly in relieving the burden of mathematical calculations. Having ascertained the loads it is not always a straightforward matter to derive the stresses; the complicated shapes that make up cylinder heads, cylinder blocks, crankcases and bases whether they are castings or fabricated parts, being frequently far from easy to analyse from a stress point of view. The computer is being brought to the aid of the designer and readers interested in advanced work in this subject can assess the state of the art from the symposium: Computers in I.C. engine design (Ref. 1).

6.12. *Bearings*

The loads taken by large end and main bearings are compounded of gas loads and inertia forces. The highest loads are frequently those due to the inertia forces. A convenient way of showing the loads on large end and main bearings is by means of polar load diagrams and one example is given in Figure 6–20. It will be appreciated that the loads vary in both magnitude and in direction.

6.13. *Oil film thickness*

Bearings which are rotating rely for lubrication on hydrodynamic principles. In a steadily loaded journal bearing, the journal takes up a position eccentric within the bearing and maintains a wedge of oil and a constant oil film thickness, see Figure 6–21. When the load changes in direction, the effect upon the oil film can be considerable. Consider Figure 6–22 where we have the journal rotating in one direction at a speed of n rev/min, the load steadily supplied in a downward direction and the outside of the bearing shell rotating in the opposite direction to the journal at an equal speed. Clearly under these conditions a film of oil cannot

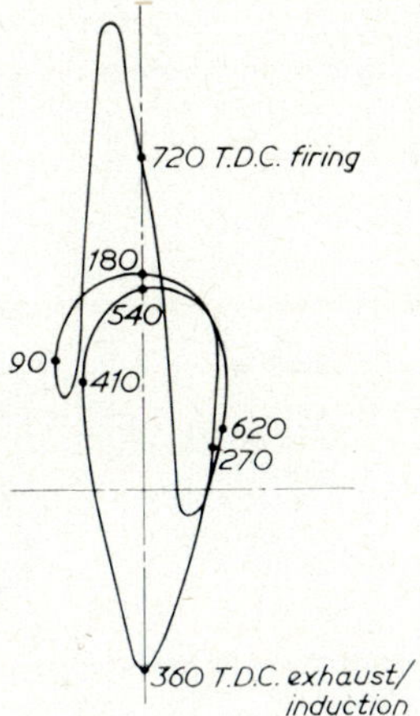

FIG. 6–20—*Polar load diagram, connecting rod bottom end bearing.*

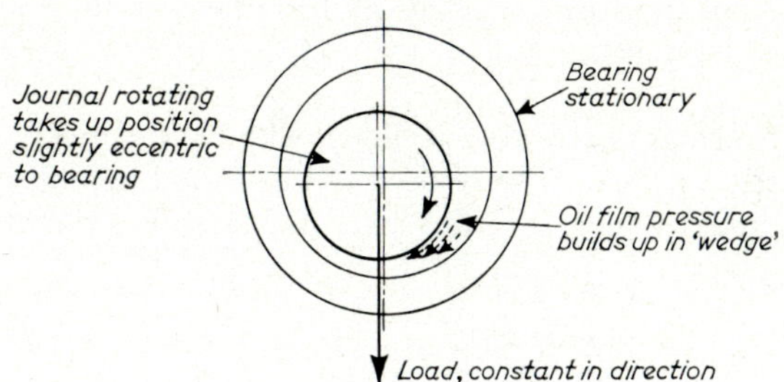

FIG. 6–21—*Journal bearing with steady load.*

possibly build up as the shell will drag oil out just as quickly as the journal tries to drag it in. This condition is exactly the same as would exist in a bearing where the shell was stationary and the load was changing its position at half the speed of rotation of the journal. During the cycle of loading of large end main bearings such a condition can be approached or may even be obtained for a number of crankshaft degrees and can lead to complete oil film failure. The condition may be modified by altering the shape of the polar load diagram perhaps by adding balance weights.

This half speed load vector is only part of the problem and the modern approach, using computer techniques, is to calculate the oil film thickness for all conditions during a complete number of revolutions for one cycle of the engine. (See Refs. 2, 3 and 4).

6.14. *Large end (bottom end) bearing*

Medium speed engines were frequently designed with large end bearings of the marine type in which the large end of the connecting rod had a palm face to which were bolted bearing housings which had been directly lined with bearing metal. Although there are still many examples of this type of bearing there has, generally, been a swing towards the automotive type which was always popular in the small high speed engines in which

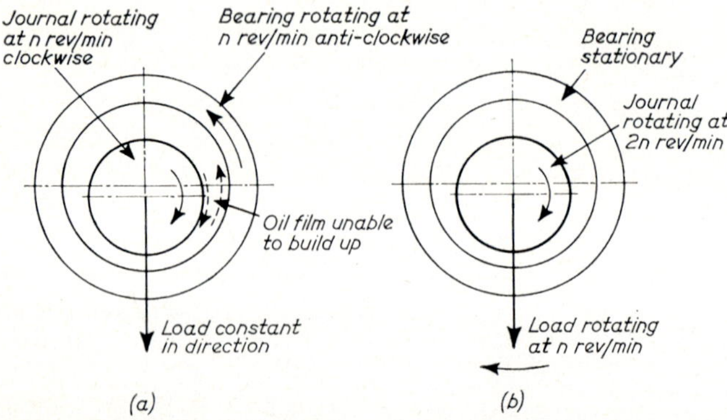

FIG. 6–22—*Rotating load, half speed load vector.*

the bearings are in the form of shells and are held to the rod by a cap, the bore to receive them being half in the rod and half in the cap, see Figure 6–23.

At first, bearings were always lined with white metal and as higher and higher bearing pressures were used it was found necessary to make this metal thinner and thinner in order that the fatigue strength of the lining should be adequate to carry the loads required.

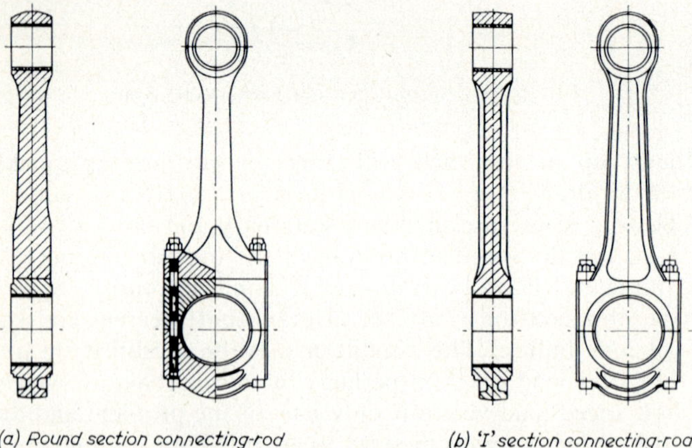

FIG. 6–23—*Connecting rod types.*

The thin wall shell bearing is a development of this requirement. It consists of a thin steel shell to which is bonded the bearing metal. This metal may be white metal or, more usually, copper lead with a white metal or similar overlay, or it may be aluminium or an aluminium tin compound.

When the bearing is assembled the thin walled shells are an interference fit in the bore which receives them. This is accomplished by having what is known as nip. Figure 6–24 illustrates the nip achieved by the shells

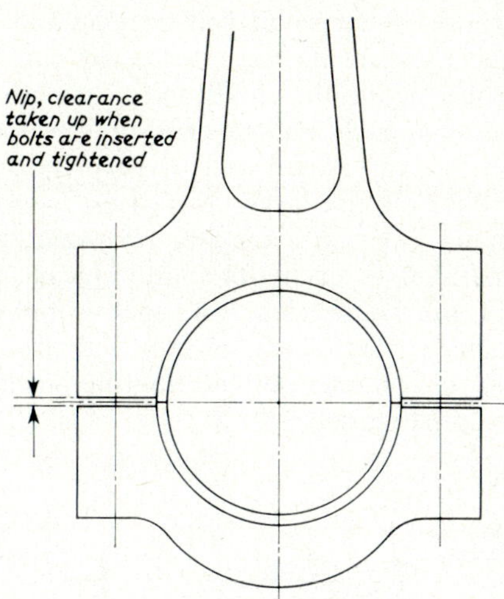

FIG. 6–24—*"Nip" of shell bearings, equivalent to an interference fit.*

being somewhat larger in circumferential length than the bore. On bolting the assembly together there is a compressive hoop stress round the bearing shells which is similar to that which occurs in a bush with an interference fit in the bore which receives it. It is easy to see that the bearing shells because of their thin flexible nature, depend for their circularity on the accuracy of the bore which receives them. They are located axially and against turning by small tangs pressed out at the butt joints. When assembling this type of bearing it is essential to see that the tangs locate in the appropriate slots in the housing without any interference and that the bore and the back of the shells are absolutely clean. Bearing failure rapidly results from the smallest particle of dirt that is trapped between the shell and the bore of the housing. Because of the highly stressed nature of the bolts and because of the frequently flexible design of the bearing housing adopted in order to achieve minimum weight and keep the inertia forces low it is essential to see that bolts are tightened to the correct tension. Maker's instructions should be followed closely. The correct tension may be achieved by means of a torque spanner or by measuring the stretch of the bolt; the latter is preferable as it is less open to error.

It is also important to see that bearings are in correct alignment as any deviation would lead to high loading on the edge of the bearing, another cause of premature failure.

6.15. *Small end (top end) bearing*

The small end bearing is quite different in its method of lubrication from the large end and main bearings. Because the journal does not rotate but oscillates through an arc, it cannot build up a hydrodynamic film and depends on the oil film being generated by squeeze pressures. It is generally recognized that in four stroke engines the reversal of load which occurs from the inertia forces applied to the piston at the end of the exhaust and beginning of the induction strokes enables oil to enter the highly loaded portion of the clearance of the small end bearing and to provide oil which is squeezed to a high pressure during the loaded part of the cycle.

With a two stroke engine the small end bearing is notoriously difficult to lubricate as there is no such load reversal or if it does occur, as on some high speed engines, it is very small and of very short duration. Small end bearings for two strokes require large clearances as much as 0·002 or 0·003 times the pin diameter. They are usually provided with a multiplicity of axial oil grooves, parallel to each other and spaced at intervals less than the arc of swing of the connecting rod, each supplied with an oil feed. Many designs have been tried: fixed bushes with fixed pins, or with floating pins; floating pins in conjunction with floating bushes, and fixed pins and saddle type bearings, see Figure 6–25.

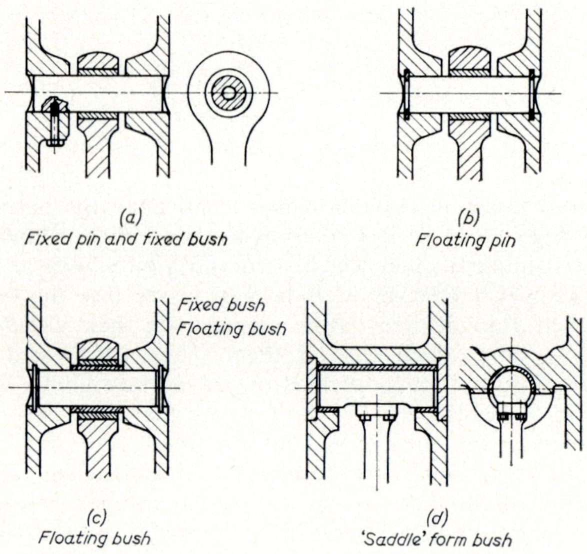

FIG. 6–25—*Some types of small end bearing designs.*

When the time comes to replace a small end bearing it is essential to see that the correct clearance is given and that the bearing is in correct alignment as, again, edge loading is a frequent cause of failure.

With all bearings the quality of the lubricating oil and especially its cleanliness is of the utmost importance. Filtration of lubrication oil is dealt with in a later chapter.

6.16. *Fatigue loads*

Practically every part of a diesel engine is subject to a fluctuating load and, consequently, a most important factor in the design of the highly stressed members is the fatigue strength of the material and of the component.

It is well known that a component subject to a fluctuating stress may fail although that stress is well below the yield point of the material. Such a failure is known as a fatigue failure. Information on fatigue strength has been gathered for many materials and is usually presented in the form of *S*-log *N* curves, where $\pm S$ is the applied direct alternating stress and *N* is the number of reversals of stress to cause failure. For ferrous metals the curve is of the general form shown in Figure 6–26. The actual values vary

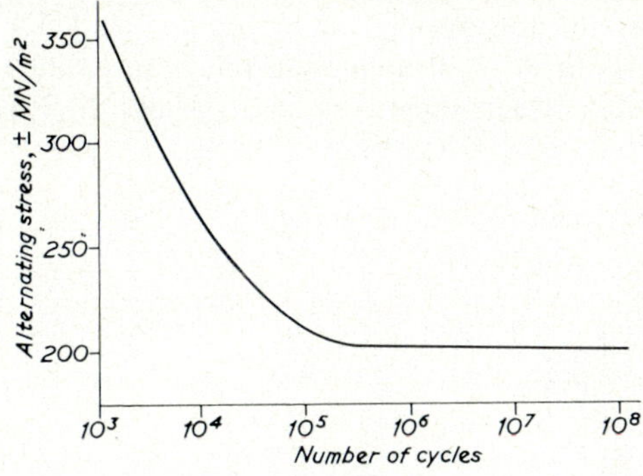

FIG. 6–26—*Typical S-N curve.*

according to the nature of the material and the size and shape of the specimen. It can be seen that there is a value for *S* below which the material will not fail however many times the stress is applied and reversed. This feature is characteristic of the materials used for the stressed members of marine engines but it is not the case with all materials. Aluminium alloys, for instance, often have *S–N* curves which do not level out and the engineering applications of such materials have to be designed for finite life.

It will be appreciated that the majority of components are subject to the maximum stress at relatively few positions; a part to which a bending moment is applied for example, carries the maximum stress on the surface at the position furthest from the neutral axis. Any notch, even the smallest scratch, in this region may account for a local raising of the stress which, if it exceeds the fatigue strength values, will cause a crack to form. Often it is in the nature of things that the stress at the root of the crack so formed will be even higher and, in consequence, the crack will propagate with increasing rapidity until complete failure. Stress raisers play a most important part in the fatigue strength of stressed components.

A large number of highly stressed parts are subject to a steady load on which a fluctuating load is superimposed. Cylinder head studs and large end bolts spring readily to mind as examples of such combined loads which are comprised of the static tightening load to which is added the dynamic working load. It is important to the designer to know what fluctuating stress can be borne by material already under a steady stress. In Figure 6–27 the abscissae represent mean steady stress and the ordinates the alternating stress. When the mean steady load is zero the alternating load that can be applied without fear of failure is the full fatigue stress, shown as the semi range of stress, A, in this figure. When the mean steady load is the UTS of the material then any further load would cause it to fracture, so at this point of the diagram the semi-range of alternating stress permissible is zero and BA the straight line joining these two points may be taken as defining the semi-range of the alternating stress that may be applied in conjunction with a steady mean load. This diagram is known as the Goodman diagram and it may be drawn in another form as shown in Figure 6–28. Where A represents the semi-range of alternating stress, the line OB represents the applied stress up to the UTS at point B. If a straight line from B to A^1 is drawn then the permissible alternating stress is precisely defined between the two lines AB and A^1B as these two lines show the total stress.

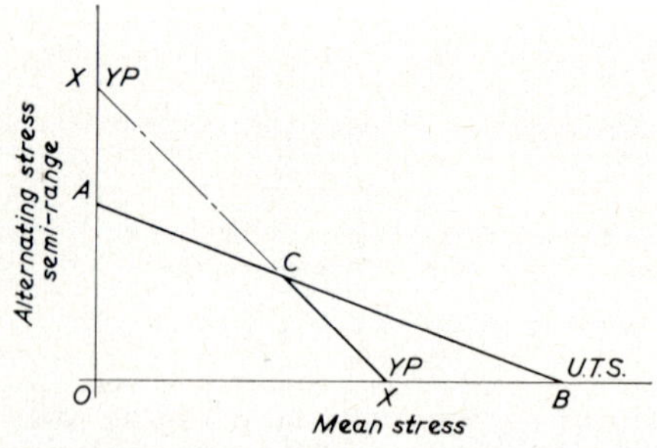

FIG. 6–27—*Goodman diagram* (1).

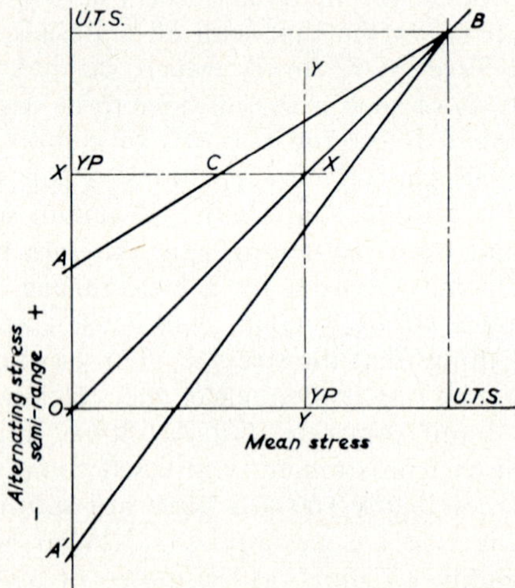

FIG. 6–28—*Goodman diagram* (2).

For a ductile material it is often desirable to limit the maximum stress to the yield stress. Lines XX and YY in Figure 6–28 indicate the limiting values of yield stress. A similar limiting line can be represented in Figure 6–27 by drawing XX from values corresponding to the yield point on both X and Y axis. The alternating stress imposed on any mean stress must then lie within the area bounded by OACX.

A great deal of practical work has been performed with the object of determining accurately the actual limiting stress above which failure will be caused. Many workers in this field have proposed a parabolic form of line in place of the straight line of the Goodman diagram, but the Goodman line is generally on the safe side and is a good practical guide. The subject matter on fatigue strength of materials is vast and interested readers are referred to *Metal Fatigue* by J. A. Pope (Ref. 5) and other works listed in the Bibliography.

When using fatigue strength information presented in the Goodman diagram form it is important to distinguish between mean stress and tightening stress in the bolt or similar part under consideration. If the dynamic load is truly alternating in character then the two have the same value, but in many practical cases the dynamic load varies from zero to a maximum in one direction only—tensile in the case of a bolt—resulting in the mean stress being greater than the tightening stress by half the dynamic stress. Referring to Figure 6–29, if a tightening load is applied corresponding to point A the maximum wholly tensile fluctuating load that can be applied is BD giving the mean level indicated by C.

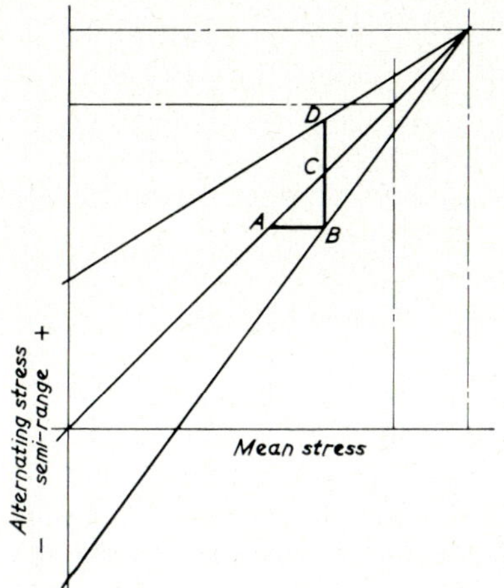

FIG. 6–29—*Tightening stress plus fluctuating tensile stress.*

6.17. *Highly stressed bolts*

Whilst the stresses in major complex parts are kept low because of lack of precise stress analysis it is essential for simple parts such as bolts which have to take up minimum space, to be designed in such a way that the stress requirements are known with considerably more exactness. In any case bolts and studs are something with which the operator and the maintenance engineer become very familiar in the course of their duties and it is important to understand the reasons for their shape, surface finish and material and to appreciate how to handle them so as to avoid disaster.

Bolts designed to carry large dynamic forces must be of adequate strength but they must also have high resilience; that is the capacity to absorb the maximum possible strain energy before reaching yield point. It is shown in text books on Strength of Materials that this condition is obtained by making them of uniform cross sectional area and as long as possible. With some bolts it is necessary to have portions of the shank enlarged in diameter, for example, the fitted portions of bottom end bolts; to obtain the greatest resilience these portions should be kept as short as possible.

A bolt that is stiff, relative to the components it holds together, will take a higher proportion of the dynamic load carried by those components than a less stiff more resilient bolt would do. This can be illustrated by the following idealized example. In Figure 6–30 part of a bearing housing is

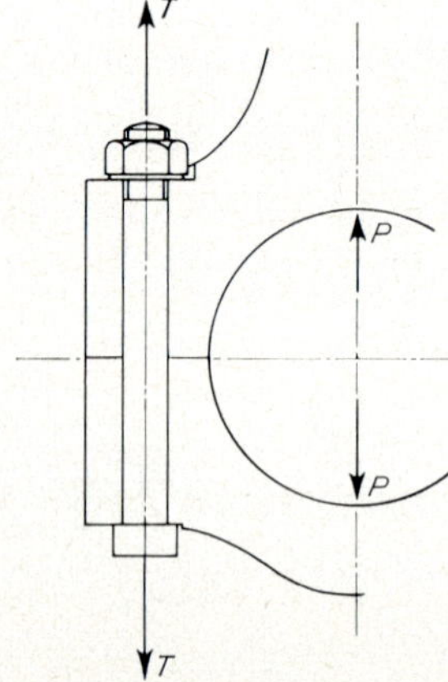

FIG. 6–30—*Bolt with dynamic load.*

represented, held together by a bolt, and subjected to an intermittent load. The diagram could equally well represent a cylinder head or other dynamically loaded component. The bolt and the housing may be assumed to obey Hooke's Law so that the expansion of one and the compression of the other is proportional to the load. When the bolt is tightened on assembly it extends and the housing compresses and in this condition the tensile load in the bolt is given by:

$$T_1 = S_b e = S_h c \qquad (6.1)$$

where S_b = stiffness of bolt (expansion per unit load)
S_h = stiffness of housing (compression per unit load)
e = extension of bolt
c = compression of housing

When the dynamic load is applied it may be assumed to act at the nut and bolt head surfaces and to extend the bolt by x. The new tension in the bolt is given by:

$$T_2 = S_b (e + x) \tag{6.2}$$

The new load on the housing is given by:

$$T_2 - P = S_h (c - x) \tag{6.3}$$

where $P =$ the dynamic load.

Subtracting equation 6.3 from equation 6.2 gives:

$$P = S_b e - S_h c + (S_b + S_h) x$$

but from equation 6.1

$$S_b e = S_h c$$

$$\therefore \quad P = (S_b + S_h) x$$

or

$$x = \frac{P}{S_b + S_h}$$

The change in the load on the bolt:

$$S_b x = \frac{S_b P}{S_b + S_h} = \frac{P}{1 + S_h/S_b} \tag{6.4}$$

Thus the amount of the dynamic load transferred to the bolt depends upon the relative stiffnesses of the housing and the bolt and a more resilient bolt will be called upon to take less load.

Another important consideration is stress concentration, and the effect of stress raisers. Where a change in diameter is unavoidable, the transition should be made as smoothly as possible, using very large radii. Screw threads represent a series of notches which, whilst not so severe as a single notch, still require consideration from the stress concentration aspect. To keep the stress in the threads within safe limits it is usual to design so that the core diameter is somewhat larger than the general cross sectional area of the shank.

Figure 6–31 shows an illustration of a connecting rod bolt with

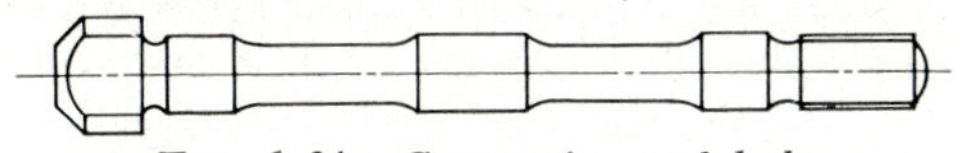

FIG. 6–31—*Connecting rod bolt.*

examples of all these features. When a dynamically stressed bolt fails because of some overload, it is almost invariably a fatigue failure which starts at some surface defect or feature. Because of this it is essential that the surface of the waisted part of the bolt which will be carrying the highest stress is finished to a high degree. In some cases the bolt is polished to a mirror finish. It will be obvious that in handling such highly stressed bolts great care must be taken not to damage their surfaces.

6.18. *Connecting rods*

The connecting rod is an example of a component subject to high mechanical loads. Axial forces result from the gas pressure and the inertia of the piston assembly modified by the side thrust arising in consequence of the angularity of the connecting rod according to the crank angle. These axial forces are, in fact, the loads on the small end bush. In addition there are transverse forces caused by the inertia effects of the mass of the rod itself and known as "whip". Both axial and transverse forces vary continuously throughout the engine cycle, but fortunately, do not attain their maximum values at the same time.

The maximum axial load is compressive and occurs at t.d.c. being due to the firing pressure modified by the inertia forces and has its highest value at low rev/min in four stroke engines. A tensile load occurs at 360° crank angle after firing, due to inertia of the piston. Its magnitude is dependent on the speed of the engine.

In two stroke engines the axial forces are often compressive throughout the cycle. If a reversal of load does occur it will be during a short interval of crank angle in the second half of the compression stroke when inertia forces increase more rapidly than the compression forces. The resulting tensile load is small. In both cases the axial load fluctuates requiring the design to be based on fatigue strength considerations. In addition a check is made on the buckling loads, regarding the rod as a strut, to ensure an adequate factor of safety against possible failure in this manner. The maximum transverse forces arising from whip occur about 75° or 80° from t.d.c. They become important in rods for high speed engines. An example of design calculations for a connecting rod is given in Ref. 6.

The cross section of the rod may be H form or round. H section rods are made as steel stampings and are very popular especially for high speed engines made in large numbers for which this method of manufacture is well suited. The thick flanges and thin web locate the material where it is required to carry the loads and form a light but strong shape for this major component of the running gear. The combination of forces on the rod results in the extreme fibres in the flanges of the H section being the most highly stressed and because of the alternating character of the stress these surfaces are sometimes polished to ensure freedom from stress raisers. It goes without saying that during maintenance operations care should be taken to ensure that polished surfaces are not damaged.

The round section of rod is favoured where production numbers are smaller. It is easily machined all over and just as easily polished to give freedom from stress raisers. A secondary function of the connecting rod in many designs is to convey the cooling oil to the pistons and this duty frequently calls for a passage of quite large diameter. Such a passage can often be more easily accommodated in a round section rod than in a rod of H section. The latter type has to be bossed to receive the drilling and this detracts from its ideal shape.

Whether the shank is H section or round it must join on to the top and bottom ends with large smooth radii or, better still, elliptical forms to spread the load evenly to the bearings, see Figure 6–23. The top end eye has thick sections of metal above and below to counter the forces from the pin which tend to split the rod, and it is important that the oil hole drilling has a smooth surface and well blending radii on its edges as any stress concentrations in this region can be disastrous. A description of the details of connecting rod design is given in Ref. 7.

The design adopted for the bottom end of the connecting rod is influenced by the conflicting requirements of the largest possible bearing surface and the ability to remove and refit the piston and connecting rod assembly.

In the interests of compact design the axial pitch length between cylinders is reduced to the minimum required to accommodate the cylinder jacket spaces or the cylinder head dimensions. After allowing for sufficient thickness of the crankwebs to provide a strong enough crankshaft the remaining length is all that is available for the main and bottom end bearings. Any increase of bearing area therefore can be obtained only by adding to the diameter of the pin or journal consequently enlarging the bearing housing.

Withdrawal of the piston and connecting rod assembly is either upward through the cylinder or downward and out through a crankcase door. Large crankpins and bearing housings coupled with a desire for engines compact in height, as well as length and breadth, place severe restrictions on a withdrawal route through the crankcase.

In the conventional design of marine type bottom end illustrated in Figure 6–32 the palm end of the rod matches the bearing housing. If this type of rod is to be drawn up through the cylinder bore the limits to bearing size are obvious. The modified design shown in Figure 6–32 in

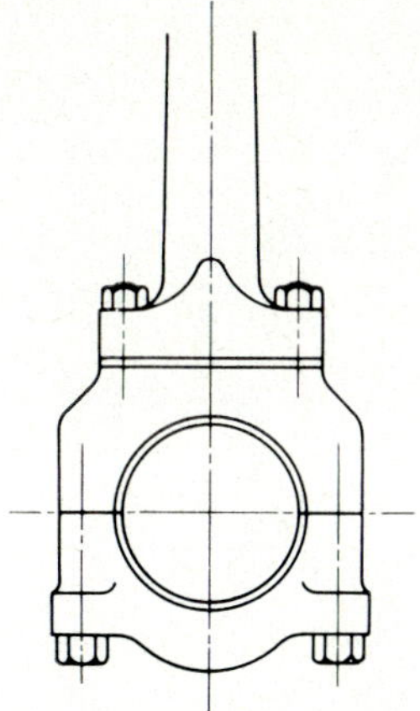

FIG. 6–32—*Bottom end with small palm.*

which the palm end is smaller than the bearing housing and is attached by separate studs, overcomes the problem to some extent.

The conventional design of "automotive" type rod has its bottom end split in a direction normal to the centre line again restricting the bearing size if the rod has to be removed through the cylinder bore. A split in an oblique direction permits the use of a bearing of larger diameter. As the loads on the cap result in shearing forces along the split some means of catering for these must be provided; a solution often adopted is that of serrated faces as shown in Figure 6–33.

For Vee form engines the most simple and straightforward arrangement is to have two rods, similar in design to those of in-line engines, side by side on the same pin. The design and production of such rods is identical and maintenance is straightforward; a cylinder on one bank may be overhauled without any need to interfere with the running gear of the

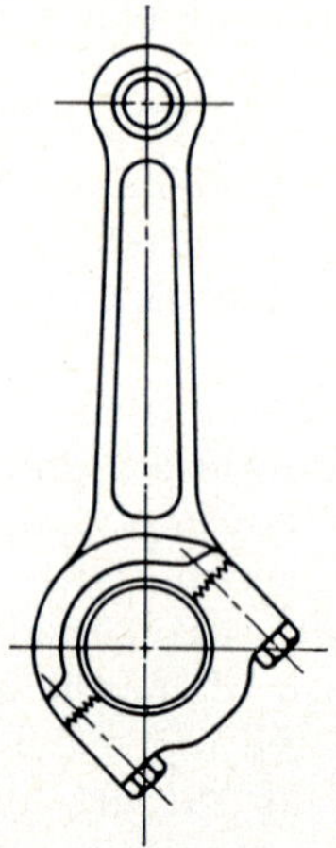

FIG. 6–33—*Connecting rod having B.E. obliquely split with serrated faces.*

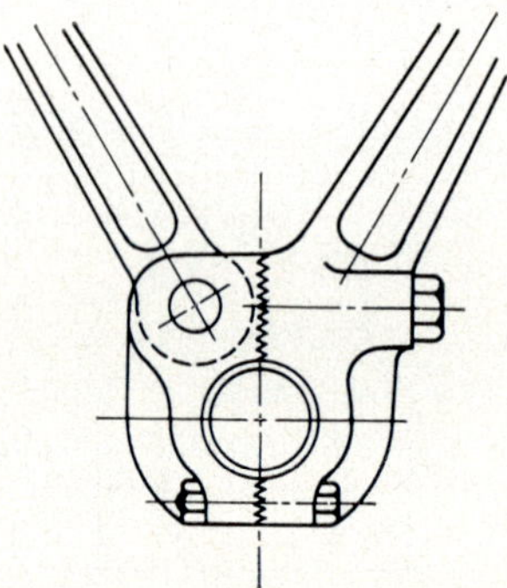

FIG. 6–34—*Articulated connected rod vertically split with serrated faces.*

corresponding cylinder of the other bank. This arrangement, however, carries a penalty of length, as the pitch of the cylinders has to be increased above that of the in-line engine.

Alternative arrangements that do not have this penalty of extra length are the fork and blade design and the articulated design. In both these designs the bottom end bearing of the master rod rotates relative to the pin but that of the slave rod oscillates. In the case of the articulated rod the motion of the piston attached to the slave rod does not correspond exactly with that of the one attached to the master rod owing to the bottom end bearing not being on the crank pin centre; this is of minor importance, however. When dismantling, the running gear of both cylinders is involved even though it may be desired to inspect one piston only.

The articulated master rod may be split horizontally if withdrawal through the crankcase is possible, but it is more usually split vertically for easier withdrawal of each half via the corresponding cylinder bore, see Figure 6–34. In both cases there are shearing forces along the faces of the split which are often made of serrated form to contain them.

REFERENCES

1. *Computers in internal combustion engine design,* Symposium. Proc. I. Mech. E., 1967–68, Vol. 182 Part 32.
2. LLOYD, T., HORSNELL, R. and MCCALLION, H., *An investigation into the performance of dynamically loaded journal bearings,* Symp. Journal Bearings for Reciprocating and Turbo Machinery, Proc. I. Mech. E., 1966–67, 181 (Pt 3B) 28.
3. MARTIN, F. A. and BOOKER, J. F., *Influence of engine inertia forces on minimum film thickness in conn-rod big-end bearings.* Proc. I. Mech. E., 1966–67, 181 Pt 1.
4. LLOYD, T. and MCCALLION, H., *A computer program for the design of reciprocating engine bearings,* Symp. "Computers in internal combustion engine design". Proc. I. Mech. E., 1967–68, Vol. 182 Part 32.
5. POPE, J. A., *Metal Fatigue,* Chapman and Hall.
6. PURDAY, H. F., *Diesel Engine Designing,* Constable and Co.
7. HENSHALL, S. H., *Just a detail—The connecting rod,* The Engineer, 6 March 1964.

CHAPTER SEVEN

Crankshafts and Shaft Alignment

7.1. *The crankthrow shape*

A basic crankthrow is shown in Figure 7–1. Its shape derives from the fundamental function it plays in the conversion of reciprocating into revolving motion. It is not a particularly good shape to carry the fluctuating stresses demanded of it and the support of the main bearings at each journal is vital to its strength and rigidity. When crankshafts are handled

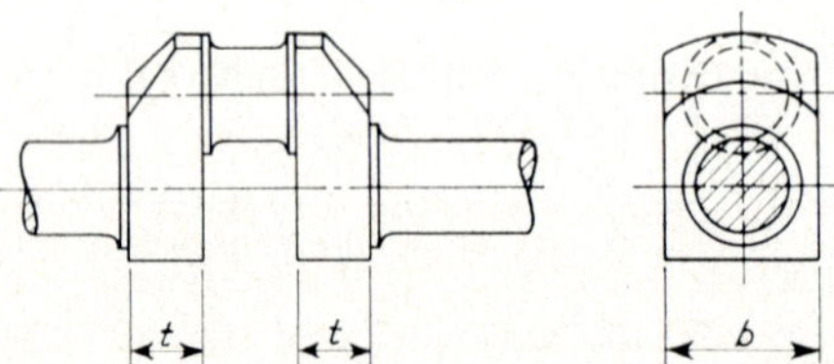

FIG. 7–1—*Crankthrow element.*

outside the engine on two supports, such as slings when lifting or stands when stored or being worked on, these supports should be placed so as to result in the minimum bending moment being applied to the shaft by its own weight. A long slender crankshaft, such as for a 16 cylinder vee engine, if supported only at its extremities can easily deflect under its own weight to an extent that the yield point of the material is reached at certain positions.

The pins and journals of the crankshafts of modern engines tend to be large in diameter and length in order to provide the maximum of bearing area to counter the large piston area and high pressures. The thickness of the webs, *t*, must, therefore, be as small as consistent with adequate strength if the engine is not to become unduly long and, in an attempt to compensate, the breadth of the web, *b*, is often enlarged, sometimes being made elliptical as in Figure 7–2. The large diameter of the pins and journals may lead to considerable overlap, lending strength to the webs.

The major stresses are due to bending of the webs and combined torsion and bending of the pins and journals.

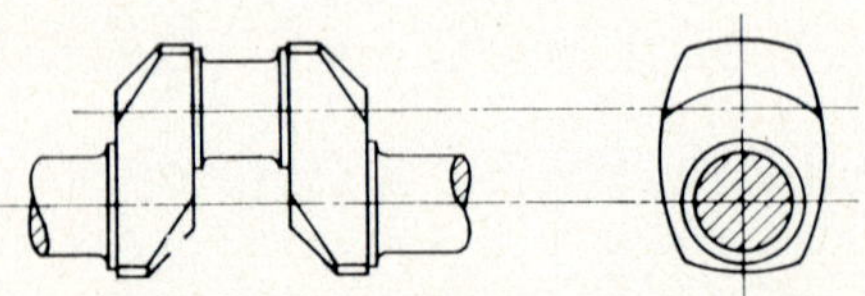

FIG. 7–2—*Elliptical webs.*

Balance weights or counter weights are sometimes fitted to reduce first order excitation or to ease inertia loads. The bolts by which they are attached, see Figure 7–3 have to withstand tensile stresses due to centrifugal force; any failure is likely to be disastrous. High speed engines frequently have crankshafts on which counterweights are formed integrally, as in Figure 7–4.

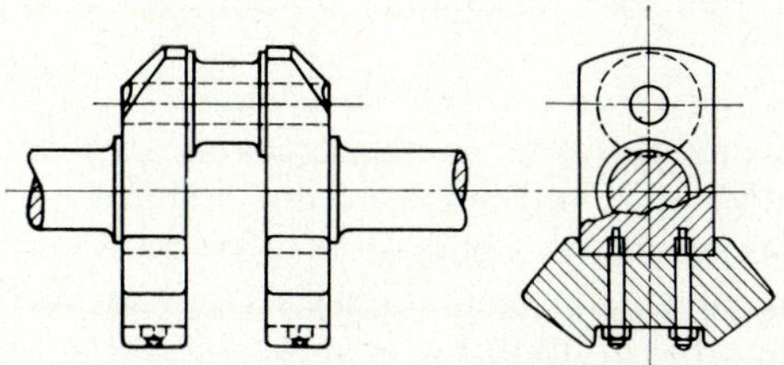

FIG. 7–3—*Counterweighted shaft.*

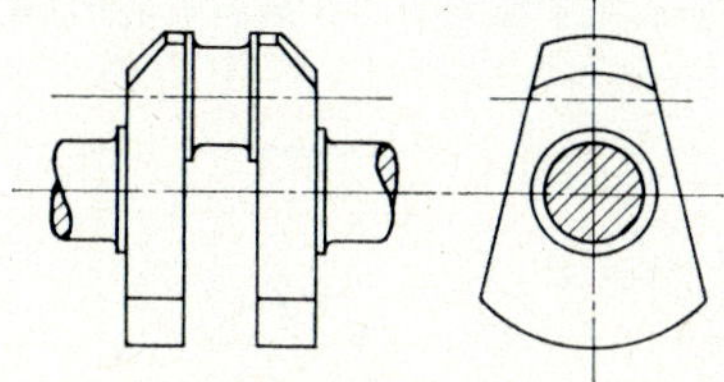

FIG. 7–4—*High speed engine crankshaft.*

To reduce out of balance effects the pins may be bored hollow as in Figure 7–3, and in lightweight engines the journals may be bored out also. Hollow pins and journals may yield an improved stress pattern.

7.2. *Crankshaft stresses*

The loads causing stress in the crankshaft arise from four sources.

1) *The working cycle.* These loads are imposed on the crankshaft by the connecting rods. They are comprised of the gas loads in combination with the inertia forces of the piston and connecting rod and are represented by the polar load diagrams of the large end bearing.

2) *Torsional vibration.* This subject is dealt with in more detail in Chapter 13. These stresses arise from the crankshaft twisting and untwisting, winding up rhythmically first in one direction and then in the other, due to vibration of the whole shafting system.

3) *Axial vibration.* These stresses arise from the crankshaft being alternately compressed and stretched along its axis in a concertina-like manner. This form of vibration is not so commonly encountered as torsional vibration.

4) *Misalignment.* If the main bearings supporting the crankshaft are not in alignment high stresses may be imposed.

In designing the crankshaft allowance must be made for the possibility of stresses arising from all of these sources and, furthermore, there must be a margin of safety to ensure that the crankshaft can survive any accidental further stressing or possible reduction of its strength during its life. A typical analysis may result in the following:

> That the sum of the stresses should not exceed 0·85 f where
> f = the fatigue strength of the crankshaft.
> Of this the cyclic working loads may account for 0·45 f
> Torsional vibration for 0·20 f and
> Axial vibration and misalignment together for 0·20 f.

7.3. *Nature of stresses in crankthrow elements*

The cyclic working loads on the crankpin can be resolved into two components: radial and tangential to the crankthrow, Figure 7–5. The radial components result in bending of the crankthrow element in the plane of the element as shown in Figure 7–6. The maximum radial component occurs when the crankthrow is vertical and the full firing pressure is present in the cylinder.

The tangential components bend the webs in the plane of the webs themselves as in Figure 7–7 and also twist the journals. The breadth of the

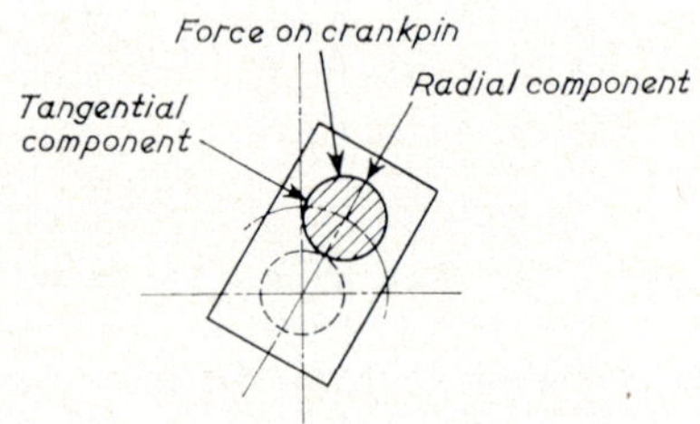

FIG. 7–5—*Force on crank pin.*

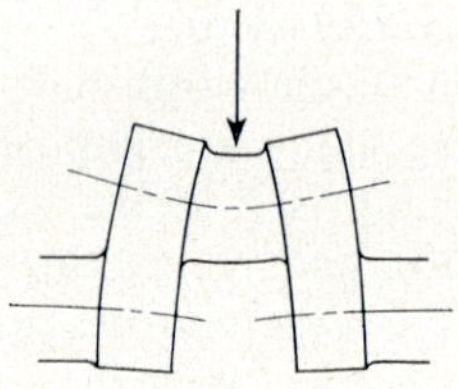

FIG. 7–6—*In plane bending from radial component.*

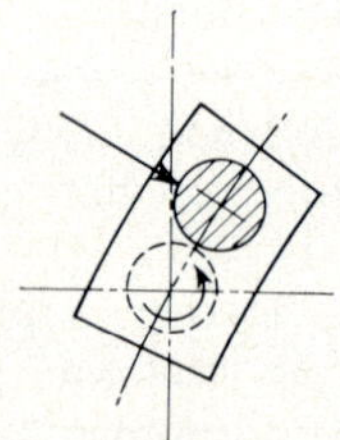

FIG. 7–7—*Tangential bending of web and torsion of journal from tangential component.*

web, dimension *b* in Figure 7–1, is comparatively large and the stresses from the bending in this direction are rarely high.

The torque transmitted along the shaft imposes torsional loading on the element under consideration. In transmitting this torque the two webs will bend tangentially as in Figure 7–7 but in opposite senses and the pin will also be strained in both torsion and bending.

Although all the power of the engine is transmitted through the last throw of the crankshaft this throw is not necessarily the highest stressed. The turning moment on each crankthrow element varies in direction during the cycle and the highest torsional loadings between adjacent elements will depend on the relative disposition of the cranks as determined by the firing order. A conventional analysis takes the crankshaft in any given angular position and summates the turning moments from the tangential components at each pin along the shaft. This is then repeated for other angular positions at small intervals to cover the complete cycle and the magnitude and position of the maximum torque ascertained.

Torsional vibration twists the elements, leading to torsional stresses in journals and pins and tangential bending of the webs. The twisting of the shaft also tends to shorten it by winding up and this may cause in-plane bending of the webs. The highest stress from torsional vibration may occur in a web or journal or pin. It is likely to be at, or near, the node.

Axial vibration results in in-plane bending of the element. Whereas torsional vibration is very common and difficult to avoid, axial vibration occurs comparatively infrequently.

Misalignment of the journals results in bending of the webs both in-plane and tangentially as the crank rotates. The in-plane bending causes the highest stresses as with the working cycle loads. The tangential bending will also result in torsional stresses in the pin.

Many of these stresses act together straining the material in combined bending and torsion. More rigorous methods of analysis are becoming possible with the advance of computer techniques.

7.4. *Regions of high stress*

Crankshaft failures, when they occur, are almost always by cracking as a result of stress concentrations at oil holes or in the fillet radii. Failure at the oil holes is caused by torsional stresses with cracks forming at 45° to the axis of the pin. To minimize the stress raising effect of an oil hole, it is essential that its edge should be smooth and well rounded with a radius about equal to the diameter of the hole.

Failures in the fillet radii may be due to bending stresses or combined bending and torsion and the cracks may progress into the web or into the pin or both. To minimize the stress concentration a large fillet radius is needed; it should be at least equal to 5% of the diameter of the pin or journal. Re-entrant fillets shown in Figure 7–10 are often used to avoid encroaching on bearing area. The highest stresses mostly occur in the fillet radius on the underside of the crankpin, see Figure 7–8.

The pins and journals of some designs of crankshafts have hardened and ground surfaces and in these cases the fillet radii are arranged so that the grinding does not affect the fillet shape. Figure 7–9. It is important that the hardened zone stops well short of the fillet or completely envelopes the fillet.

A process that is occasionally applied is that of cold rolling the fillet radius to increase the fatigue strength by producing a residual compressive stress at the surface.

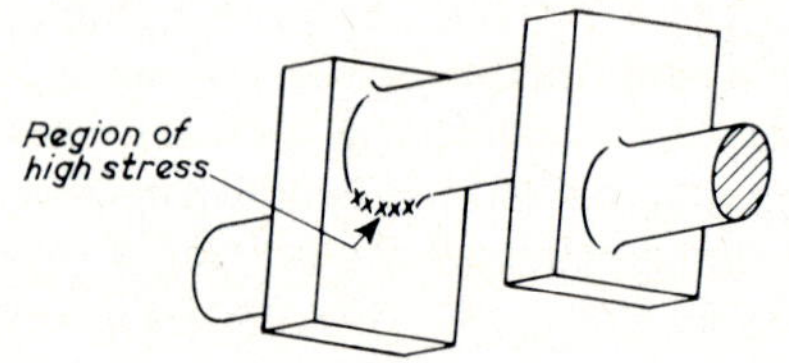

FIG. 7–8—*Region of high stress in crankpin fillet.*

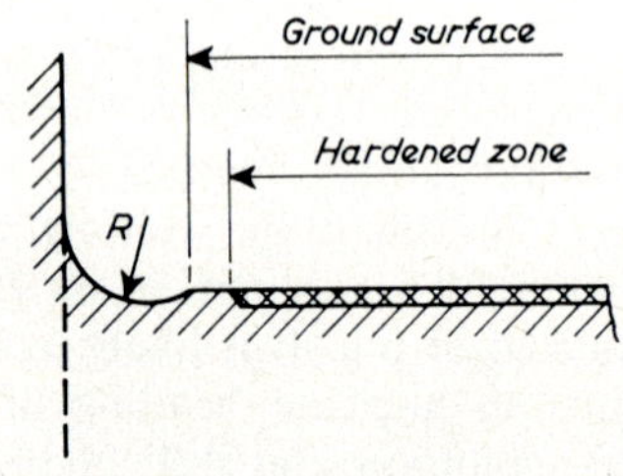

FIG. 7–9—*Fillet of surface hardened and ground pin.*

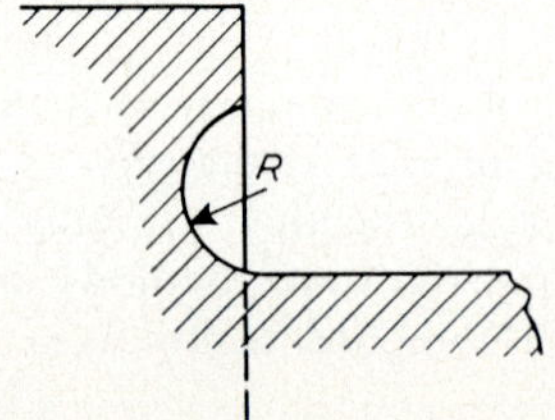

FIG. 7–10—*Re-entrant fillet.*

7.5. *Materials*

The materials for crankshafts are chosen from a comparatively small range of carbon and alloy steels. It cannot be over emphasized that as all the loads imposed on a crankshaft are fluctuating in nature, the property most required is that of high fatigue strength and this is more important than a high UTS. Many highly alloyed steels, although having a high UTS, show little improvement in fatigue strength over steel having less alloy content.

A further requirement in modern high output engines is that the running surfaces of the shaft shall be hard enough to work in conjunction with bearing materials having a high fatigue strength to withstand high bearing pressures. Fortunately, hardness and strength tend to keep in step.

Twenty-eight ton (433 MN/m^2) 25 carbon steel was, at one time, widely used with white metal bearings but the high loads of modern engines required harder bearing materials and harder shafts.

The 40 carbon steels having a brinell of about 220, are easy to forge and work. They have a UTS of 37–45 tons (about 575 to 700 MN/m^2) and are suitable for use with white metal thin wall bearings and are sometimes used with copper lead bearings having an overlay of white metal, or with aluminium-tin bearings.

The 1% chromium-molybdenum and 3% nickel steels have a higher tensile strength of about 45 tons (approx. 700 MN/m^2) and correspondingly higher hardness; they are generally employed for induction or flame hardened crankshafts; the surfaces thus reaching 480–490 brinell for use with copper-lead or aluminium-tin bearings.

Two and a half per cent nickel-chromium-molybdenum steel of 55–65 tons (about 850 to 1,000 MN/m^2) UTS develops its strength and hardness of about 300 brinell by a simple heat treatment and is a suitable material for the crankshafts of medium speed engines.

Chromium-molybdenum steels can be nitrided to give a surface hardness of 900 to 1,000 VPN. They are popular for smaller high speed engines, the shafts being made by die forging.

7.6. *Forgings*

Crankshafts for medium speed engines are conventionally forged by making a slab from the billet, reducing it to form journals and twisting the blocks so formed to the appropriate angles to become the crankthrow elements. As can be seen from Figure 7–11 the material which was at the

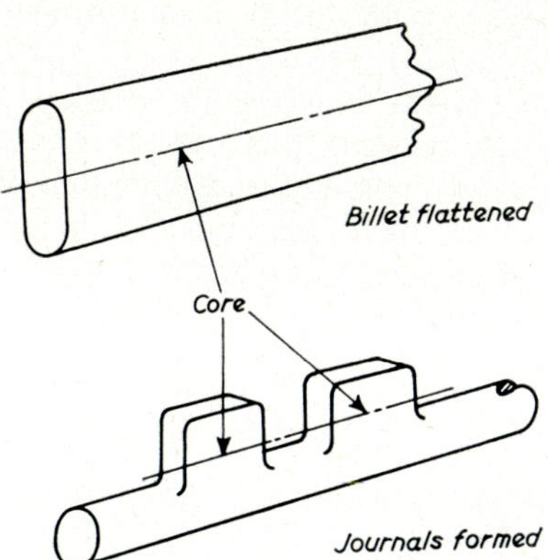

FIG. 7–11—*Crankshaft forging conventional methods.*

core of the original billet appears at the underside of the crankpins when machining is completed. The least satisfactory material is thus situated at the point where stress concentrations are highest; this influences the fatigue strength for the crankshaft and a much lower value has to be assumed than might otherwise be taken for the material.

Methods have been devised for forging crankthrow elements between sets of dies to produce what is termed continuous grain flow. By these processes material is distributed so that the grain structure follows a path round the crankthrow in a manner which contributes to the strength. The material which was originally at the core of the billet is displaced so that it remains in the centre of the pins, webs and journals. One of the chief advantages of producing crankshafts by continuous grain flow methods is that good material appears at all the surfaces that are highly stressed. See Figure 7–12.

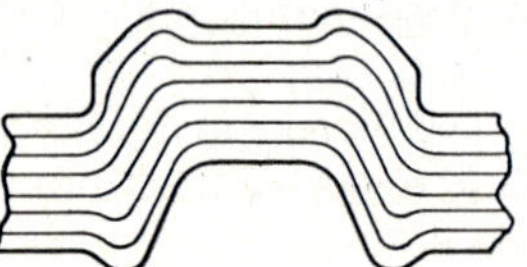

FIG. 7–12—*Continuous grain flow forging.*

Crankshafts for small engines may be forged completely between closed dies which tends to produce a desirable pattern of grain flow.

7.7. *Reconditioning*

During the life of a crankshaft, some wear on the pins and journals is likely to be experienced. These cylindrical surfaces tend to wear to an oval shape and eventually, a limit may be reached, perhaps when a bearing ceases to perform satisfactorily because of the effect of the worn shape upon the oil film. Reconditioning of the crankshaft then becomes advisable to produce truly cylindrical surfaces again, to be used in conjunction with undersize bearings. Because of this requirement it is not uncommon for crankshafts to be designed with an allowance on the pins and journals for such reconditioning.

The webs are usually the most highly stressed part of the crankshaft and, although it is often possible to recondition a shaft by regrinding of pins or journals, it is rarely possible to do anything with damaged webs.

The importance of the fillet shapes cannot be overstated in connexion with the re-grinding of pins or journals and it is essential that the new surfaces are blended in with large radius fillets into the existing surfaces. On no account must a sharp corner or step remain. Figure 7–13. It is also

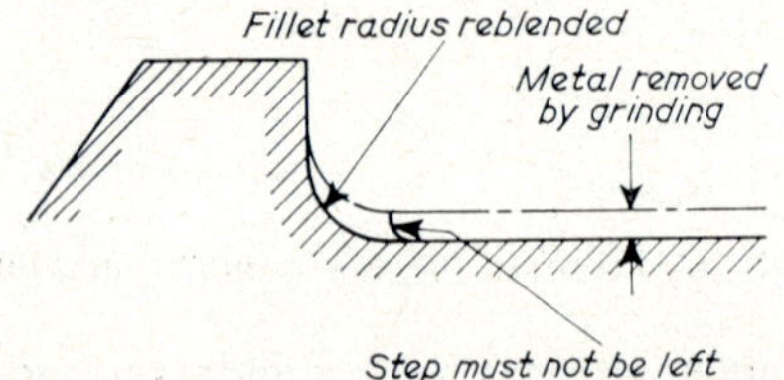

FIG. 7–13—*Attention to fillet radii after regrinding.*

essential to round off any sharp edges that may be produced at the oil hole. If fillet rolling has been applied to the original shaft it is obviously essential to repeat the process on the reformed fillet of the reconditioned shaft.

Crankshafts that have hardened surfaces for the pins and journals have obvious limits on the amount of re-grinding that may be performed. Only a small amount of material may be removed if the hardened case is not to be destroyed altogether. Nitrided shafts have a very thin hardened zone and it is usually impossible to re-grind them without removing it completely. However, it is possible to re-harden by nitriding after grinding.

Local accidental damage to the surface of pin or journal such as may be caused by a knock, can be tolerated if it is not near to a fillet radius and if the edges are made well rounded and smooth.

7.8. *Coatings*

Sometimes attempts to recondition crankshafts are made by rebuilding the surfaces of pins and journals. A popular means of doing this is to grind the surface cylindrically and restore it to diameter by chromium plating. The chromium plate provides a hard surface and one or two manufacturers have used it on new shafts because of this property. However, chromium plating can be dangerous when applied to some materials which are notch sensitive as fatigue failures can start from the interface between the parent metal and the chromium deposit. Before re-claiming shafts by this method the manufacturers should be consulted in the matter.

Metal spraying is sometimes proposed as a method of reclaiming shafts. Again, this has dangers from fatigue conditions originating at the interface, and requires close collaboration with the manufacturers to avoid unnecessary risk.

7.9. *Engine frames*

If the crankshaft is to perform satisfactorily every care must be taken to see that it is properly in alignment: this means that the main bearings supporting it at each journal must be correctly aligned, and this in turn depends upon the engine frame and the supports or seating for it.

Engine frames may be made of cast iron or fabricated from steel

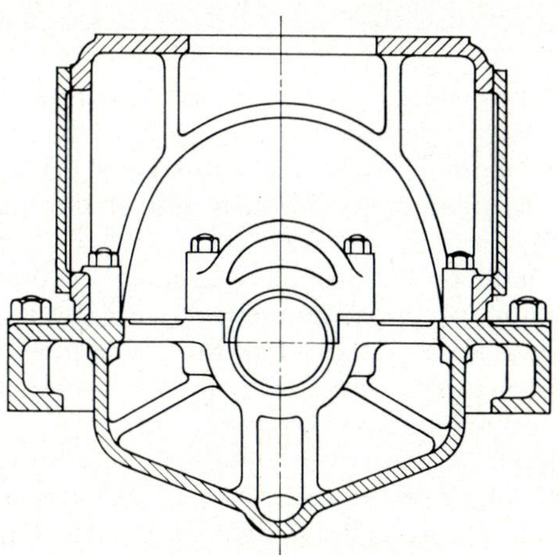

FIG. 7–14—*Bedplate.*

plate or from a combination of steel castings and plates. The designs fall into three main classes: bedplate, underslung and tunnel form.

The bedplate, Figure 7–14, is the conventional form for medium speed engines and has been used for many years. Its chief assets are easy access to the crankchambers and convenience of production for larger engines. Because of its comparatively shallow depth it is usually given continuous support.

A modern variant has been to deepen the bedplate to produce a U-form shape in which the crank chamber doors are incorporated. This gives a much more rigid bedplate better able to maintain its alignment and support the shaft but it is more difficult to machine.

With all bedplate designs the reactions between the bearings and the gas forces on the cylinder heads have to pass through a joint between the bedplate and the column of the engine. Frequently, through-bolts are used to carry these loads in order to avoid heavy sections in the frame, but even so, the oil tightness of joints between the column and bedplate depends on careful maintenance of bolt tightness.

The underslung design, Figure 7–15, has been conventionally used with

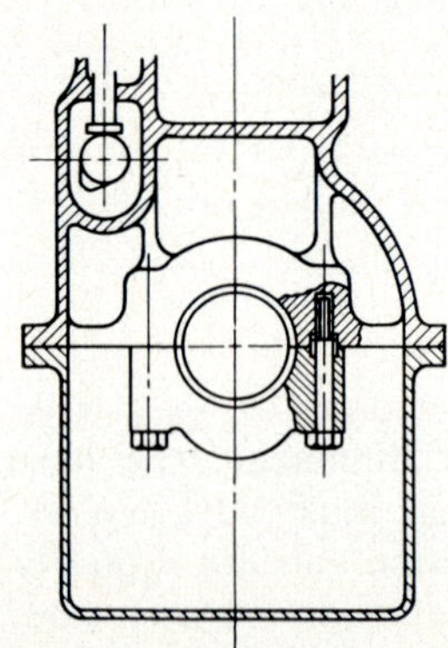

FIG. 7–15—*Underslung main bearing high speed engine.*

smaller engines. It provides a very rigid construction but is difficult of access. Many small high speed engines rely on removing the engine from its mountings and inverting it in order to obtain access to the crank chamber. In marine service this method is obviously out of the question for all but the smallest engines.

A number of modern designs have applied an underslung principle to large engines combining the advantages of a rigid construction, with gas forces and bearing reactions contained within a one piece frame work, with adequate access through the sides for bearing inspection. Figure 7–16 shows a fabricated frame of this form.

The greater depth of the frame obtained by combining the cylinder casing column and crankcase together gives great benefits in rigidity and frames of this design are frequently used with point supports only. Point supports give comparative freedom from distortion due to working of the

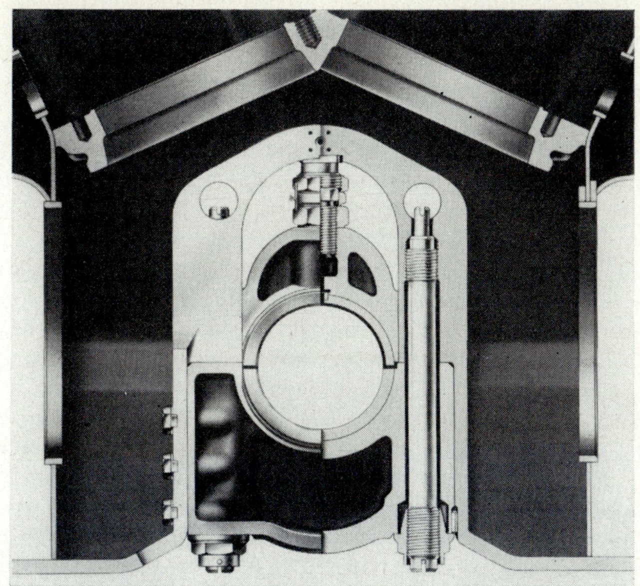

Fig. 7–16—Underslung shaft with removable bearing caps.

seatings but require a frame design rigid enough to maintain its shape without continued support.

Tunnel shaft designs provide the most rigid frames and are commonly used in conjunction with point supports. They have a minimum of joints to be maintained oil tight and gas forces and bearing reactions are contained within the one member. For the design shown in Figure 7–17 bearing caps

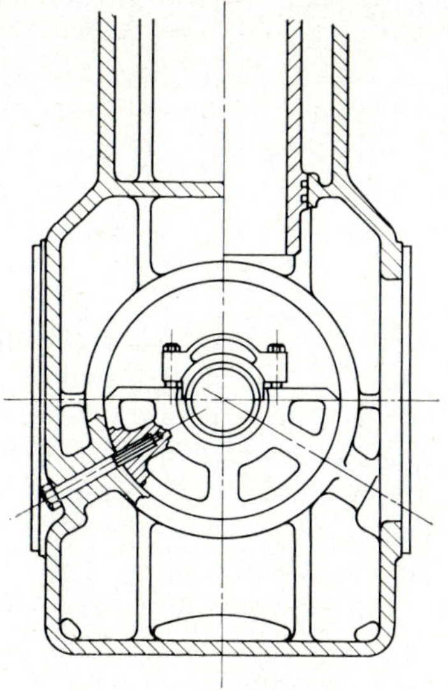

FIG. 7–17—*Tunnel shaft.*

are provided and bearing inspection is carried out in a conventional manner. Tunnel shaft engines require axial space for the withdrawal of the crankshaft.

7.10. *Alignment*

Correct alignment calls for two main considerations: that each piece of machinery (*e.g.* engine, gearbox) is supported without distortion, and that it is held in correct attitude to the neighbouring components of the installation.

As well as giving proper attention to these matters when building the vessel it is also most desirable to carry out regular checks throughout its life to detect at an early stage any deterioration such as the onset of abnormal bearing wear and to correct any resulting misalignment.

All modern medium speed and high speed engines have their main bearings bored accurately in-line in the frame. When erecting the engine in the work shop or installing it in the vessel it is essential that the frame is mounted and supported so that its shape is exactly the same as when it was machined. Obviously the more rigid designs of frame are easier to deal with in this respect in that they can be supported at a fewer number of points without causing distortion.

External reference marks or surfaces on frames can be of the greatest assistance in ensuring that they are not distorted when placed into position on the seating. A series of horizontal and vertical faces are the best arrangement as they permit the easy mounting of optical levelling equipment.

The most frequently used method of checking that the bearings are accurately in-line is by crank web deflexions. A deflexion micrometer dial gauge is placed between each pair of webs in turn and readings taken with the crank in four angular positions at 90° to each other, top, bottom, port and starboard. This is performed for each of the crankthrows and from a study of the readings it can be deduced which bearings are high or low or horizontally displaced and corrections made to the assembly or supporting of the frame. Usually distortion occurs only in the vertical direction being due to the effects of self weight of the engine. The engine builder will have laid down the acceptable limits for deflexion gauge readings. For medium speed engines a maximum deflexion of the order of $\pm 0{\cdot}025$ mm is generally specified.

If one main bearing is lower than the others the crankshaft may be sufficiently rigid to bridge across this bearing without the deflexion readings being large enough to reveal the situation. However, it is essential that bearings in this condition are detected as the shaft will certainly deflect when the high forces of the operating cycle come upon it. The usual method is to force the shaft journals down on to the main bearings by taking up the clearances in the caps with packing soft enough not to damage the shaft; sometimes special caps are used for this purpose, the upper shells having been removed.

With larger engines it is common practice for the manufacturer to provide a bridge gauge of the form shown in Figure 7–18, stamped with a set of readings giving the relative height of each crankshaft journal as measured by feelers in the position shown when the engine is new and run in. Any increase of these readings is a measure of bearing wear.

Abnormally heavy wear of one bearing may not always be evident from bridge gauge readings as the sag of the shaft over the span of two crankthrow elements may be less than the amount by which the bearing is worn. A reading near to that given by the natural sag for the double span is an indication of heavy wear which should be confirmed by removing the shell in question and measuring its thickness.

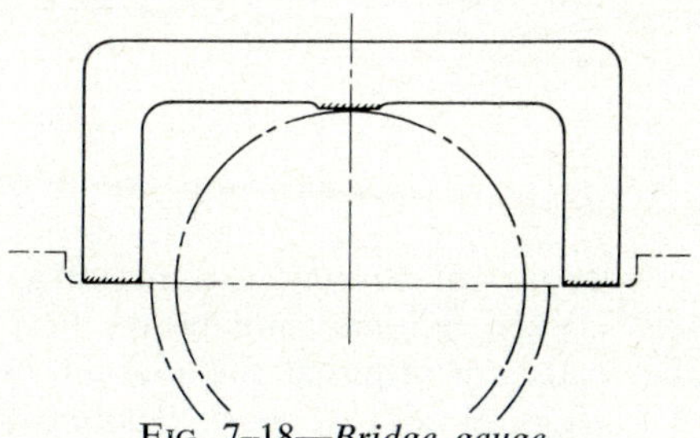

FIG. 7–18—*Bridge gauge.*

Large shafts sag more than small shafts under their own weight and for smaller engines bridge gauge readings are less valuable as an indication of uneven wear between bearings.

Should a large overhung mass such as a heavy flywheel or the driving half of a fluid or electro-magnetic coupling be attached to the crankshaft it will influence the deflexion readings, sometimes considerably, depending on the rigidity of the shaft and the disposition of the bearings. Ideally, the best course of action is to remove the mass and ensure that there is no distortion of the engine frame by checking the deflexions of the unloaded crank. In practice the effect of known overhung masses is usually known to the engine builder and he can advise on the deflexion readings to be expected.

7.11. *Correct attitude*

The problem of the overhung mass mentioned in the preceding paragraph is one example of the difficulties that arise in aligning or re-aligning a complete machinery installation. The first and most important matter to be settled is whether the shaft system should follow a straight line. In many cases the best line does not do so. Considerations of bearing loading in relation to the various masses supported by the shaft and considerations of hull deflexions under various conditions are taken into account by the shipbuilder and Naval Architect when determining the line that the bearings and their supports should follow. When replacing shaft bearings or machinery, the original manner of alignment must be ascertained as a first step. To assume a straight horizontal line irrespective of whether the ship is afloat or in dry dock, is courting disaster.

Taking the overhung mass as an example, a state of affairs is represented in Figure 7–19 which shows a mass at the end of a crankshaft supported in bearings A, B and C. At (a) the bearing A is in-line with B and C and the shaft bends under the influence of the mass and produces a deflexion reading in the crank element between B and C. By raising bearing A as shown at (b) the bending moment on the crankshaft between B and C is reduced and the deflexion reading can be brought to zero by adjusting the level of A. There is, of course, a higher load imposed on the cap of bearing B. By this means the crankshaft is not so highly stressed

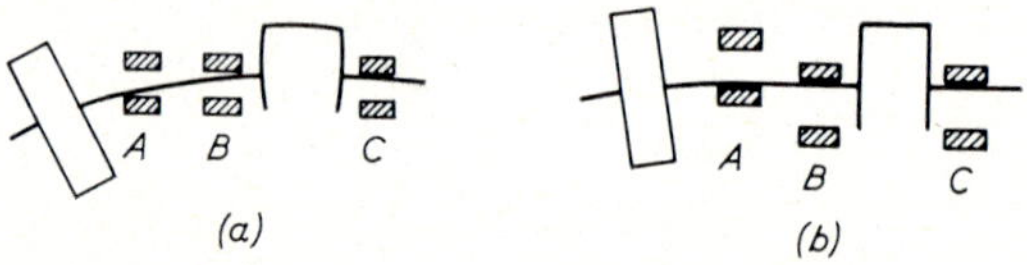

FIG. 7–19—*Diagram of bearing alignment with overhung mass.*

when in the static condition, but bearing A is no longer on the same horizontal line as B and C. Even this may not be the best solution, for when the dynamic working loads are imposed the journals will be forced down to the bottom of bearings B and C periodically with bending in the opposite sense. In fact, the correct height for bearing A may have been the subject of lengthy investigation by the engine builder. This is one illustration of one of the factors to be considered in settling the line for the shafting to follow.

If a mass is overhung only temporarily during the lining up and installation processes, then the mass itself should be supported on temporary jacks and its height adjusted to relieve the shaft of stress and so position it in a straight line to match the supporting shaft which is to be joined to it at the other side. This is illustrated in Figure 7–20. When

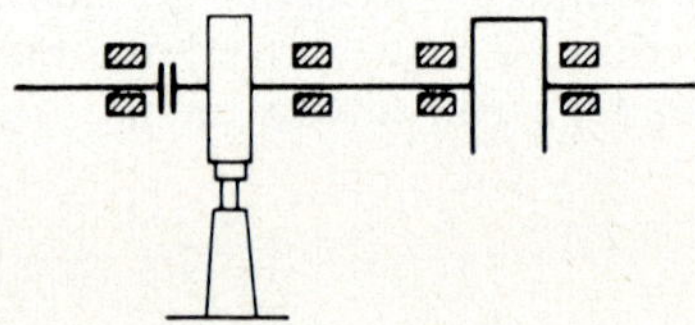

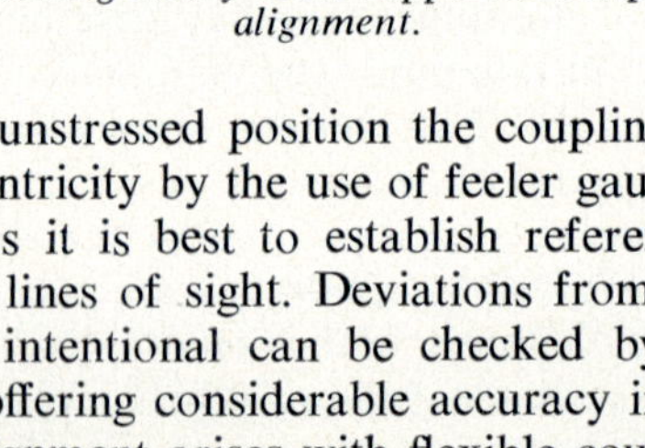

FIG. 7–20—*Diagram showing heavy mass supported temporarily during coupling alignment.*

the shafts are in the unstressed position the couplings can be checked for parallelism and concentricity by the use of feeler gauges and straight edges.

For most systems it is best to establish reference datum lines using taut wires or optical lines of sight. Deviations from these reference lines, both accidental and intentional can be checked by direct measurement, the optical methods offering considerable accuracy in this respect.

A problem in alignment arises with flexible couplings. By their design they are intended to cope with malalignment and, therefore, there is no rigid relationship between input and output shafts on which reliance may be placed. The methods to be adopted differ according to the design of coupling and the advice of the manufacturer should be sought in each case. Quite often the use of special setting gauges or other equipment is needed and a precise set of instructions is required to be followed.

Because a coupling is advertised as tolerating a certain amount of misalignment, angular or lateral, it does not mean that this can be regarded as a margin for error in aligning the installation. The best service will result from correct and true alignment leaving the scope of the coupling available to cope with the displacements that occur during operation.

Some couplings are designed primarily to accommodate angular and lateral displacement arising from temperature changes or deflexions of the machinery and seating. Others may be chosen entirely from considerations of torsional flexibility in order to provide acceptable torsional vibration characteristics and may tolerate only very small deviations from parallel and concentric input and output shafts.

7.12. *Axial positioning*

The thrust collar transferring the main propulsive force to the hull provides a fixed reference relative to the hull for the axial position of the machinery shafting system. The thrust is usually located immediately aft of the gearbox or immediately aft of the engine in cases where no gearbox is fitted. It often forms part of the gearbox, being embodied in the casing. Shafts rigidly connected to the thrust shaft must be free to accommodate relative expansion under temperature or other effects. This includes engine crankshafts.

Double helical gears impose fixed axial relationships on the shafts which carry them, but single helical and straight spur gears do not. The input shafts of gearboxes with the latter form of teeth therefore, require axial location.

Many flexible couplings are designed to be incapable of transmitting axial loads in order that they may accommodate axial displacements as well as those in other directions. Wherever they are used appropriate axial location of shafting is required.

Fluid couplings exert a thrust tending to separate the two halves axially. Their casings embody thrust carrying bearings to take account of this. These bearings determine the relative axial positions of input and output shaft systems.

Axial location of the engine crankshaft, as opposed to the accommodation of large axial thrusts, is usually achieved by allowing limited clearance at the sides of one particular journal bearing. Flanges (in some instances, crankwebs) on each side of the one journal in question bear against flanges or rings on the side of the bearing shell, cap or housing see Figure 7–21. Engines are designed so that the axial location can be included or not, according to the requirements of the system. It is important that only one location of this nature exists on each shaft assembly and very important that such a location should not be in existence on the same shaft assembly as a thrust bearing.

7.13. *Seatings and mountings*

The satisfactory alignment of the installation and its maintenance under all operating conditions depends a great deal upon the seatings for the various components. Not only do they have to sustain the weight of the machinery but they must hold it in a reasonably rigid relationship against accelerations in all directions and under occasional adverse conditions of list and trim; in addition they must absorb the thrust and torque reactions and the effects of unbalanced forces and couples without giving rise to vibration.

The mounting for the thrust block, for example, must be designed to convey the thrust responsible for propelling the ship from the shafting to the hull. This requires not only strength but also rigidity if axial positioning of the components is not to be affected. Thrust is not always steady and lack of rigidity can lead to axial vibration being set up.

The torque reaction of the engine is equal and opposite to the turning moment it applies to the crankshaft. As the turning moment on the shaft

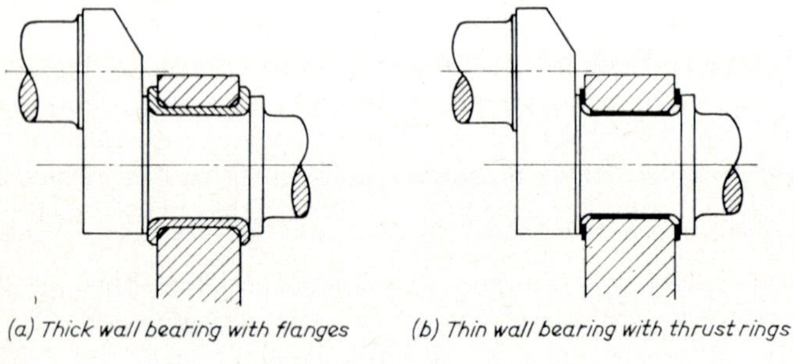

(a) Thick wall bearing with flanges (b) Thin wall bearing with thrust rings

FIG. 7–21—*Axial location.*

is not uniform the torque reaction is not uniform either. It may be regarded as arising from the side thrusts of the pistons being transmitted to the frame of the engine and tending to rock it about the crankshaft centre line in opposite sense to the fluctuating turning moment applied by the pistons to the crankshaft. Because of the fluctuating nature of the torque reaction, vibrations can arise and again it is essential that the seating and mountings should be designed from considerations of rigidity as well as strength.

A gearbox which effects a reduction in speed between input and output also increases the torque and in consequence has a torque reaction. The flywheel and couplings will have smoothed out most of the torque fluctuations before the input gear but it is prudent to expect some cyclic variation in torque reaction at the gearbox seating.

The unbalanced forces and couples, which are described in Chapter 6 react on the engine seating in the vertical and horizontal planes and may be a source of vibration if the design is inadequate.

In large installations the seatings for the thrust block, gearbox and engine may be separate, but in most cases the thrust block is incorporated in the gearbox so requiring the gearbox seating to be capable of transmitting the thrust. Small high speed engines are often of a form in which gearboxes with thrust bearing and engine are integral and it is the engine bearers through which the thrust is transmitted to the hull.

Seatings generally take the form of strong longitudinal girders forming part of the structure of the hull and braced laterally.

Not infrequently lubricating oil tanks are built into the seatings for engines and gearboxes. During operation the lubricating oil becomes heated raising the temperature of the portions of the seating forming these tanks. Unless carefully designed, the resulting expansion can lead to undesirable distortion of the seatings which is reflected in misalignment of the machinery. Ideally, the design should be such that no changes in shape can be transferred to cause distortion of any equipment, but this is rarely possible to achieve and it is a wise precaution to check the alignment when the machinery is warm on the occasion of the trials and to carry out any necessary corrections.

Location of the machinery and the engines in the horizontal direction is usually by means of fitted bolts which may be responsible for transmitting thrust forces and these are usually situated at one end only of engines or gearboxes so that relative expansion of machinery and seatings can be accommodated.

Small high speed engines having underslung or tunnel shaft designs of frame usually have a small number of mounting points and are easily levelled. Larger engines requiring a continuous mounting are, in practice, supported by chocks placed at close intervals.

Efforts are sometimes made to prevent vibration from the machinery being transmitted to the hull by using resilient mountings to support it. It is shown in Chapter 12 that, in order to isolate the vibrations of the engine from the hull, the resilient mountings have to have great flexibility. In marine applications where the weight of the engine may act in different directions at different times the static deflexion of such mountings must receive careful consideration. It is necessary to provide some form of limit stops to prevent excessive movement in lateral and fore and aft directions as well as in the vertical direction.

The movement of an engine on flexible mountings can be considerable giving rise to a need for corresponding flexibility in the shafting. Cardan shafts are sometimes used to meet the requirements. Also all piping connexions to the engine must be flexible.

The centre of oscillation of an engine on flexible mountings is often reasonably close to the shaft centre. Whilst this eases the problems of the shafting alignment it can lead to large oscillatory movements of the engine itself.

In twin engine installations there may be advantages in mounting both engines rigidly on a common base or "raft" which is supported on flexible mountings. The amplitudes of vibration are generally smaller but the centre of oscillation is obviously not near to the crankshaft centre line of either engine. The alignment problems that arise from this are solved if the gearbox also is mounted on the raft, but this requires a separate thrust block on a seating rigidly attached to the hull structure and a flexible coupling between the gearbox output and the thrust shaft.

BIBLIOGRAPHY

ARCHER, DR. S., *Some factors influencing the life of marine crankshafts,* Trans. I. Mar. E., 1964, Vol. 76.

HOSHINO, J. and ARAI, J., *Strength analysis of diesel engine crankshafts,* Trans. I. Mar. E., Indian Division Supplement, Nov. 1966.

HOSHINO, J., *Strength design of diesel engine crankshafts,* J.S.M.E. Bulletin, Vol. 11, No. 43, 1968.

LANGBALLE, M., *Investigations into the stressing of crankshafts for large diesel engines,* Trans. I. Mar. E., 1966, Vol. 78, No. 12.

CHAPTER EIGHT

Some important details

8.1. *Piston ring sealing action*

Piston rings play a vital role in the successful operation of the diesel engine. They are amongst the least accessible of the components and are designed nowadays to run for upwards of twenty thousand hours between overhauls. It is imperative, therefore, that on the occasions when they are exposed they should be most carefully inspected for correct functioning. Renewals or reconditioning of pistons and rings must be carried out thoroughly bearing in mind the long period of service likely before the next opportunity for overhaul.

The action by which the rings seal the gas in the cylinder is seen in Figure 8–1. The pressure of the gas in the clearance spaces forces the

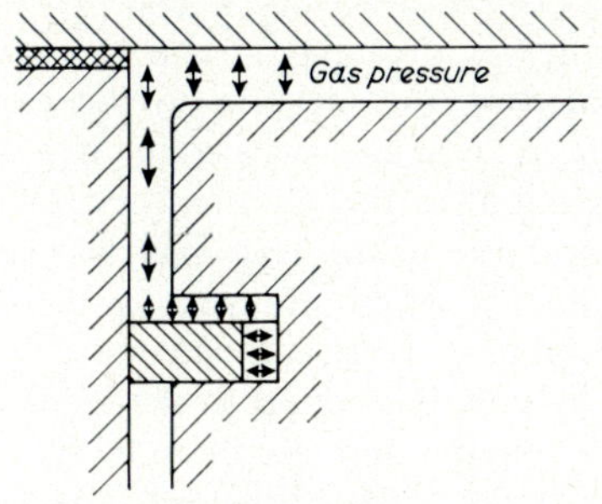

FIG. 8–1—*Piston ring sealing action.*

ring down on to the side of the groove and outwards on to the cylinder wall. Contact at these faces must be gas tight demanding smooth mating faces all the way round the ring. To make it possible to assemble the rings on the piston and to enable them to conform to the cylinder bore they have the familiar split or gap. This gap provides a leakage path for the combustion gas so that one piston ring alone is insufficient to seal adequately. A few diesel engines employ just two compression rings but usually there are three or four. Sometimes the lowest ring in the pack performs the dual function of gas sealing and oil control. The top ring bears the brunt of the sealing task, it sustains the greatest pressure drop across it and it operates at the highest temperature.

8.2. *Piston ring lubrication*

Lubrication of both sealing faces of each ring is essential. The face in contact with the cylinder liner wall is rubbed up and down it at high speeds and under high pressure when at the top of the stroke. There is also relative movement between the lower face of the ring and the groove as the piston moves across from one side of the cylinder to the other when the direction of thrust changes as the crank passes through the dead centre positions. As the cylinder bore wears more at the top of the stroke, where the wall pressure of the ring due to the gas is highest, than at the bottom the ring diameter is continually changing during operation and it does this to a greater degree as the engine gets older. This also contributes to the relative motion between the ring and its groove. If lubrication is to be maintained the top ring must be situated in a region where the temperature of the groove is sufficiently low to avoid excessive oxidization of the lubricating oil. Generally speaking it must be below 200°C. The depth of the top land or junk is determined by this consideration and this accounts for it being considerably larger than the lands between the rings. The operating conditions for the top ring will obviously be less arduous if the clearance between the junk and the cylinder liner wall can be kept to a minimum when the engine is on full load. A modern technique is to turn a fine thread on this portion of the piston; the hot clearance may then be made very small and in the event of the piston touching the cylinder bore the metal has a place to which it can retreat without causing damage.

8.3. *Gap clearance*

Wear of piston rings is greatest on the cylindrical face and is most easily assessed by measuring the increase in the gap dimension when the ring is placed squarely in a gauge consisting of a flat rigid ring with an inside diameter exactly equal to the nominal bore of the cylinder. If such a ring gauge is not available the piston ring may be placed inside the cylinder liner bore right at the bottom where the wear on the cylinder bore is usually negligible, taking care that it is positioned squarely. The dimensions of the gap when the ring is new and when at the limit of wear are individual to the make and size of engine and the manufacturer's instructions should be followed implicitly. It is clearly desirable that the ring gap should be as small as possible and it is equally important that it should never close completely. If the butts of the ring were to make contact with each other due to thermal expansion during operation the face would be forced into heavy contact with the cylinder bore and severe scuffing would ensue causing the ring to overheat, the lubrication to be destroyed and the situation aggravated by itself to the point of disaster.

8.4. *Side clearance*

Axial or side clearance of the ring in its groove is essential to ensure that it is free to move relative to the piston in order to maintain contact with the cylinder bore. The clearance is best kept to the minimum that will serve this end for a number of reasons. A large clearance permits the ring to hammer the land below it increasing the stress in the land and encouraging wear of the ring and the groove. Groove wear can be a problem with light alloy pistons.

Piston rings tend to pump oil from the lower, well lubricated side to the upper, combustion chamber side where it is burnt. The compression rings are lubricated by this process but if it is excessive a high lubricating oil consumption will result together with carbon formation in the groove which interferes with the correct operation of the rings and is conducive to high rates of wear. During downward motion of the piston during the induction stroke the gas pressure on the ring is so low that it takes up a position at the top of the groove as shown in Figure 8–2. The oil film on

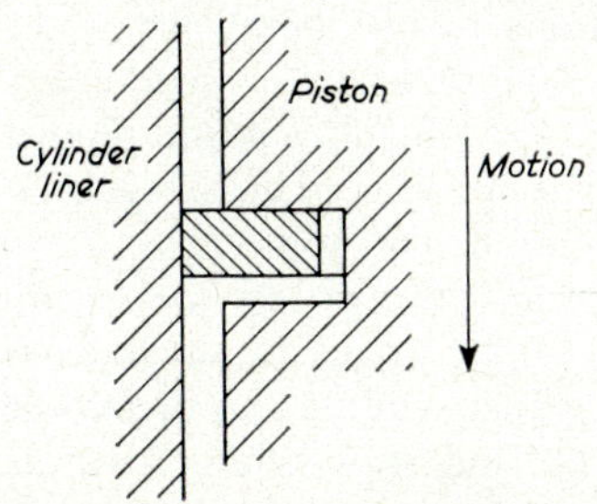

FIG. 8–2—*Piston ring oil pumping action.*

the cylinder wall will be comparatively thick below the ring as a result of splash from the crank chamber. The ring scrapes this oil film into the clearance space below and behind itself and then when the piston begins its upward travel and the compression of the charge commences the ring is forced down on to the bottom of the groove and the oil is transferred around the back of the ring to its upper side. Excessive axial clearance increases the stroke of the ring in its groove with consequently larger amounts of oil being transferred and consumed.

8.5. *Back clearance*

Any gases occupying the clearance spaces round the ring are compressed during the high pressure part of the cycle by hot combustion gases entering the spaces. Large clearance spaces allow greater quantities of hot gas to enter and to raise the temperature of the groove with oxidization of the lubricating oil to form larger amounts of undesirable carbonaceous deposits. When the ring is pushed to the back of the groove its cylindrical face should lie below the lands on either side by a small amount which will be equivalent to the radial or back clearance when working. This back clearance is necessary to avoid possible side loads from the piston being imposed on the ring. The back clearance must also be as small as possible once this condition is satisfied as a large back clearance is conducive to oil pumping and it enlarges the spaces into which hot gases can compress and expand to overheat the ring and groove and oxidize the oil leading to carbon packing behind the ring.

Gap, side and back clearance all increase as the ring wears and together with groove axial wear and cylinder liner wear account for much of the deterioration in the performance of an engine over its life and in between overhauls.

8.6. *Ring material*

Piston rings need to be made of material which is compatible with being rubbed up and down the cylinder bore without scuffing, which is hard wearing and which is strong enough to withstand the strains of conforming to the groove and bore geometry under the influence of the gas pressures. Flake graphite cast iron is generally used alloyed with manganese, chromium and up to one per cent modybdenum to provide the necessary strength.

Almost invariably the top ring is chromium plated on the cylindrical face, the obvious exceptions to this being when chromium plated cylinder bores are used. This is a point to be remembered if worn cylinder liners are reclaimed by boring out and chromium plating. The chromium plate offers a hard face resistant to wear. The surface is not too good from the point of view of oil retention and sometimes the periphery of the ring is given small circumferential grooves to assist lubrication. In any case chamfering or rounding of the edges of the face is necessary to promote lubricating conditions in which scuffing can be avoided. Figure 8–3, a section through a chromium plated ring, shows these features.

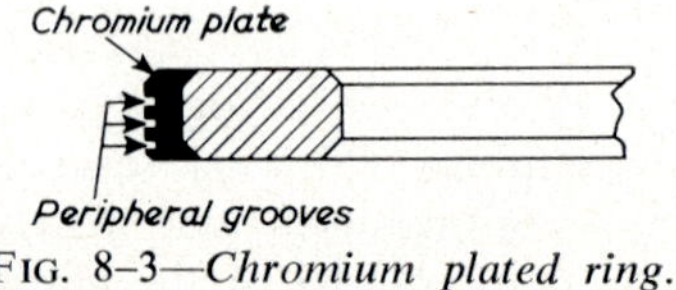

FIG. 8–3—*Chromium plated ring.*

8.7. *Running in*

Despite considerable efforts in both design and manufacture the shapes of the cylinder bore and the ring when working are not perfectly circular and sealing is only finally achieved by a process of running in. This is accomplished by running the engine at considerably less than full load for a period to allow the rings to wear slightly and bed to the cylinder bore. Application of full load to a newly assembled ring pack would result in high pressure hot gases blowing past the rings at positions where contact was light or not quite established causing local overheating, destroying the lubrication with resultant scuffing and even channeling or guttering of the ring face. If this is permitted a condition will result which will never heal however much careful running in is attempted subsequently.

In production testing and particularly after overhauls in service it is highly desirable that the necessary run-in period should be of as short a duration as possible. This is somewhat incompatible with the provision of

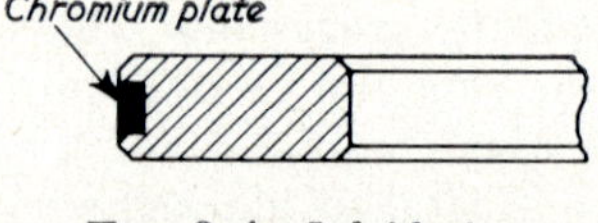

FIG. 8–4—*Inlaid ring.*

hard wearing long lasting surfaces for normal service. One device to overcome this problem is the use of inlaid rings. These may be of the design shown in Figures 8–4 where the chromium plated region is in the centre of the face leaving cast iron edges initially slightly proud which rapidly wear to a shape conforming to the cylinder bore. Another form is shown in Figure 8–5 where thin bronze inserts are let into the surface of a chromium plated ring, again initially these are slightly proud and designed to run in rapidly.

A different but popular approach is illustrated by Figure 8–6. The face of the ring is very slightly tapered so that initial contact is made at one edge; this edge rapidly beds in to form a band over which contact with the bore is established. As running in progresses the band widens until contact is completed over the full width of the ring. The engine is usually ready to accept full load long before this final stage of contact over the full width

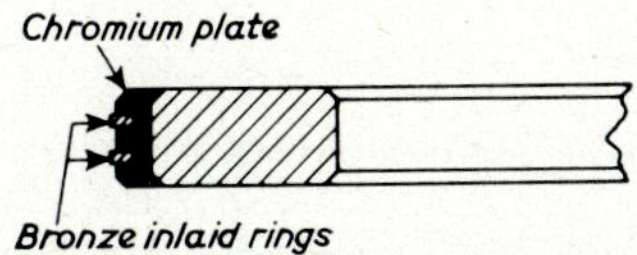

FIG. 8–5—*Chromium plated ring with bronze inlays.*

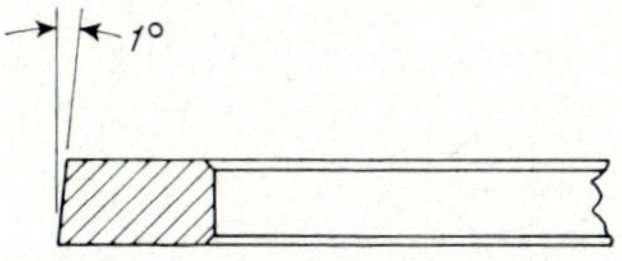

FIG. 8–6—*Taper faced ring.*

is reached. When assembling taper faced rings of this kind on a piston they should be placed with the contacting edge at the bottom or high lubricating oil consumption will result. The taper on the face is very slight and is difficult to distinguish so that generally ring manufacturers mark such rings to show which side should go uppermost.

An important factor in the running in process is the surface condition of the cylinder bore. A surface that is too smooth delays the running in intolerably. New cylinder liners have considerable attention paid by the manufacturer to the type of finish in the bore to give as true a surface as possible yet with oil retaining spaces and sufficient roughness to promote running in. Typically surface roughness is in the region of 0·4 to 1·2 micro metres for cylinders up to 250 mm bore and about 0·8 to 1·5 micro metres for sizes above this. When a re-ringed piston is fitted into an existing bore it is good practice to break the glaze on the bore surface by honing. A more satisfactory running in will be attained.

8.8. *Groove wear*

Axial wear of piston ring grooves is not usually a problem with cast iron pistons or two piece pistons with steel crowns, but it does occur with light alloy pistons. In fact it is often a major consideration in the choice of the aluminium-silicon alloy from which such pistons are made, a compromise having to be reached between wearing properties, thermal conductivity and castability. Some manufacturers offer replacement piston rings of above standard width so that worn top ring grooves may be recut at overhaul times to extend the life of the piston. When grooves are recut the engine makers advice should be sought; the land under the top groove is highly loaded by the gas pressure on the ring and is frequently made of greater depth than the other lands to withstand these forces. Obviously reduction of the width of this land when recutting the groove will detract from its strength. On the other hand enlarging the groove too far upwards may result in it being too hot. Furthermore it is the practice of some manufacturers to machine the sides of the ring grooves at an angle as in Figure 8–7 when the piston is cold so that they become square to the bore at working temperatures. This is to avoid the situation shown in Figure 8–8

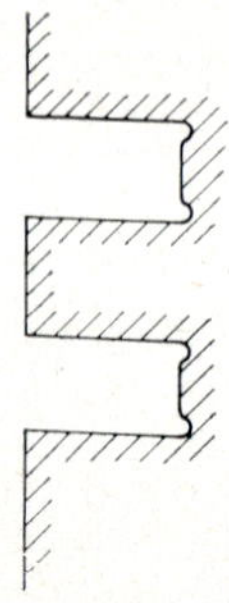

FIG. 8–7—*Angled grooves—cold.*

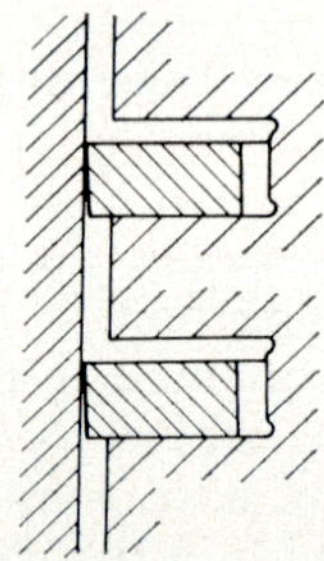

FIG. 8–8—*Rings in non-angled grooves—hot.*

which results in the ring scraping lubricating oil upwards and also being highly stressed because of unsymmetrical bending by the gas forces. A similar condition arises if the land is made too thin to provide sufficiently rigid support for the ring.

Groove wear in light alloy pistons is often countered by the use of ferrous inserts which embrace the top or the top and second grooves and which are cast in during manufacture. High expansion cast iron such as ni-resist is generally used and it is bonded to the aluminium alloy by proprietory processes.

8.9. *Rings for two stroke cycle engines*

Piston rings for two stroke engines have a more arduous time because of having to cross the ports in the cylinder wall. As a ring crosses a port the unsupported section will be blown outwards by the gas pressure behind the ring. It is often the case that two stroke engine rings are made of greater axial width than rings in four stroke engines of the same size. A simple calculation will show that the deflexion of an unsupported segment of the ring is not affected by its width as the increase in force from the additional area presented to the gas pressure is countered exactly by the increase in second moment of area of the ring due to the increased axial dimension. A ring that is blown out into the ports will, however, be snagged at the port ends as it returns to the full support of the cylinder liner. This snagging bends the ring axially and it is in resisting this treatment that the additional width provides a stronger construction. In this respect the second ring is often treated more roughly than the top ring. This is because the pressure between the rings lags behind that in the cylinder. When the pressure rises in the cylinder during combustion it acts first on the top ring and leaks through into the space between the top and behind the second ring and then on to the space between the second and behind the third ring. As the gases in the cylinder expand when the pistons descend their pressure falls and may eventually reduce below the pressure behind the second ring. Leakage then takes place in the reverse direction from the space between the first and second rings into the cylinder and at the time that the rings cross the ports the pressure behind the second ring is greater than that behind the first ring. Note also that the second ring reaches the ports before the first ring does, when all the pressures are higher. The action is illustrated in Figure 8–9 which shows

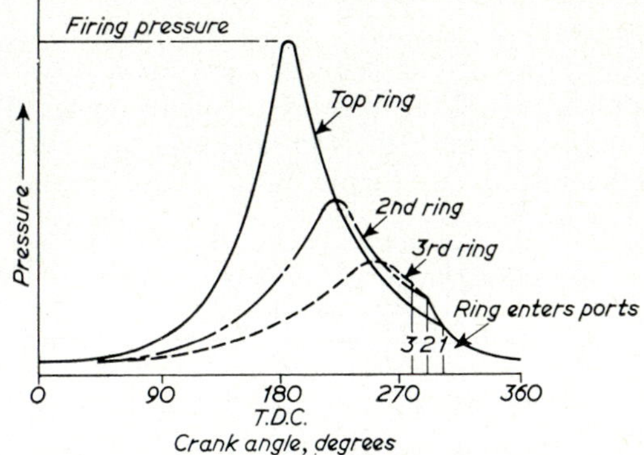

FIG. 8–9—*Pressure behind piston rings two-stroke engine.*

the pressure behind the rings and in the cylinder related to the position of the piston.

The horns of the ring, if they come opposite a gap, will be blown farther out because of their being cantilevered instead of bridging the port. They are, therefore, far more susceptible to breakage, both from pressure behind the ring and from snagging at the port ends. To prevent this many engines employ pegged or pinned rings. The ring is prevented from rotating by a peg as shown in Figure 8–10 which is located in line with a port bar to prevent the ring rotating. The peg design is important as it receives considerable buffeting from the ring during operation of the engine. It should be well secured to the piston preferably being embedded in the side of the ring groove and being squared at the sides with a corresponding gap in the ring so that contact is made over an area and not at a point. The peg should be located at the upper side of the groove so

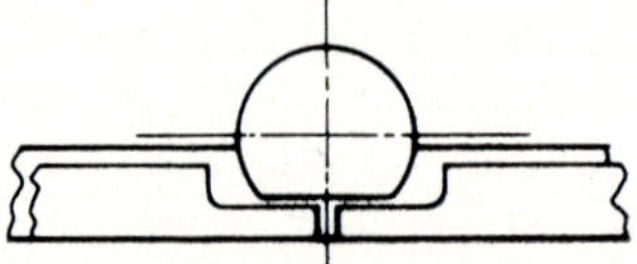

FIG. 8–10—*Pegged ring.*

that the lower face of the ring may be extended to the usual gap dimension; in this way the maximum sealing of the gas is obtained. To minimize the buffeting of the peg by the ring it should be located on or near to the centre of the thrust sides of the piston. Pegs situated in the plane of the cylinders receive shock loads during each cycle as the piston slams across at the dead centre positions.

An alternative means of preventing rotation of the ring is to manufacture it with a small projection on the inside which enters a hole in the back of the groove. The fit of the projection in the hole must not prevent the ring from seating on the lower side of the groove.

If at all possible it is best not to peg the rings but this demands narrow widths of the exhaust ports. Rings that run over inlet ports only can usually survive without pegging.

The top rings of two stroke engines are arduously worked because of the high thermal loading of the pistons in such engines. In the past, fire rings have been used to protect the top land and to secure positive cut off at the ports by the piston, see Figure 8–11. They were almost impossible

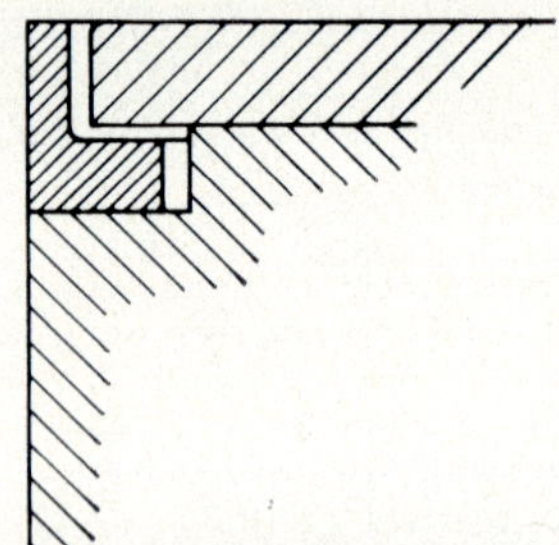

FIG. 8–11—*Fire ring.*

to lubricate correctly, particularly under varying load conditions and are rarely used in engines today. Taper section or wedge section rings are sometimes used to combat the onset of sticking. These rings and the corresponding grooves may have tapers on both sides or on the upper side only as shown in Figure 8-12. They permit an increase of 10 or 15°C in the groove temperature before the onset of sticking but the rings are highly stressed particularly those with tapered lower sides as they are inevitably twisted when forced down on to the side of the groove and are required to change their attitude of twist as the piston changes thrust sides at the top of the stroke when the gas pressure on the ring is highest.

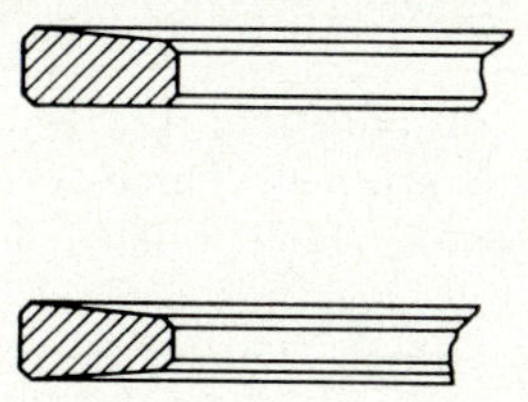

FIG. 8–12—*Taper section ring.*

8.10. *Piston shape*

The piston trunk functions as a cross head taking the side thrust from the connecting rod whilst sliding up and down the cylinder wall. At working temperatures it must approximate to a uniform cylindrical shape on the thrust surfaces to give even contact over a large area maintaining the lubricating oil film separating it from the cylinder liner bore.

The junk and ring lands need to have the minimum clearance from the cylinder bore in order to protect the rings and to give the maximum possible support to them. This portion of the piston need not touch the cylinder wall but some light contact is not unusual.

These requirements result in the shape of a typical one piece piston such as is shown by the curve in Figure 8–13. The trunk or skirt is barrel

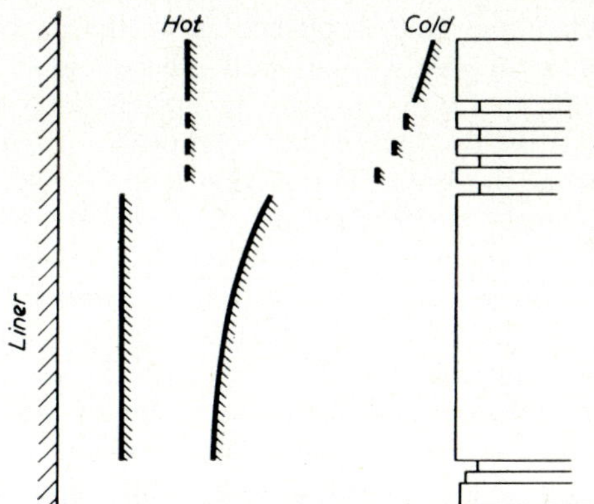

FIG. 8–13—*Piston clearances.*

shaped and the ring belt region is tapered as well as being stepped back. Expansion of the piston skirt is greater along the axis of the gudgeon pin than it is in the direction normal to it and the skirt may be formed into an elliptical shape in plan to accommodate this distortion when hot or there may be relieved areas on either side of the skirt around the region of the gudgeon pin bosses.

8.11. *Piston lubrication*

The piston trunk must be well lubricated for it to perform its cross head duties with a minimum of friction. On the other hand the compression rings need to be fed with the least possible amount of oil that will lubricate their sides and faces as any excess will be burnt with a resulting increase in both lubricating oil consumption and the creation of undesirable carbonaceous deposits in the grooves. There is an abundance of lubricating oil splashed on to the cylinder bore walls from the crank case and the problem is to make sure that just the right proportion of this oil reaches first the trunk and then the rings. This flow of oil to the trunk and compression rings of the piston is controlled by other piston rings specially designed for this purpose.

8.12. *Oil control rings*

The simplest oil control ring is a plain ring bevelled on the face to provide a narrow edge in contact with the cylinder bore. This type of ring is termed a single edged oil control ring and its form is shown in Figure 8–14. It acts on the oil in two ways; as a seal by means of a high wall

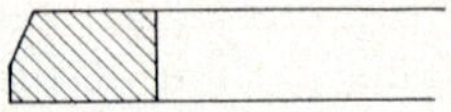

FIG. 8–14—*Single edge bevelled oil control ring.*

pressure and as a directing influence on the passage of oil as a result of its bevelled shape. When the ring is pushed upwards the bevelled face rides up over the film of oil on the cylinder wall. The wedge of oil that is formed easily generates sufficient film pressure to spring the ring inwards whilst on the downward stroke this influence is completely absent and the oil film is scraped down from the bore. This action leads to the term "oil scraper ring" but there is little doubt that on the upstroke some oil is pushed upwards by the ring and on the down stroke some of the oil film is left on the bore. There is a strong balance in favour of downward passing of the oil. The action of taper faced compression rings described in article 8.7 is just the same and hence the care that must be taken to see that such rings are assembled the right way up as their shape is not so obvious as that of the single edged bevelled scraper. The greater taper of the oil control ring ensures that it rides over the oil film correctly and that its high wall pressure is maintained as it wears. The purpose of the design shown in Figure 8–15, a "hooked" scraper ring, is to maintain the face dimension and

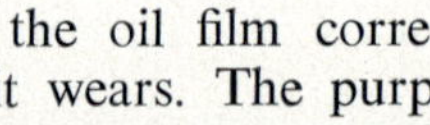

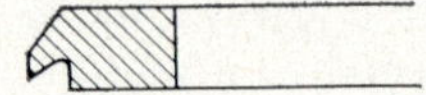

FIG. 8–15—*Single edged hooked oil control ring.*

hence the wall pressure, more accurately as wear takes place; it also has a more severe scraping action.

The wall pressure is the intensity of pressure that the ring exerts against the bore. It depends on the strain of the ring when in position and the area in contact with the bore. A high wall pressure is exerted by a strong section ring with a narrow face in contact with the bore.

Unlike the compression rings the oil control rings do not have the gas pressure to force them against the wall and have to rely on their own strength to generate the wall pressure necessary to keep the oil film to the required thickness. The wall pressure of a simple ring is restricted by the radial depth of the ring, which is controlled by the need for it to be capable of passing over the piston diameter when sprung outwards to its limit, in order that it may be assembled in its groove.

Various designs of spring backed rings are aimed at providing higher wall pressures. One is shown in Figure 8–16. They also have another most

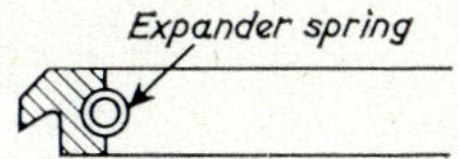

FIG. 8–16—*Spring backed conformable oil control ring.*

important advantage which is that the ring itself can be made thin and flexible so that it can conform to a worn or distorted bore that is no longer circular. The conformable portion of this ring is pressed against the bore by a backing spring and it is this spring which generates the high wall pressure.

Some oil control rings rely entirely on wall pressure whilst others combine wall pressure with directional tapers. The contact edge may be arranged in the middle of the ring section as in Figure 8–16, rather than at the bottom as in the simple bevelled ring of Figure 8–14. This is done to maintain a symmetrical attitude under the bending forces imposed when assembled and so ensure a better wear pattern throughout the life of the ring. Many rings are made with two scraping edges and also have escape routes for the oil. Several of the various types are illustrated in Figure 8–17.

The design of the piston will have embodied in it escape routes for the unwanted oil. A stepped or bevelled space is usually provided immediately below the ring in which the oil can collect; this is shown in Figure 8–18. Drain holes convey the oil from this space and also from the back of the groove. The siting of these holes demands considerable care in design to ensure that the oil does in fact flow from the ring towards the inside of the piston and not in the other direction.

8.13. *Operation at an angle to the vertical*

Any engine used for marine purposes must be capable of continuing to operate at any power up to its rated power without being affected by the motion of the vessel during rolling or pitching and also it must continue to perform satisfactorily if the vessel takes on a list. The mechanical construction of high and medium speed reciprocating engines is well suited to these demands as the high rev/min, necessarily involves consideration of high acceleration forces which are usually well in excess of those resulting from the motion of the vessel during normal operation at sea. It is unusual for the mechanical parts of such engines to depend on gravity for either their operation or their securing and the running of the engine at

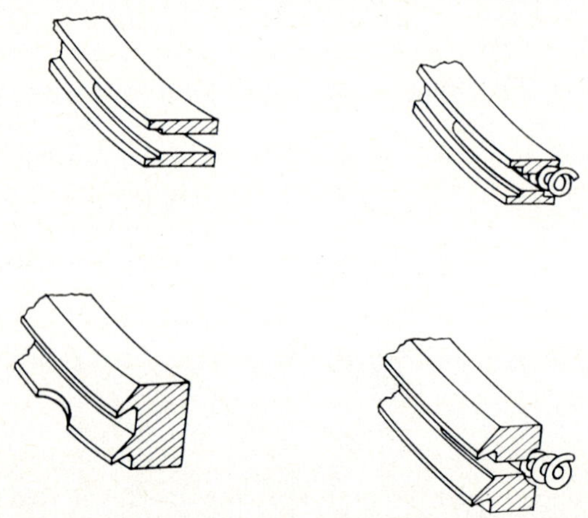

FIG. 8–17—*Examples of double edged oil control rings.*

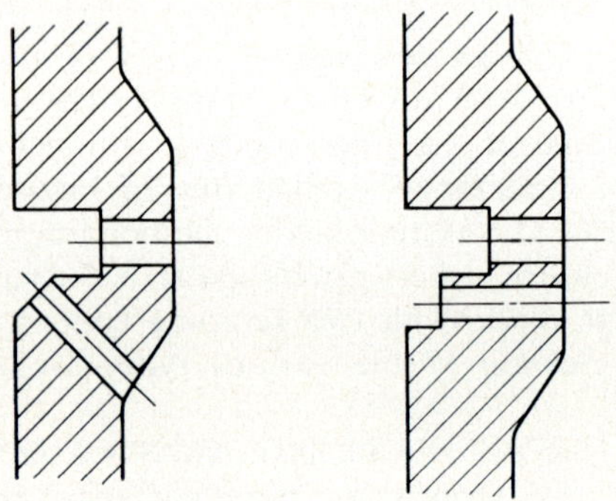

FIG. 8–18—*Drainage paths from scraper ring grooves.*

considerable angles from the vertical presents little or no problems on this score. However, if it is proposed to use on shipboard an engine which has been developed primarily for use on land these points are worth checking.

The control of the liquids in the engine are a different matter and they frequently define the limiting angles at which satisfactory operation is possible. As no engine will run for more than a very short time without lubricating oil feeds to its bearings and to its pistons if they are oil cooled lubrication generally becomes the most important criterion. Lloyd's Register of Shipping lays down in its rules that the lubrication arrangements for engines, including drainage, should remain efficient with the ship inclined from the vertical at any angle up to 15° transversely and when pitching 10° longitudinally and when rolling up to $22\frac{1}{2}$° from the vertical.

Most engines are installed with their crankshaft axes fore and aft, parallel or nearly parallel to the axis of the hull, but occasionally auxiliary engines may be positioned athwartships. Should this position be proposed then it is a wise precaution to examine the lubricating oil arrangements carefully as many engines which happily withstand tilting through large angles about the crankshaft centre line are limited in the amount of inclination they can tolerate in the longitudinal plane.

Drainage paths from various parts of the engine must function satisfactorily at all inclinations up to the desired limit in order that oil pump suctions should remain covered. Oil that does not drain down correctly may result in the level of oil at the pressure pump suction being insufficient to prevent ingress of air leading to failure of the oil supply to the bearings. In practically every engine design there are pockets, lubricating oil troughs and the like, where owing to the inclination the oil level will change from the normal and in many instances this will have a negligible effect, the amount of oil withheld from the sump being very small. The changed level may result in seepage through joints or small apertures not normally below oil level and this may be a nuisance but tolerable. If, however, there is any tendency for large quantities of oil to drain outside the engine then this situation will clearly give rise to a dangerous set of circumstances as this oil will then not be available for recirculation through the bearings and the pistons.

The oil mist sealing arrangements at openings through which shafts emerge from casings, for example the aperture for the crankshaft, are not always liquid oiltight, the seals at many of these positions taking the form of labyrinths or sometimes oil thrower disks or oil scraping arrangements. Although wet sump engines are designed with deep chambers they may, nevertheless, be limited in the angle to which they can be inclined in the longitudinal plane by the oil level reaching openings such as these.

A further limitation that may arise with wet sump engines tilted longitudinally is that the crank and bottom end bearing of one or more cylinders at one end or the other may dip into the oil causing excessive splash and over oiling of the cylinders concerned. This over oiling can lead to fairly rapid deterioration in performance in some cases. It may also be responsible for increasing the rate of consumption of lubricating oil considerably so requiring more frequent checks and replenishment than when operating in a normal position.

In some designs cams or gear wheels are lubricated by dipping into a trough into which lubricating oil is fed or into which it drains. A change in angular position may result in under lubrication of some mechanisms and over lubrication of others. Often the condition is tolerable for a reasonable period of time to meet with emergency situations but can become dangerous if running is continued for long periods.

If at all possible oil level gauges and dip sticks in oil sumps and tanks should be situated where changes in angular position do not result in misleading readings regarding the quantities of oil contained or the head covering pump suctions. If such siting is not possible then the effect of any list of the ship on the level readings must be ascertained and replenishment practices modified accordingly during periods of operation under such conditions.

Small high speed engines with integral reverse reduction gears are generally of wet sump design. It is not unusual for this type of engine to be installed at considerable angles of rake and they are usually designed to suit this practice. The manufacturers are well aware of the limiting angles of operation and their advice should be sought if abnormal installation positions are intended to be used or if unusual operating conditions are foreseen.

8.14. *Crankcase explosions*

Although crankcase explosions occur only rarely their effects can be both drastic and dangerous. An understanding of how such explosions take place indicates the precautions that may be taken to minimize the chances of their occurrence and the safeguards to ensure that there is little danger to men and machinery in the event of one happening.

For an explosion to occur there must be in the crankcase a mixture of oil and air in a ratio that is within the range of inflammability and in addition there must be a source of high temperature energy sufficient to initiate combustion. This latter is called a hot spot.

The splashing of the lubricating oil inside the crankcase breaks it up into droplets or globules of widely varying size distributed in varying density throughout the crank chamber. The overall mixture strength is usually very weak, *i.e.*, too high a proportion of air to support combustion, although it will obviously vary from one part of the crankcase to another. For combustion to begin oil must first evaporate from the surface of some of the droplets and then continue to rise in temperature until combustion is initiated. As the flashpoint of lubricating oil is of the order of 215° (420°F) this is not likely to happen, however finely divided the oil may be, unless some other agency is present. Once started combustion can spread very quickly through this mixture of oil droplets finely divided and dispersed in air. But to start the process the presence of a hot spot is essential.

The hot spots which cause trouble can occur over a wide variety of components. The cause may be failure to maintain sufficient lubrication to a bearing but not only the large end bearings, main bearings, small ends and gudgeon pins are vulnerable; large explosions have been the result of failures of lubrication to jockey sprockets and similar small parts. Sometimes the hot spot arises from the rubbing together of parts never intended to come into contact, for example unexpected axial thrusts along the crankshaft may bring a web into contact with the side of a main bearing. Hot gases blowing past the pistons may provide sparks sufficient to cause an explosion but this is rather unusual, as although the temperature of a spark is high the quantity of heat it carries is low. Incipient piston seizure on the other hand can easily provide the conditions leading to a hot spot. Also perforation of trunk piston crowns has been known to occur and to be the cause of crankcase explosions.

8.15. *Pressures resulting from crankcase explosions*

When a hot spot occurs its presence both vaporizes the oil with resultant increase in the richness of the mixture and initiates the combustion. Once the flame is started it will spread more or less rapidly according to the strength of the mixture around it. The speed of propagation of the flame may be as low as a few inches per second or it may reach thousands of feet per second or even higher. If conditions are favourable to it the flame front will progress with increasing speed building up a pressure ahead of itself which also tends to increase the temperature of the oil and air mixture by compression leading to increased acceleration. In extreme conditions it is possible that detonation may occur.

However the evidence from most explosions is that the pressures experienced are fairly low, probably of the order of 0·3 bars, but even these low pressures when applied to the large surfaces which form the crankcase enclosure represent forces quite sufficient to cause considerable damage.

Because the flame front gathers in intensity of pressure as it progresses, the effect in a long crankcase can be severe and engines with small crankcases probably experience less dangerous pressures if an explosion occurs. In any case, the crankcases of small engines can be made considerably stronger than those of large engines without incurring any penalty of weight.

Two crank chambers connected together are obviously an increased hazard in respect of flame travelling from one to another gathering pressure as it progresses. Such a position should be guarded against in twin engined or multi engined installations. An explosion in one engine has been known to be followed by an explosion in another engine close by. It is possible, but unlikely, that the flames emerging from one engine may heat up the crankcase of the other very rapidly until flashpoint is reached.

A most important precaution to be taken is to avoid direct connexions between the crankcases. If lubricating oil drains lead to a common tank they should enter by pipes which are submerged below the oil level.

8.16. *Secondary explosions*

If the pressure wave reaches an opening through which it can escape to the atmosphere the pressure pulse is immediately followed by a suction pulse of lower magnitude but greater duration. This suction pulse can be responsible for drawing in a charge of fresh air to take the place of that which has been burned by the initial explosion. A secondary explosion may ensue and it has been known for this to be far more violent than the first.

8.17. *Flame*

The opening through which the original pressure wave escapes unfortunately may be made by the pressure itself blowing off or fracturing to pieces a crankcase door or similar cover. One of the dangers that occurs when a pressure pulse escapes in this way is that of flame. The issuing of the pressure wave through an opening causes it to expel a mixture of lubricating oil and air, some of which is burning and some not yet burnt. This mixture continues to burn as long as its temperature is sufficiently high, causing a large tongue of flame to issue from the hole.

8.18. *Explosion relief valves*

Nevertheless the escape of the pressure wave immediately relieves the pressure inside the crank chamber and it is desirable to allow this to occur in order to keep the pressure low provided that it can be managed with safety, without danger from flame and without intake of a fresh charge of air which may lead to a violent secondary explosion. These requirements form the basis of one of the most essential safeguards: the explosion relief valve.

Relief valves are designed to incorporate the following features. They must open very rapidly if the crankcase pressure rises by a significant amount above atmospheric pressure and this may be very small, *e.g.* 0·1 bar. The size of the valve and the rate of opening must present in a very small space of time a wide escape route for the gases causing the pressure rise. The suction pulse following the pressure pulse must close the valve equally rapidly to prevent fresh air entering the crank case and leading to a secondary explosion. The valve should embody a flame trap, *i.e.*, a means of rapidly cooling the gases that issue through the open valve in order to quench the flame and reduce the risks of burning and fire. It is also usual to provide a shield arranged to direct any flame that may emerge down the side of the engine which is less dangerous than allowing it to be projected horizontally. Additionally when closed during normal running which the valves are throughout the life of the engine they must seal the oil mist in the crankcase without allowing unsightly seepage.

The BICERA crankcase explosion valve developed by the British Internal Combustion Engine Research Association exhibits all these features. It consists of a light disk shaped valve held on its seat by a spring designed to give a low opening pressure. A rubber seal round the edge of the valve disk prevents oil leakage during normal operation. A low explosion pressure causes rapid opening of a large passage and a return to atmospheric pressure causes the valve to reseat just as quickly. The flame trap consists of several layers of gauze forming a dome shaped cover situated over that side of the valve within the crankcase where the splash of oil from the crankshaft maintains it in a wetted condition. The wetted gauze is far more effective in quenching the flame than is dry gauze. The open area through the gauze is well in excess of the area through the open valve, of the order of 30% extra. On the outside of the valve there is a shield to deflect downwards any flame that may penetrate the gauze.

8.19. *Size and number of valves*

The crankcases of small engines are often sufficiently rigid to contain safely the pressure from any explosions that may develop in them but above a certain size of engine the provision of relief valves is imperative. Classification societies lay down rules for the number and size of valves that are to be provided. They require all engines of 200 mm bore and above to have them and the larger the engine the greater is their number and size.

A large number of smaller valves is preferable to a small number of larger valves but the minimum size of valve that is allowable is one in which the passageway is about 76 mm (3 in) diameter or of equivalent area.

8.20. *Smoke detection*

As the vaporization of the lubricating oil in sufficient quantity to approach the inflammability limits usually produces dense smoke a number of warning instruments have been based upon recognizing this condition. They work on the principle of taking samples from various points in the crank chamber at regular small intervals of time. The samples are passed before a source of light and the transmission of the light through them compared with that of transmission through a similar sample of clean air. The comparison is carried out automatically by photoelectric means and any abnormal level of smoke is arranged to trigger off an alarm system.

8.21. *Crankcase ventilation*

Investigations over the years have established that the general mixture strength throughout an engine crankcase is invariably a weak mixture in normal running conditions. There is therefore no danger from and some safety to be obtained by having a positive ventilation for the crankcase. This may take the form of an exhaust fan or may be by means of a small bore pipe connexion from the crankcase to a low pressure point on the air intake of the engine. In both cases the pressure level in the crankcase is maintained at about 25 mm of water below the atmospheric pressure. This minimizes oil leaks and seepage by creating a small pressure gradient towards the inside of the crank chamber and is particularly effective at the inevitable small openings such as the annular spaces round the crankshaft seals.

8.22. *Inspection of crankcases*

Although the normal mixture strength of oil and air inside the crank-ease is a weak one there are occasions when abnormal conditions may be suspected and when these arise great care must be exercised in investigating the situation. Should an explosion have occurred and the crankcase remained intact it would be foolhardy to allow fresh air to enter what may be a rich mixture. Should the crankcase be suspected of containing a rich mixture the engine should be shut down and allowed to cool before opening any crankcase doors or carrying out any action which may allow air to enter the oil rich and possibly high temperature zone. When eventually opened no naked light should be used in inspecting the interior of the crank chamber.

CHAPTER NINE

Fuels and Lubricants

9.1. *The fuel requirements of diesel engines*

The early designs of diesel engines were intended to run on a wide variety of fuels but as development progressed it was soon seen that rapid improvements followed from the use of relatively specialized fuels. Research into fuels progressed at the same time and was directed towards developing the most suitable fuels for the more critical engine designs.

High speed engines require top grade distillates having a high degree of cleanliness. The very short time available in each cycle for the combustion of fuel demands a sufficiently high volatility so that some of the fuel in each droplet entering the cylinder will vaporize rapidly which it must do before combustion can commence. The slower speed engines are not so critical in this respect and cheaper but heavier fuels can be used.

Large slow speed diesel engines found themselves in competition with steam turbines and were developed to burn the heavy viscous fuels used as boiler fuels. Most ocean-going motor ships use residual fuels or blended fuels containing a high proportion of residuals. Medium speed engines when installed in similar vessels are naturally more attractive if they, too, can burn these heavy fuels. In consequence a good deal of development over recent years has been devoted to making medium speed diesel engines capable of running on heavy fuel.

The cheaper heavy fuels are, not surprisingly, found to contain more foreign matter, sediment and water than distillate fuels. Before they can be passed through the fuel injection equipment of any diesel engine either larger bore or medium speed, they must be cleaned by centrifuging or filtration or a combination of both. To accomplish this, heating the fuel is necessary in order that its viscosity can be reduced to a value at which it can be pumped easily. The capital cost of all this equipment or its depreciation must be set against a saving in the cost of fuel alone. In addition because of the quality of the fuel the engines will be found to require more maintenance. The use of heavy fuel is as much a question of economics as of engineering; in many cases a net saving in running costs can be shown.

At the present stage of development it is generally the case that engines that run faster than 1000 rev/min cannot satisfactorily burn heavy fuel as the time required for combustion is too short.

9.2. *Fuel specifications*

Fuels have been divided into grades according to specification in B.S. 2869–1967. Extracts from the specifications of various grades are shown in Table 9–1. The fuels that fall into class A are all distillates. Fuels

TABLE 9.1

Fuel Specification and type		*Viscosity*							
		Kinematic cs. max.	*S.R.1 (1) at 100°F max.*	*Water Content % vol. max.*	*Sediment % wt. max.*	*Conradson Carbon Residue % wt. max.*	*Cetane Number min.*	*Sulphur Content % wt. max.*	*Ash Content % wt. max.*
B.S. 2869: 1967									
	Class	**at 100°F**							
Engine Fuels	A1	6·0	41	0·05	0·01	0·2	50	0·5	0·01
	A2	6·0	41	0·05	0·01	0·2	45	1·0	0·01
	B1	14	65	0·10	0·02	0·2	35	1·5	0·01
	B2	14	65	0·25	0·05	1·5	—	1·8	0·02
		at 180°F							
Burner Fuels	E	12·5	200	0·50	0·15	—	—	3·5	0·10
	F	30	1000	0·75	0·25	—	—	4·0	0·15
	G	70	3500	1·0	0·25	—	—	4·5	0·20
	H	115	7000	1·0	0·25	—	—	5·0	0·20

falling into class B usually contain some residuals and the other classes contain higher proportions of residuals and may be wholly residual.

Distillates are easily recognized by the fact that they are transparent although they range from colourless fluids to those having a dark brown or red colour. Residuals are black and opaque and a small addition of a residual fuel to a distillate will render it also black and opaque. All lower cost fuels contain residual fractions.

9.3. *Chemical composition*

All diesel engine fuels are mixtures of various hydrocarbons. Their chemical composition is dealt with in Chapter 4 together with the chemical properties that have effects on the combustion process.

9.4. *Specific gravity*

The specific gravity of the fuel is principally used to convert weights into volumes and vice versa. In association with the volatility range it can yield information on the types of hydrocarbons present in the fuel. Although the residual fuels are termed heavy this is not really a reference to their specific gravity but more to that of viscosity.

In defining specific gravity it is necessary to specify the temperature of both fuel and water to which it refers. In Britain both are usually 60°F and specific gravity is said to be at 60°F/60°F. European countries usually express it as the density of the fuel at 15°C compared to that of water at 4°C.

In the U.S.A. gravity is often quoted in degrees A.P.I. which are related to the specific gravity by the following relationship:

$$\text{Degrees A.P.I.} = \frac{141{\cdot}5}{\text{S.G at } 60°\text{F}/60°\text{F}} - 131{\cdot}5$$

9.5. *Viscosity*

Fuels are generally graded by viscosity as can be seen in Table 9–1. Those termed heavy are in the range from 250 seconds Redwood No. 1 to 9,000 seconds Redwood No. 1. Viscosity is a measure of the resistance to flow of the fluid. More precisely it is a measure of the tangential force between layers of a liquid which resist movements of one layer over the next one. If the surface areas, the distance between them and the rate of movement are defined in metric units the resultant force is the absolute unit of viscosity, known as the "poise". Kinematic viscosity is the ratio of absolute viscosity to the specific gravity of the fluid at the temperature of the viscosity measurement. Its unit is the "stoke" but it is more usually quoted in "centistokes". These are the scientific measurements of viscosity. In practice viscosity is usually measured by the time taken for a given quantity of the fluid to flow out of an instrument called a viscometer. The viscosity is then expressed in "seconds". In the United Kingdom the Redwood Viscometer is the most commonly used, in the United States the Saybolt and on the Continent the Engler Viscometer.

9.6. *Calorific value*

The calorific value of a fuel is the amount of heat obtained when it is burned completely. There is a gross and a nett calorific value. One of the products of combustion of hydrocarbons is water vapour and the gross calorific values include the heat given up by this water vapour when condensing. The nett calorific values do not include this factor. This is dealt with in further detail in Chapter 4. As the engine cycle after combustion is always above the temperature at which water vapour will condense it is the nett calorific values which are appropriate to comparisons of engine performance. Obviously an engine producing a given power will require a greater amount of fuel if the calorific value is low than it will if the calorific value is high.

9.7. *Cetane number*

In the Chapter on combustion attention was drawn to the ignition delay period. The cetane number of fuel is a measure of the ignition quality. the higher the cetane number the less the delay. Cetane number is the percentage of cetane in a mixture of cetane and α–methylnaphthaline, having the same ignition qualities as a test fuel in the same engine at the same conditions.

9.8. *Diesel index*

Cetane number is not easy to determine as it requires careful and skilled engine testing. Diesel index is often used as a substitute. It is obtained by calculation from the aniline point and the specific gravity. The aniline point is the lowest temperature at which a fuel is completely miscible with an equal volume of aniline. It gives an indication of the chemical composition of the fuel.

The diesel index is then obtained by the multiplication of the aniline number and the gravity in degrees a.p.i., divided by 100. For the same fuels the diesel index is usually a few points higher than the cetane number.

9.9. *Volatility*

All fuels are mixtures of hydrocarbons. A pure hydrocarbon will boil at a given temperature according to the pressure at which it is maintained, but a mixture will boil over a range of temperature. If a fuel is distilled and the amount recovered at successive temperature intervals recorded, a picture of its volatility is obtained.

9.10. *Flash point*

The flash point of a fuel is determined largely as a safety measure. It is the temperature at which its vapour gives off an inflammable mixture with air. It is also an indication of its volatility in that the higher the volatility the lower the flash point.

9.11. *Cloud point*

The cloud point of a fuel is the temperature at which wax begins to solidify out of solution. It is an indication of the temperature above which it must be maintained to avoid clogging filters.

9.12. *Pour point*

The pour point is an indication of the lowest temperature at which the fuel can be pumped.

9.13. *Carbon residue*

Diesel fuels can form carbonaceous deposits when strongly heated. The carbon residue is a measure of their tendency to carbon formation when used in an engine.

9.14. *Sulphur*

Crude oil contains sulphur in a variety of forms. In distillate fuels a good deal of the sulphur is extracted but residual fuels contain higher percentages.

9.15. *Ash*

The ash content of the fuel is the amount of material which remains as a residue after heating in air under oxidizing conditions at a high temperature. High ash content can be detrimental to high speed engines but low speed engines are not so sensitive to it.

9.16. *Sediment*

Sediment is material present in a fuel which is insoluble in it and which is large enough to be retained by a filter. It is not identical with ash as it may contain material which can be burnt.

9.17. *Water*

Water is sometimes present in diesel fuel particularly in the heavy residual fuels and may interfere with combustion.

9.18. *Fuel cleaning, heating and viscosity control*

The very fine clearances essential in the fuel injection equipment necessitate the cleaning of fuels so that damage does not occur. A filter must be included in the system just before the fuels approach the fuel injection pumps. This filter is a micropack type having cloth, felt or paper elements and capable of filtering at least below 15 microns even for the larger medium speed engines and possibly down to two microns for high speed engines. When distillate fuels are used a simple filter but of adequate area having fine filtration of this order is sufficient.

In the case of heavy fuels, however, much more elaborate cleaning processes must be provided. It is usual to provide settling tanks in which the fuel can remain for a period to allow the water which may be contained in it to separate out. The most popular type of cleaning equipment following this is centrifuges used first as purifiers to extract the remaining water and the heavier particles of foreign matter, followed by clarifiers in which the smaller particles of foreign matter are extracted. The fuel is then pumped to the service tank.

In order to convey the fuel through this equipment transfer pumps and booster pumps are necessary. If the fuel is at all viscous then it must be heated so that it can be pumped and centrifuged, and also its viscosity reduced sufficiently for the fuel injection equipment to deal satisfactorily with it in introducing it to the cylinder. For centrifuging its viscosity must be reduced to the order of 100 to 150 seconds Redwood No. 1, whilst for the injection equipment some engines may tolerate it at a viscosity as high as 100 seconds Redwood No. 1, whereas others may require a lower viscosity for example 80.

To control these viscosities it is usual to use thermostatic methods. Figure 9–1 indicates the relationship between temperature and viscosity. If the fuel bunkered is regularly of one kind a simple thermostat will suffice. If the fuel is varied in quality then thermostats must have variable setting facilities so that they can be adjusted to the type of fuel in use. More sophisticated equipment involves the use of viscostats which continuously measure the viscosity and control the temperature accordingly.

In addition the rails supplying the fuel injection pumps must be arranged to form a loop so that the hot fuel can continuously be circulated and all these pipes including the fuel pumps and the high pressure pipes to the injectors must be heated by tracing with steam or electricity or in some such manner. All this is in addition to the system for distillate fuel which must also be available so that the engine can be changed over on to it before shutting down for long periods during which maintenance can be carried out, see Figure 10–8.

9.19. *Effects on fuel injection nozzles*

If fuel injection nozzles are operated at too high a temperature they tend to form carbon trumpets around the holes in the nozzles. These trumpets interfere with the injection spray pattern and reduce the combustion efficiency, so aggravating the temperature problem. This tendency has been associated with fuels having a high carbon residue figure. As residual fuels often have high carbon residues and as they are heated in order to reduce the viscosity sufficiently for them to be handled properly by the fuel injection equipment they are liable to result in both a high temperature at the nozzle and a fuel prone to trumpet formation. Whereas an uncooled distillate fuel passing through the nozzle has a considerable cooling effect, injectors for use with heavy fuel must be cooled by either water or light oil. Incidentally, it is important not to overcool as cold corrosion can result, see Chapter 4.

9.20. *Effects on exhaust valves*

One of the most important features of medium speed engines running on residual fuels is that of the exhaust valve condition and life. Residual fuels often contain small quantities of sodium and vanadium which during combustion form salts and compounds which have melting points not far removed from the temperatures attained by the exhaust gases. These molten compounds can adhere to exhaust valve seats forming deposits which are corrosive and in addition prevent the valve from seating properly. Under these conditions the valve is often insufficiently cooled so that the deposits build up at an even greater rate. Eventually some of them break away leaving gaps through which the hot high pressure gases rush and the result can be severe guttering of the valve.

Blended fuels are just as difficult as residuals in this respect as many

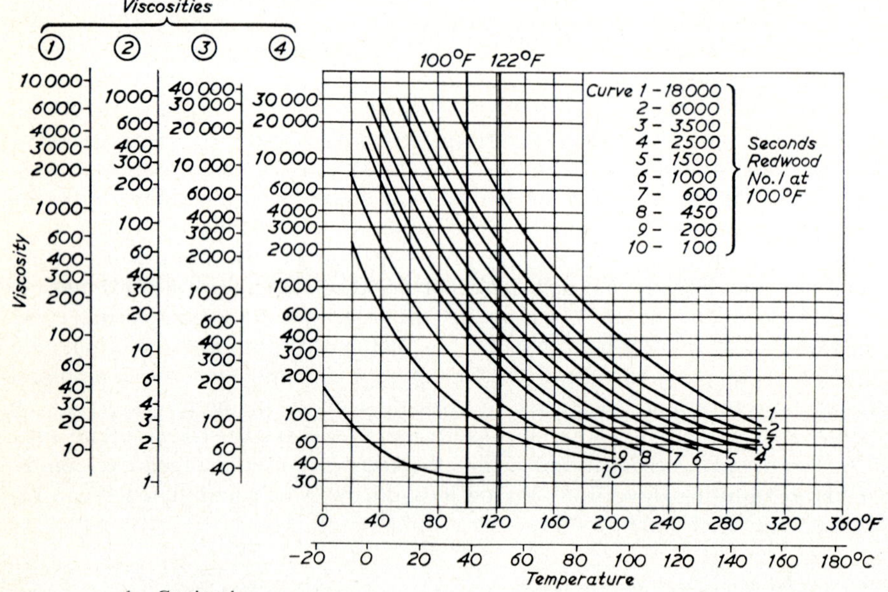

1. Centistrokes
2. Engler degrees
3. Saybolt Universal seconds
4. Redwood No. I seconds

Note: Viscosities Redwood II approximately equals one tenth Redwood I
Viscosities Saybolt Furol approximately equals one tenth Saybolt Universal

FIG. 9–1—*Fuel oil viscosity/temperature curves.*

of them are the result of the addition of quite small percentages of distillate to heavy residual fuels so they may, and frequently do, contain significant amounts of sodium and vanadium even though they are comparatively low in viscosity.

By keeping the temperature of the valve seat below about 550°C (1,020°F) the adhesion of the vanadium and sodium compounds is reduced to an extent that a rapid build up of deposits does not occur. Limiting the exhaust valve temperature in this way may also limit the output which the engine is capable of giving on heavy fuel. Attention has therefore been given in the development of medium speed engines to keeping the exhaust valves cool, a matter discussed in Chapter 5 on Thermal Loads.

9.21 *Wear, abrasion and corrosion*

The high sulphur content of residual fuels can play a large part in the problem of cylinder wear in that it leads to corrosive conditions. Wear is usually thought of as being the result of abrasion and this is, in a large measure, true. When corrosive influences are also present the abrasive processes remove the products of corrosion even more easily than they remove the parent metal and at the same time they expose fresh surfaces to the corrosive agents. In consequence the conjoint action leads to extremely rapid rates of removal of material.

Abrasion during normal operation can result from foreign matter induced in the air, a problem to which auxiliary engines are frequently exposed when loading certain cargoes, or from hard carbonaceous particles resulting from combustion or incomplete combustion of fuel or lubricating oil or both. Abnormal conditions in the engine can also lead to excessive wear. A faulty injector for example may be the cause of poor combustion and the formation of hard carbon particles or of piston distortion or of removal of the lubricating oil from the cylinder wall by washing it with fuel. Faults in the cooling of cylinders or of pistons may also lead to distortion or malfunctioning of the piston rings.

Generally speaking excessive wear in marine engines is very often the result of corrosive influences. The sulphur in the fuel burns to form either sulphur dioxide (SO_2) or sulphur trioxide (SO_3). Most of the sulphur goes to form sulphur dioxide, which is fortunate as it is relatively non-corrosive. The sulphur trioxide, however, combines with the water vapour produced by the combustion of hydrogen to form sulphuric acid upon condensation. In spite of the relatively high temperatures in the diesel engine cylinder condensation can occur. The water cooling of the diesel engine cylinders results in the skin temperature of the liner wall being less than 260°C (500°F). Sulphur trioxide even in very small proportions raises the dew-point or condensation temperature of the combustion gases considerably. The pressures in the cylinder are, of course, quite high during part of the cycle and the dewpoint of all gases increases with pressure. As a result of the combined effects of the pressure and the presence of sulphur trioxide it is extremely likely that condensation of water vapour will occur leading to formation of sulphuric acid.

Both the sulphur oxides and sulphuric acid attack lubricating oil films and if these are destroyed the conjoint action of abrasion and corrosion is greatly increased. Sulphur compounds also combine with the carbonaceous deposits formed from the partial combustion of hydrocarbons with the result that these can be quite hard as compared with those formed from fuels which are low in sulphur. In this way the abrasive particles are multiplied when using high sulphur fuels.

Hard deposits are frequently found on the top land of the piston, particularly in small high speed engines; they can become excessive and take up clearance space and when this occurs they become hard and brittle and break up, the particles causing abrasion of the liner and the piston rings. They also form in or work their way into piston ring grooves causing packing or sticking of the rings.

Jacket temperatures should always be kept as high as practicable in order to avoid the cool conditions at the inner wall of the liner which would lead to corrosion. This is especially important in small engines where the liners are thin. These high jacket temperatures should not be obtained by restricting the water flow which is a very dangerous proceeding, but by circulating the full amount of water and using recirculation or similar means to keep up the temperature. Almost all engines these days use closed fresh water jacket cooling systems in which operation in this manner is easily carried out. A temperature rise of about 10°C between inlet and outlet is reasonable. Operation at this level will also reduce the risk of cracking of cylinder heads or liners.

The material of the liners obviously plays a part in the matter of wear. Usually cast iron alloys are used, in some cases incorporating vanadium and titanium.

In small engines chromium plated liners have found increasing favour and they are also used in several medium speed engines. Wear rates with such liners are usually considerably less than for plain iron liners in the same engines. The cost of chromium plating has to be taken into consideration and for large liners it has restricted the use of this means of combating wear. The success of its application is no doubt due to the fact that it is resistant to acid attack.

An alternative is the chromium plating of the face of the top piston rings. This is obviously cheaper than the chromium plating of a liner bore and is extensively used particularly in the larger engines.

9.22. *Piston deposits*

The lubricating oil used in an engine must be chosen carefully with regard to the type of fuel on which the engine is expected to run. The choice of lubricating oil will materially affect the condition of the piston rings and liner. A common source of trouble, liner wear and other difficulties is the sticking or sluggishness of piston rings. Carbonaceous deposits which form on the piston are the result of oxidization to a greater or lesser degree of both fuel and lubricating oil. Those at the top of the piston and round the top land or junk are mainly from the fuel whilst those in the ring belt and on the skirt are mainly from the lubricating oil. The deposits of the kind found at the top of the piston are loosely termed "carbon" although in fact they comprise partially burnt carbonaceous and asphaltic matter, ash and sulphur compounds originating some from the fuel and some from the lubricant as well as from foreign matter taken in with the induction air. In the ring zone this accumulation gradually takes up clearance at the sides and back of the rings, causing them to become stuck in their grooves. They then cease to seal properly, leading to lack of compression resulting in deterioration of combustion which in turn imposes even more severe conditions on the failing rings; ultimately the hot gases blow by the rings contaminating the crank case oil and in extreme cases burning the rings and piston.

Similar deposits build up on the ports of two stroke cycle engines and the rapidity with which they accumulate here can depend a good deal on the nature of the load cycle and the operating conditions.

The thin deposits that are brown to black in colour that form on the piston skirt termed "varnish" are caused mainly by low temperature oxidization of lubricating oil. Unless they become excessive they are usually not harmful.

9.23. *Lubricating oils*

Modern lubricating oils for diesel engines contain additives to assist them to perform satisfactorily in the arduous conditions required for lubrication of pistons and rings. Most of these oils contain oxidization inhibitors to render them more stable under these conditions and arrest the deterioration of the oil. Of course any oil will oxidize if the temperature is high enough and it is finely divided in the presence of air. In consequence it is most important that the groove which receives the top most piston ring and which is at the highest temperature at which lubrication is required should still be at a low enough temperature for the lubricating oil to survive. 200°C is usually reckoned to be the limiting temperature in medium speed engines; high speed engines run somewhat hotter.

Ring zone deposits have yielded to oils containing detergent additives which cleanse the rings and grooves. These additives are both detergent and dispersive in nature and keep the carbon suspended in a finely divided state in which condition it passes through the engine without harming it. Detergent oils are classed according to their performance in certain standard engine tests. B.S. 1905–1952 lays down standards for the quality and performance of additive type oils for diesel engines. These tests which are based on the American MIL–0–2104 tests are made on a Caterpillar engine. This engine imposes particularly difficult conditions on lubricating oils when run at high rating and with high temperatures. Oils with a high degree of detergency meet more stringent tests and are termed supplement 1, series 2 or series 3 in order of increasingly higher performance.

9.24. *Alkaline lubricating oils*

Oils having a high degree of detergency are usually highly alkaline. This alkalinity helps considerably in combating corrosive wear of the cylinder bores and corrosion of the crankshaft and bearings in the crankcase, the latter resulting from contamination of the crankcase oil by the acidic products of combustion. It has been recognized of recent years that many medium speed diesel engines require the high alkalinity but do not need the high detergency and oils having alkalinity approximately equivalent to that of a series 3 oil (and sometimes even higher) but with a detergency level corresponding to a supplement 1 oil are now being developed. These oils promise to be cheaper than series 3 oils.

The burning of heavy fuels with high sulphur content was first attempted in slow speed cross head engines and difficulties were experienced in a number of cases with corrosion of crankshafts as a result of the contamination of the crankcase oil by strong acids formed by the combustion of the fuel. The problem was solved by introducing diaphragms which separated the cylinders from the crankcase and running gear, the piston rods sliding through glands or scraper boxes in the diaphragms. The cylinders of cross head engines are lubricated by a metered feed quite separate from the oil used in the crankcase and any sludge formed as a result of products of combustion mixing with this oil was thus prevented from finding its way into the crankcase. As medium speed engines are almost invariably of trunk piston design separation of cylinders and crankcase is not possible, and understandably it was feared that if engines of this type were run on heavy fuel they would suffer from crankshaft corrosion. The fact that there has been no problem of this nature is due in a large measure to the development and use of highly alkaline oils. Other factors have undoubtedly played a part such as efficient sealing by the piston ring pack and more even temperature of the crankcase at a sufficiently high level to avoid condensation of moisture.

9.25. *Neutralization value and total base number*

The neutralization value of an oil is a measure of its acidity. It is the number of milligrams of potassium hydroxide (KOH) required to neutralize the acid in one gram of the oil. For a lubricating oil which is alkaline, by virtue of the additives in it, the number of milligrams of KOH equivalent to the amount of acid required to neutralize one gram of it is a measure of its alkalinity and is termed the total base number, (TBN).

Lubricating oils oxidize in service and become more acidic. A rise in the neutralization value of a straight mineral oil provides an indication of the degree of oxidization taking place. Alkaline oils are used in circumstances where contamination by acidic products of combustion is likely and the rate of decrease of total base number is a measure of the rate of contamination.

9.26. *Lubricating oil filtration and cleaning*

In its passage round the engine the lubricating oil in addition to its functions of lubricating and cooling also cleanses the working parts of the engine. It does this by washing particles of foreign matter, detritus from wear and carbonaceous deposits from various parts of the engine. As it would be harmful to recirculate this dirty oil it is cleaned by filtration.

Regular attention to filters to keep them clean is important if the lubricating oil is to do its job properly and not cause damage to the engine. Two periods during the life of an engine are of particular importance. One is when the engine is first put on the test bed when metal cuttings, foundry sand and manufacturing foreign matter have to be removed from the engine and the lubrication system. The other important occasion is the trials of the ship when the lubricating oil system may be run for the first time and foreign matter which is in the system will be washed by the oil into the filters. It is usual practice to flush out systems in an endeavour to clean them before the engine is run but vibration of the engine when it first starts to operate is often responsible for shaking free dirt which has not been dislodged by normal flushing methods and careful attention to filters during this period is often rewarded by more satisfactory operation of the engines during their subsequent service.

Filters range from a coarse mesh of wire gauze restraining only comparatively large particles to micropack types restraining all particles above a few microns in size. The finer the degree of filtration the larger is the surface area required to accommodate a given flow. Because of this inescapable fact filters of reasonable size that deal with the full flow of oil to the engine, cannot be expected to give the finest degree of filtration and similarly filters that give very fine degrees of filtration can deal with only a portion of the oil on a by-pass system.

The simplest form of filter is a wire mesh gauze strainer. Although the gauze can be of the finest the particles arrested by it are still comparatively large. This type of filter is frequently used at the intake of the lubricating oil pumps to prevent the entry of foreign matter that may accidentally have found its way into the engine or larger fragments of metal that would cause damage to the pumps themselves. Before the oil enters the engine it requires finer filtration than this.

Edge type filters are used to give a much finer degree of filtration. They may consist of a number of plates of metal or plastic spaced apart by small shims or they may be constructed by winding hard wire often of rectangular or other special cross section round an open drum. In practically all cases the oil is arranged to pass from the outside of a stack of plates or the drum to the inside, the particles filtered being left on the outer edges. Many of these filters include cleaning devices some of which are mechanical, having a scraping action, and others are hydraulic using back-flushing of the oil itself. These filters are usually capable of filtering down to about 0·05 mm. A few are capable of filtering to a much finer degree than this. One example consists of sticks of plastic washers threaded on rods. The washers have small triangular section grooves cut across their flat surfaces which form filtering passages when the washers are stacked together. The grooves can be of different sizes and can give filtration as fine as 5 microns. By arranging a number of these stacks within a single casing the overall size of the filter is kept reasonably small for a comparatively large flow as the filtering surface area is large compared to the overal volume. Cleaning is carried out by flushing back each of the sticks of washers in turn and as there are six, eight or more of these sticks in a single filter this can be done without interruption of the oil supply.

Surface filters in which the filtering medium consists of felt or paper packs are claimed to give filtration to a very fine degree. The element is usually of cylindrical form and in the case of the felt medium is often corrugated to obtain a larger surface area within a given filter casing size. Again the oil usually passes from the outside to the inside and the foreign matter is caught by the felt or paper whilst the clean oil is allowed to pass through. This type of filter element must be renewed from time to time. Attempts to clean these elements by washing usually result in some particles being left which eventually channel their way through leaving passages through which oil will flow more readily without being filtered.

In most of these filters a by-pass is incorporated which allows unfiltered oil to pass to the engine if the pressure drop across the filter exceeds a predetermined figure by reason of the oil being cold and viscous or the filter becoming choked.

Some of the paper pack type filters have a very fine degree of filtration indeed, less than 2 microns. The rate of filtering with such filters is necessarily very slow if the filter is to be of reasonable size and this type is usually used on a by-pass system.

Sometimes oil is cleaned on a batch system in which it is completely removed from the engine and put through a cleaning process after which it is ready for use once again. Centrifuges are widely used for cleaning oil in this way. The oil is introduced to a bowl or drum rotating at high speed. Foreign particles usually being of higher specific gravity than the oil are carried to the periphery of the drum allowing the clean oil to be extracted from the centre. If the oil contains water it is separated from the oil by the same action. Some centrifuges are arranged so that the solid matter is continuously removed; others require periodic dismantling and cleaning. It was frequently the practice at one time to introduce water into the oil in order to dissolve impurities which were insoluble in the oil but soluble in water and so to remove them along with the water. This practice was advantageous in the case of non-inhibited oils but it cannot be carried out with modern lubricants as many of the additives used are soluble in water and would be removed by the process.

Centrifuges are also used on by-pass systems and there are small centrifuges operated by the oil pressure itself which are capable of dealing with comparatively small flows but are very simple and cheap to use.

9.27. *Tests on used lubricating oils*

Routine tests on used lubricating oils not only give an indication of whether the oil is fit for further use, but also frequently provide valuable information on the operation of the engine, the manner and rate of deterioration often being more informative than the degree.

Checks should be made on the viscosity of used lubricating oil. The viscosity will be increased by oxidization and an increase of say 10% would normally indicate a bad state of affairs. As well as renewing the lubricant an attempt should be made to find out what operating conditions are causing heavy oxidization.

If the viscosity is reduced it is an indication that there may be dilution of the lubricant by fuel oil. The viscosity test is not a good one for dilution when residual fuels are used as the viscosity of such fuels may well be equal to that of the lubricating oil.

A test of the flash point, however, will reveal dilution whether it is due to residual fuel or distillate fuel. The closed flash point for an SAE 30 oil is in the region of 215°C (420°F), that for fuel is about 93°C (200°F), even if it is of the heaviest residual type. The difference between the two is quite adequate for a test on the flash point to show when quite a small amount of fuel has been added to the lubricating oil.

The alkalinity of lubricating oil is measured by its total base number and it is important when running on residual fuels with a high sulphur content to keep a regular check that the crankcase contents remain at a sufficiently highly alkaline value.

Traces of water in the lubricant are a useful indication of leaks internally which might otherwise not be noticed.

A test to determine the amount of insoluble matter is one of the most important on used lubricants. The solids present in the oil which are insoluble in heptane and benzine frequently consist of atmospheric dust or metallic wear products from the engine or carbon or soot from incomplete combustion, all of which can give valuable information as to the operating condition of the engine. Spectrographic analysis is being used more and more often for the routine determination of the presence of metal in used lubricants. It can rapidly reveal a sudden increase in the rate of wear and often gives an indication of the part which is wearing.

BIBLIOGRAPHY

British Standard Specification for Petroleum Fuels for oil engines and burners, B.S. 2869–1967.

HUGHES, J. R., *Storage and handling of petroleum liquids: practice and law,* Griffin, London, 1967.

GOODGER, E. M., *Petroleum and performance,* Butterworth Scientific Publications, London, 1953.

Diesel fuels and lubricants, Esso Petroleum Company Ltd., London, 1960.

CHAPTER TEN

Flow Systems

10.1. *Cylinder jacket cooling, sea water*

The simplest water cooling system is shown in Figure 10–1(a), using sea water passing through the jackets of the engine. In order to provide adequate circulation a pump is necessary and in a system as simple as this it is usually engine driven. The capacity of the pump will be chosen to provide a sufficient quantity of water for cooling at the full output of the

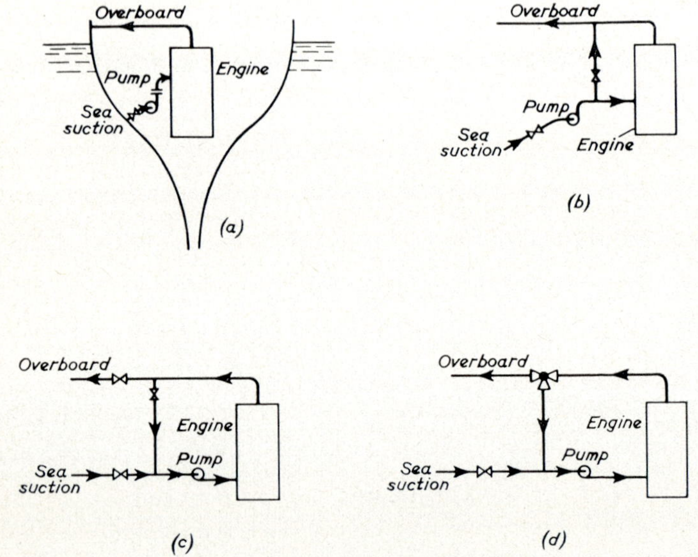

FIG. 10–1—*Simple sea water cooling systems.*

engine. At reduced speeds it will give a reduced quantity which is a desirable trend but in order to cater for variations in the sea water temperature or the engine load it will be necessary to have some further regulation of the quantity.

The most common form of regulation is by the addition of a by-pass. One way of arranging this is to take it from a point at the pump delivery to a point at outlet from the engine as shown in Figure 10–1(b). A valve in the by-pass will control the flow of water by varying the resistance of the by-pass in comparison to the constant resistance of the engine. This arrangement however, has certain disadvantages, the chief one being that the temperature rise through the engine is always a maximum, whilst the quantity of water flowing through is varied.

Much better results in the operation of the engine are obtained if the flow of water is kept to the maximum and the temperature rise through the engine to as small a figure as possible according to load. This can be arranged by a by-pass from a point at the outlet of the engine to a point at the intake of the pump as shown in Figure 10–1(c). By this means a quantity of the water is recirculated and so the temperature at the inlet to the engine jackets can remain high, the quantity of water flowing through the jackets always being the maximum capacity of the pump. In order to regulate this system two valves are shown, one in the by-pass and one on the overboard discharge. By adjusting the amount by which these valves are opened the temperature at which the heat is extracted from the engine can be varied. In practice matters should be arranged so that it is not possible to close both valves completely, especially if the pump is a positive displacement pump, as this could lead to disastrously high pressures in the engine jacket.

The arrangement can be improved and made to look after itself by introducing a thermostatic three-way control valve at the point B where the by-pass is joined to the engine discharge pipe. This arrangement is shown in Figure 10–1(d). The thermostatic valve will consist of a temperature sensing element, the output from which can be arranged to divert the flow of water along the by-pass or over-board as required to maintain a reasonably constant temperature at the engine outlet. The operation of thermostatic valves is described in Chapter 15. For sea water operation the outlet temperature must not be allowed to exceed 55°C, otherwise salt deposits will occur in the cooling system. As will be seen later very much the same thermostatic control arrangements are used with fresh water systems but in that case it is much better to run the engine at higher temperatures, for example, 70°C or 80°C according to the make of engine. Thermostatic valves cannot snap open immediately the set temperature is reached but require 2 or 3°C to operate from the by-pass fully closed to the by-pass fully open.

Simple sea water cooling systems of the type described were very popular some years ago but today are found only with installations of comparatively low horsepower in small craft. Sea water is not a suitable medium for cooling the jackets of modern highly developed engines as it contains corrosive elements, sediment and other matter likely to cause damage either by corrosion or deposition.

10.2. *Cylinder jacket cooling, fresh water*

A fresh water system is basically the same as that shown in Figure 10–1(d) with the exception that the hot water instead of being discharged overboard is passed through a heat exchanger and returned to the pump suction, the heat exchanger being circulated by sea water supplied from a second pump. The arrangement is shown in Figure 10–2(a).

In small craft the heat in the fresh water from the engine jackets may be removed by keel cooling or skin cooling systems in which the fresh water flows through ducts in, or alongside the keel, or of which the skin of the vessel forms a part. With these systems the heat is passed to the sea without sea water being brought inboard.

Because of the recirculation of the fresh water the system is completely closed and this requires the introduction of another feature; the header tank. The closed fresh water system has a certain capacity which is filled with the volume of water. As the water temperature rises it will overfill the capacity of the system, or if air is introduced into the system it will become overfull. The introduction of air into the system commonly occurs at pumps during the suction stroke of plunger pumps or along the spindle in the case of centrifugal and other rotary pumps. Although the leakage may be very small it can accumulate and make a significant difference in the volume. By including a tank open to the atmosphere and placed at a higher level than any other point in the system these changes in volume of the contents are accommodated by a change in the level of the free surface of the water in the tank.

10.3. *Venting*

In Figure 10–2(a) the tank is shown connected to the engine outlet at point P. It could be connected at point Q at the pump delivery, or point R at the pump suction. These alternative connexions have an influence on the behaviour of vent pipes. Vent pipes are necessary where there are local high points in the system in order that air which accumulates in such

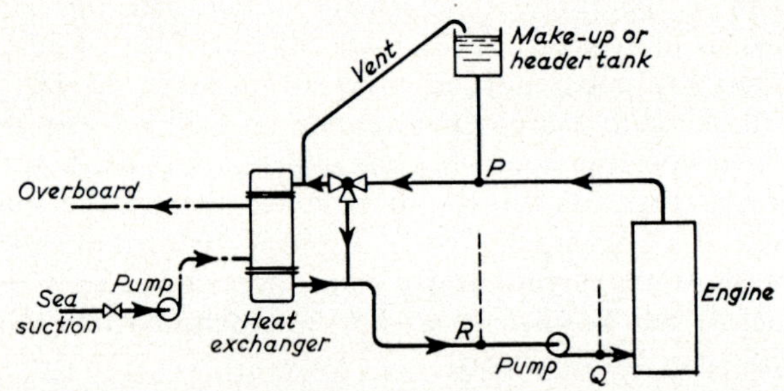

(a) PRIMARY FLUID CONTROL

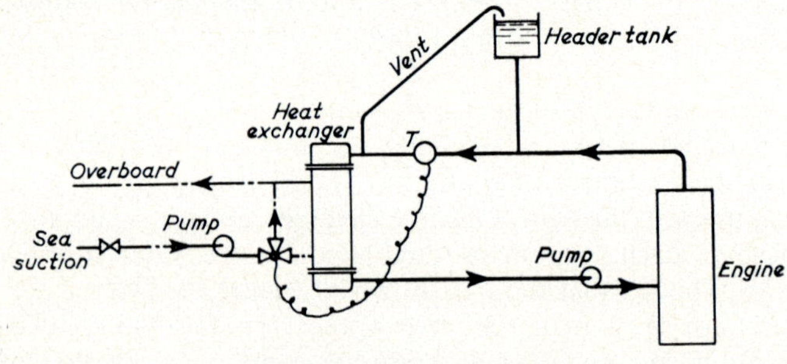

(b) SECONDARY FLUID CONTROL

FIG. 10–2—*Fresh water cooling.*

positions can be extracted. They may be merely vent cocks which can be opened until all the air is out and then closed again or they may be permanent vents, such as the one illustrated, which are taken up to the header tank and discharged into it above the normal water level. The circulation of water round the system by the pump involves a pressure drop. The pressure is highest at point Q at the pump discharge and falls as the water progresses round the circuit being lowest at point R, the pump intake. The absolute level of pressure is determined by the head provided by the header tank. If the header tank is connected to point Q then at no point in the system can the pressure exceed that given by the head of water at point Q. If, however, the header tank is connected at point R then the pressure at Q and at a number of other points early in the system can exceed that provided by the header tank. Any air vents taken from such points of higher pressure such as the one shown, will run full of water most of the time. This water does not pass through the heat exchanger and therefore there is a reduction in the cooling capacity of the system. This is not necessarily a bad thing unless a large number of such vents for air extraction are added to an existing system without regard to the effect on the cooling capacity.

10.4. *Primary and secondary fluid control*

In a system such as this which uses two fluids for cooling; the fluid cooling the engine itself, in this case the fresh water, is known as the primary fluid; and the fluid to which the heat is transferred, in this case the sea water, is known as the secondary fluid. A control arrangement with the by-pass in the fresh water system, as shown in Figure 10–2(a), is known as primary fluid control. An arrangement can be made, such as shown in Figure 10–2(b), where the by-pass is in the sea water circuit, which is called secondary fluid control. It should be noted that both cases are aimed at control of the temperature of the primary fluid at outlet from the engine and therefore the temperature sensing part of the thermostatic valves must be located at the engine outlet in both cases. With secondary fluid control the temperature sensing part and the valve itself are, therefore, separate and must be connected by electrical, pneumatic, or other means. Secondary fluid control uses more sophisticated equipment, but is advantageous in that the amount of sea water circulated through the cooler is reduced in circumstances where a large flow could be harmful. For example, when the engine is operating at low power in harbour or estuarial waters which are usually high in foreign matter content and are of a corrosive nature, lesser amounts are circulated through the cooler.

10.5. *Water pumps*

The two water pumps shown may be engine driven or may be separately driven, or one may be driven from the engine and the other driven by electric motor. Whatever the arrangement it is usual to provide cross connexions so that one pump can take over the duty of the other in an emergency, cooling the engine by means of sea water alone. However, unless the emergency is a severe one, it is desirable to avoid introducing sea water into the fresh water systems and it is common practice to provide stand-by pumps for both fresh and sea water pumps. Not infrequently the bilge and ballast pump is arranged to take over the duties of the sea water pump if necessary and the general service pump piped in so that it can take the place of the fresh water pump should it be required.

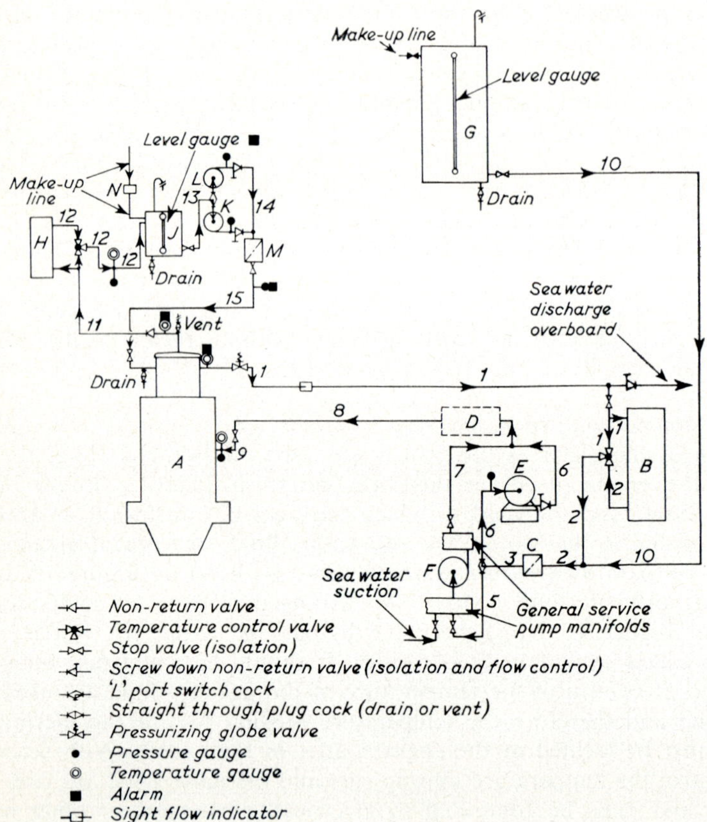

A Main Engine
B Jacket Water Heat Exchanger
C Jacket Water Strainer
D Jacket Water Heater
E Jacket Water Pump
F General Service Pump (Stand-by)
G Make Up Tank
H Exhaust Valve and Injector Cooling Water Cooler
J Exhaust Valve and Injector Cooling Water Header Tank
K Exhaust Valve and Injector Cooling Water Pump
L Exhaust Valve and Injector Cooling Water Pump (Stand-by)
M Exhaust Valve and Injector Cooling Water Strainer
N Exhaust Valve and Injector Cooling Water De-Ionizer

1 Engine to Cooler
2 Cooler to Strainer via Temp. Control Valve
3 Strainer to Switchcock
4 Switchcock to Pump
5 Switchcock to stand-by Pump
6 Pump to heater
7 Stand-by Pump to Heater
8 Heater to Engine Stop Valve
9 Stop Valve to Engine
10 Make-up Line Tank to Pipe Ref. 3, Valve Gauge and Injector Cooling System comprising:
11 Engine to Heat Exchanger
12 Heat Exchanger to Header Tank
13 Header Tank to Pumps
14 Pumps to Strainer
15 Strainer to Engine

FIG. 10–3a—*Fresh water and deionized water systems single engine motor driven pumps (manned engine room).*

10.6. *Injector cooling water*

In addition to water for cooling the engine jackets some highly developed engines require water of a high degree of purity for cooling other components such as injectors and exhaust valve cages or the exhaust valves themselves. Because these parts usually have very small internal passages the water used for cooling is often confined to a separate circuit which embodies fine filters, the water itself being distilled or carefully treated before being introduced.

10.7. *Complete water cooling system*

Figure 10–3 shows fresh and distilled water systems together with the sea water system for a single engine. Points to be noted are the alternative sea water intakes at high and low level and the filters which are necessary to exclude the larger particles of foreign matter. As well as serving the jacket water heat exchanger the sea water also passes through the lubricating oil cooler, the combustion air intercoolers and the gearbox lubricating oil cooler. It may well serve other machinery in the engine room also and the size of the sea water pump and the piping is determined often by considering not only the main engine but other machinery as well.

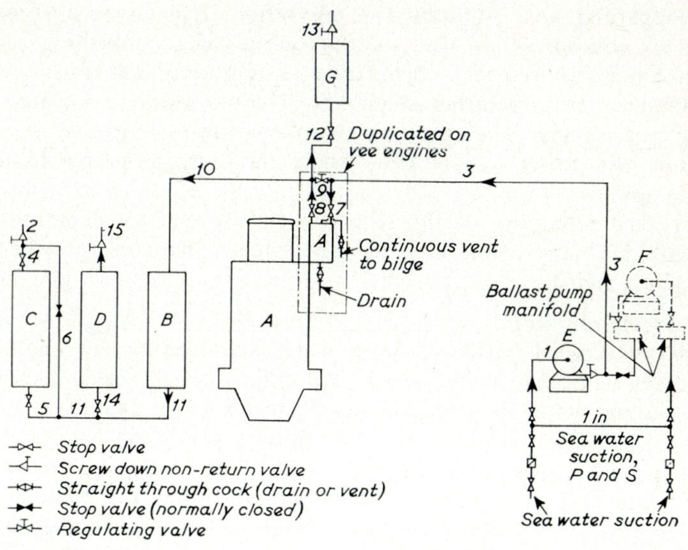

A Main Engine
B Oil Cooler
C Jacket Water Cooler
D Gearbox Oil
E Exhaust Valve and Injector Water Cooler
F Sea Water Pump
G Stand-by Pump (Ballast)

1 Sea Water Suction Main
2 Sea Water Discharge Overboard
3 Pump Delivery Main
4 Jacket Water Cooler Outlet
5 Jacket Water Cooler Inlet
6 Jacket Water Cooler Bypass
7 Inlet to Intercooler "A"
8 Outlet from Intercooler "A"
9 Intercooler Bypass
10 Inlet to Oil Cooler
11 Outlet from Oil Cooler
12 Exhaust Valve and Injector Cooler Inlet
13 Exhaust Valve and Injector Cooler Overboard Discharge
14 Gearbox Oil Cooler Inlet
15 Gearbox Oil Cooler Overboard Discharge

FIG. 10–3b—*Sea water system single engine motor driven pumps (manned engine room).*

10.8. *Twin engine jacket water, combined and separate systems*

For a twin engine installation the fresh water system may be combined to serve both engines or it may be two entirely separate systems, one for each engine. The choice of system depends on various circumstances and the following are some of the factors to which consideration must be given.

A combined system generally follows the lines of that for a single engine, but branches to serve the two engines at entry to the jackets and recombines at exit from the engines. Valves are provided in the branches to regulate an even flow to the two engines and to isolate either one from the system when required. Only one heat exchanger is needed but of a capacity sufficient for both engines.

If the full power of the installation is to be available in the event of failure of the circulating pump a standby pump of at least equal capacity must be provided for the combined system. With separate systems it is most unlikely that both circulating pumps would fail at the same time so that one standby pump, half the size of that required for the combined system, will suffice for the maintaining of full power.

The power from one engine alone is often regarded as adequate for the propulsion of the ship in emergency in which case, with separate systems there is no need for any standby pump as one complete system can be regarded as a standby for the other. This saves a considerable amount of equipment as well as the pump. The connecting pipes and valves are not required and, in the case of remote or automatic control of the machinery, the operating solenoids or motors and associated control gear are not needed. The resulting twin separate systems are less complicated and use fewer components than the corresponding basic single engine system.

Increased reliability of the ship can be claimed for separate systems as the failure of any component in one system cannot prevent the running of the other engine.

10.9. *Circulation rate*

The amount of jacket cooling water required by an engine varies considerably according to its design. An engine which rejects almost all its heat to cooling water may require say 50 to 60 m^3/hr per 1,000 kW brake power (8 to 10 gallons per brake horsepower per hour), whilst an engine which uses oil for piston cooling and rejects considerable quantities of heat by this route may require only 30 m^3/hr per 1,000 kW brake power (5 gallons of water per brake horsepower per hour). The sea water pump will have to be sized according to the amounts of heat which are rejected not only by the main engine but by other components as well.

10.10. *Heat exchangers*

The type of heat exchanger most used in marine work is of tubular form as shown in Figure 10–4. The principles of construction and operation are the same for the two applications of cooling jacket water and cooling lubricating oil. A stack of thin walled tubes is supported by spacer plates within a cylindrical casing, the ends of the tubes communicate with header boxes, the secondary fluid (sea water) enters by one of these and

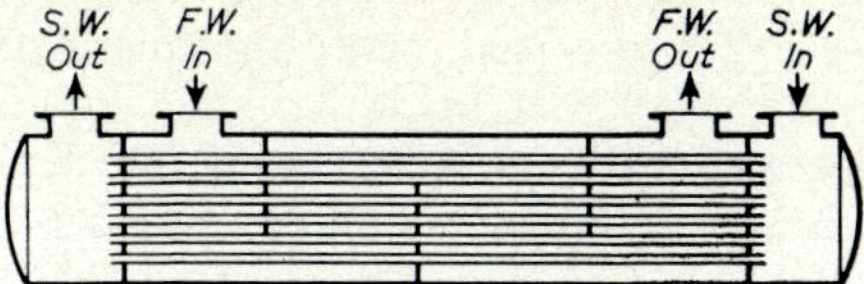

FIG. 10–4—*Heat exchanger.*

leaves by the other after flowing through the tubes. The primary fluid (fresh water or lubricating oil) flows through the cylindrical casing over the outside of the tubes giving up heat to the sea water in the process. The header boxes can be removed and the tube stack withdrawn for cleaning.

The rate at which heat is transferred from the primary to the secondary fluid depends upon the total surface area of the tubes, their radial thickness, the conductivity of the material, the heat transfer coefficients from primary fluid to tube surface and from tube surface to secondary fluid depending on the velocity of the fluids for various paths and directions of flow, and the logarithmic mean temperature difference between the fluids. The theoretical relationships between these factors are derived in textbooks on Applied Thermodynamics. The size of a heat exchanger to perform the task of keeping the temperature of the primary fluid within the required limits may be found from these relationships but in practice considerable margin of capacity is given to safeguard against deterioration due to the arduous circumstances that inevitably accompany marine service.

The chief troubles that beset heat exchangers are fouling and corrosion. The former detracts directly from the heat transmission; the latter usually results in temporary repairs being made by plugging the ends of the offending tubes diminishing the heat transmission surfaces of the cooler. Effects of this nature, coupled with the deterioration of pumps and restrictions arising in various fittings as they age so reducing flow rates explain the need for amply sized heat exchangers.

Regular inspection and cleaning of heat exchangers is essential in the interests of maintaining the efficiency of cooling systems. Sea water is the chief source of debris and deposits and for this reason it is usual to pass it through the tubes as the bores can be cleaned by mechanical means using rods and brushes, rotating and power driven if necessary. If deposits of hard scale are formed then chemical cleaning is necessary. Deposits on the outside of the tubes from fresh water are by no means unknown and chemical cleaning is the only possible way to remove them. The method used is to soak the tube stack in a weak acid solution containing corrosion inhibitors. It should only be carried out by the manufacturers or by specialists acting under their guidance, as the utmost care must be taken and only approved solvents used to avoid the dangers of setting up corrosion or blocking parts of the cooler by uncontrolled precipitation of deposits.

Scale formation can occur on both fresh water and sea water sides and is the result of thermal breakdown of bicarbonates and other salts in solution in hard water. When it occurs on the sea water side it is generally the result of running at too high sea water temperatures. The effects are often cumulative; because of diminished heat transmission the tubes become hotter and the boundary layer of sea water then increases in temperature leading to an accelerated rate of build up of the deposits. On the fresh water side it is the result of unsuitable make up water.

10.11. *Corrosion*

Corrosion is electro-chemical in nature. The processes involved differ for different metals and several different processes may operate on the same metal but three features are commonly present. At certain areas of the metal surface the metal dissolves away; these areas have a negative electric potential and are called anodes. At different areas the potential is positive and a complementary action occurs: these areas are called cathodes. For these processes to take place it is necessary for oxygen to be present in the water, except in the case of a form of corrosion aided by sulphate reducing bacteria.

The attack can take several forms and it may be initiated in one form and then progress by another. There may be general wastage in which all the material surfaces deteriorate in a fairly uniform manner often with excessive deposition of corrosion products. Graphitization is a form of general wastage to which cast iron is subject, the iron being removed leaving behind the carbon in the shape of the original casting. The material remaining is soft and can be carved away by a penknife. Brasses are subject to wastage in which the zinc is dissolved leaving the copper behind by a process known as dezincification.

Corrosion may also take the form of pitting in which the anode areas are small. Concentrated attack of this form often proceeds rapidly and the pits may quickly penetrate a metal wall. Crevice corrosion may occur under loose scale or at the bottom of a pit. A crevice forms a back water in the cooling system in which, as a result of the corrosion processes, conditions may become quite different from those in the main bulk of fluid and often result in rapid dissolving of the metal.

Where two different metals immersed in water are in contact a potential difference will exist between them, the more "noble" metal having a higher electrical potential than the more "base" metal. The latter will become anodic and tend to corrode whilst the former becomes cathodic and thereby protected. This is termed bimetallic or galvanic corrosion.

High water velocities can result in impingement attack, particularly if the water contains abrasive particles, *e.g.*, sand. The products of corrosion are eroded away exposing the metal to further corrosion which is thus encouraged to proceed rapidly.

In certain conditions vibration leads to the formation of vapour bubbles which collapse with severe effect leading to cavitation corrosion. This form of attack occurs on the water side of cylinder liners as a result of vibration caused by hammer blows from the motion of the piston. It can also occur on pump impellers.

As the processes of corrosion are dependent upon oxygen being present in the water much can be done to prevent it by excluding air from the system wherever possible. Pet cocks are often provided, particularly with reciprocating pumps to avoid water hammer; they should be adjusted to admit the minimum amount of air to achieve this purpose and should certainly not be used as a means of regulating the pump output. Vent cocks should be used regularly and high points in the system fitted with permanently open vents where possible. Casings should not be left partially full of water, particularly sea water.

10.12. *Water treatment*

Water treatment is recognized as being an essential safeguard for expensive modern marine machinery. It is well known practice to add chemicals termed inhibitors to the cooling water in closed systems with the object of preventing both corrosion and scale formation. Most inhibitors work by stifling the electro-chemical actions by providing conditions in which the corrosion products become insoluble in the water. These corrosion products then tend to form a protective film on the metal. This inhibiting action may take place at the anode or at the cathode or at both. Most inhibitors operate at both regions to a greater or lesser extent but the terms anodic or cathodic inhibitors are frequently used to indicate the major effect.

Inhibitors are also classed as "safe" or "dangerous". Safe inhibitors reduce total corrosion without any tendency to increased attack on unprotected areas. Dangerous inhibitors, if used in insufficient concentration, may leave small areas unprotected and promote intensified rates of attack at these points. Anodic inhibitors mostly fall into the dangerous class whilst cathodic inhibitors are safe, but there are exceptions in both cases.

The action of an inhibitor depends a great deal on the environment in which it is used; corrosion may be prevented in one set of conditions but increased in another. The choice of an inhibitor is influenced by the type of system, the composition of the water, its temperature and rate of flow, the composition of the metal surfaces and the presence of dissimilar metals. The condition of the system is of importance and whether a water treatment has been used previously. In Table 10.1 some of the more commonly used inhibitors are listed.

Commercially available inhibitors intended for use in diesel engine cooling systems often have several constituents, as a mixture of two inhibitors operating by different processes often produces a better result than either on its own. They are chosen to protect the wide variety of metals of which most systems are composed, which includes cast iron, steel, copper, brass and solder; without attacking the rubbers and non-metallic materials used for joints and seals.

As an alternative to chemical additives, soluble oils are used, preventing corrosion by their propensity to form a protective film over the internal surfaces of the cooling system. If used in excess of the proper concentration there is a danger of too thick an oil film being deposited with consequent interference with the heat transfer. Certain blends of tannins

TABLE 10.1

Behaviour of some typical inhibitors towards materials found in engine cooling systems

Additive	*See Note*	*Materials*					
		Cast Iron	*Carbon and Alloy Steels*	*Copper and Copper Alloys*	*Solders*	*Alum. Alloys*	*Rubbers*
Sodium Nitrite	1	P	P	NA	C	P	NA
Sodium Chromate	2	P	P	P	P	P	NA
Sodium Benzoate	3	NA	P	NA	P	NA	NA
Sodium Silicate	4	V	V	NA	V	P	NA
Sodium Borate	5	P	D	P	V	V	NA
Sodium Phosphate	6	P	P	P	P	V	NA
Triethanolamine Phosphate	7	—	P	C	—	P	NA
Benzotriazole	8	NA	NA	P	—	—	NA
Sodium Mercaptobenzthiazole	9	NA	NA	P	—	—	NA
Soluble oil	10	P	P	P	P	P	V
Tannins	11	P	P	P	—	—	NA

Key:

- P Protective, definitely inhibitive.
- C Corrosive.
- NA No action, neither protecting nor corroding to any appreciable degree.
- V Variable behaviour, depending on circumstances.
- — No data.

Notes for TABLE 10.1

1. Ferrous metal inhibitor, operates at pH 8·5 to 9, the alkalinity being provided by alkalis 4, 5, 6. Attack on solder countered by 3.4.5.6.
2. Operates at pH just above neutral.
3. Included in formulations to protect solder.
4. May be used by itself or in conjunction with 1.5. Is an inhibitor for solder and aluminium alloys provided pH does not exceed about 11.
5. Protects ferrous metals by virtue of its alkalinity; high pH's may be dangerous to solder or aluminium alloys. Borate specified may be metasilicate or borax.
6. Usually di-sodium salt. May be corrosive to aluminium alloys when heat is transferred.
7. With 9 developed for aluminium alloy engines, but now employed in cooling circuits containing ferrous metals. NaMBT counters corrosion of copper alloys by the triethanolamine.

8.9. Inhibitors specific for controlling copper corrosion, and present in many formulations.

10. Operates at about pH7.
11. Operates at about pH7.

have also been used with success in diesel engine cooling systems; they also act by forming a protective film.

The water selected for use in a closed system should preferably be distilled or de-ionized before the inhibitors are added. At the very least water having a low hardness should be chosen. It is also of obvious importance to determine accurately the capacity of the cooling system so that the correct strength of solution is used.

10.13. *Lubricating oil systems*

The lubricating oil must be circulated outside the engine in order to cool it and to filter it. Some of the smaller high speed engines operate on a "wet sump" system in which the oil is contained in an enlarged sump forming part of the engine. There is one pump provided, often engine driven, which draws the oil from the sump, passes it outside the engine through coolers and filters and then returns it to the lubricating oil passages within the engine.

Larger engines are usually designed to have a "dry sump" and a separate drain tank is provided in which the oil is retained. The oil may flow to the drain tank by gravity and this is a fairly common arrangement, the tank being situated in the structure of the vessel below the engine.

Alternatively if this situation of the tank is not possible, the oil may be taken from the engine sump by a scavenge pump and delivered to the drain tank. The engine lubricating oil pressure pump draws its oil from the drain tank and passes it through cooler and filter before returning to the engine in the same way as with the wet sump engine. If a scavenge pump is employed it is essential that it should exceed the capacity of the pressure pump by sufficient margin to keep a low level of oil in the engine under all conditions of list and trim of the vessel. The pumps can be engine driven or separately driven and in either case there are often cross connexions arranged so that in an emergency a dry sump engine can be run as a wet sump engine on one pump only. If the pumps are engine driven it will be necessary to include also a priming pump to ensure that the lubricating oil system is full and oil circulating before the engine is started. Obviously if the pumps are separately driven this is achieved in any case. Figure 10–5 illustrates the simple principles of a dry sump system. In this figure the filter is situated between the cooler and the engine; the oil passing through the filter immediately before entering the engine lubricating oil passages. Some operators consider it preferable to filter the oil before it passes through the cooler as the hotter and less viscous oil is more easily cleaned. Ideally two filters should be installed but the increased cost is difficult to justify.

The oil temperature must be regulated so that its viscosity is kept within bounds. For this purpose there is a by-pass on the cooler and in Figure 10–5 this is shown on the lubricating oil or primary fluid side.

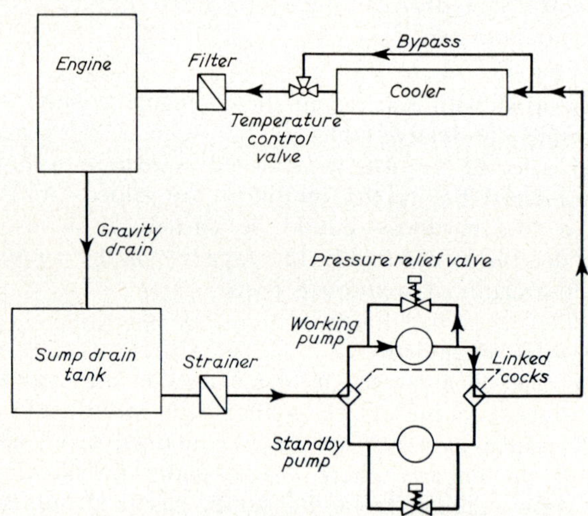

FIG. 10–5—*Lubricating oil basic system.*

Secondary fluid control can be used in the lubricating oil system just as it was in the jacket water systems and has an additional advantage in that the full flow of lubricating oil is maintained at all times, so avoiding cold slugging in the cooler. This is an occurrence in which some of the tubes

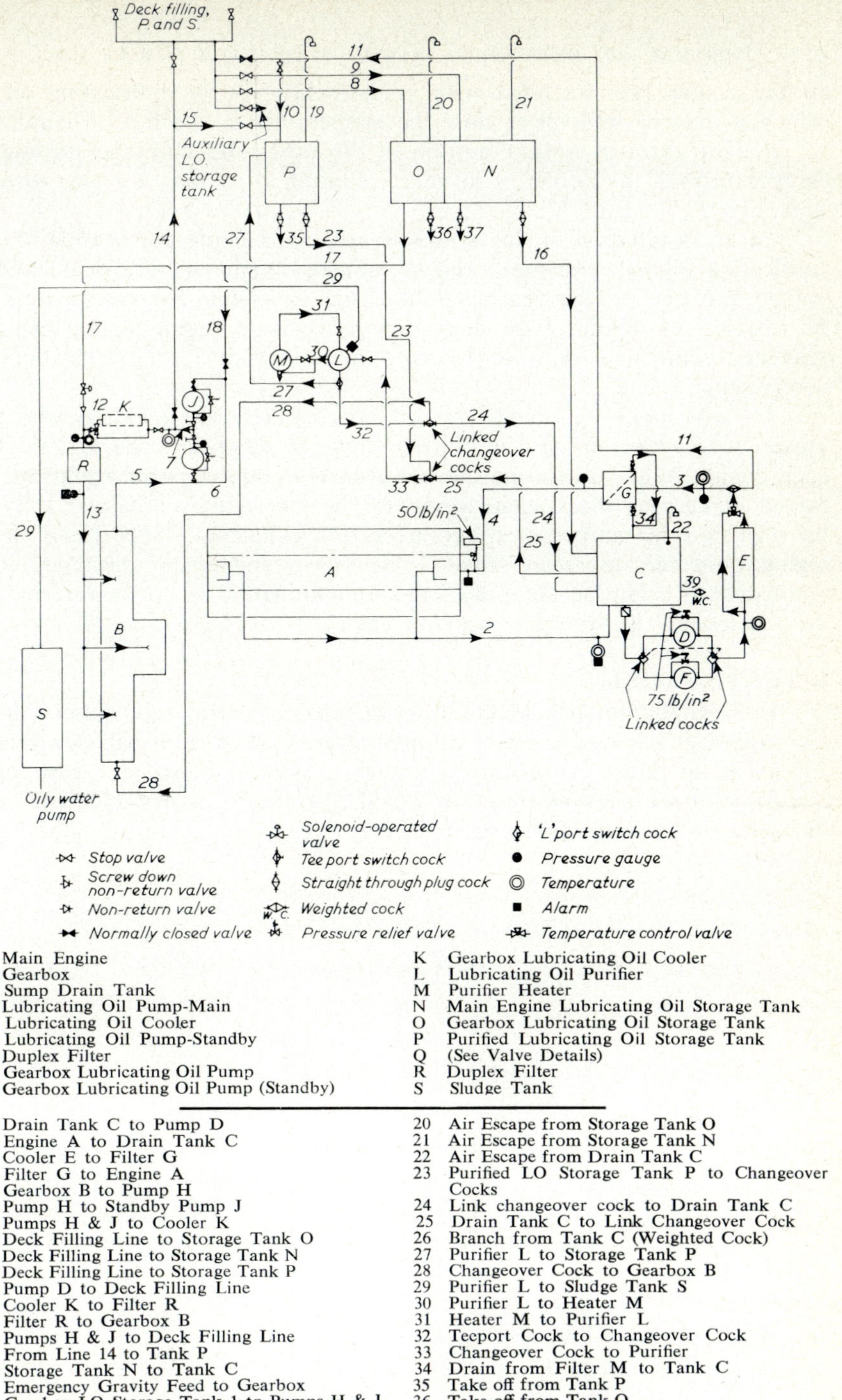

A Main Engine
B Gearbox
C Sump Drain Tank
D Lubricating Oil Pump-Main
E Lubricating Oil Cooler
F Lubricating Oil Pump-Standby
G Duplex Filter
H Gearbox Lubricating Oil Pump
J Gearbox Lubricating Oil Pump (Standby)
K Gearbox Lubricating Oil Cooler
L Lubricating Oil Purifier
M Purifier Heater
N Main Engine Lubricating Oil Storage Tank
O Gearbox Lubricating Oil Storage Tank
P Purified Lubricating Oil Storage Tank
Q (See Valve Details)
R Duplex Filter
S Sludge Tank

---

1 Drain Tank C to Pump D
2 Engine A to Drain Tank C
3 Cooler E to Filter G
4 Filter G to Engine A
5 Gearbox B to Pump H
6 Pump H to Standby Pump J
7 Pumps H & J to Cooler K
8 Deck Filling Line to Storage Tank O
9 Deck Filling Line to Storage Tank N
10 Deck Filling Line to Storage Tank P
11 Pump D to Deck Filling Line
12 Cooler K to Filter R
13 Filter R to Gearbox B
14 Pumps H & J to Deck Filling Line
15 From Line 14 to Tank P
16 Storage Tank N to Tank C
17 Emergency Gravity Feed to Gearbox
18 Gearbox LO Storage Tank 1 to Pumps H & J
19 Air Escape from Storage Tank P
20 Air Escape from Storage Tank O
21 Air Escape from Storage Tank N
22 Air Escape from Drain Tank C
23 Purified LO Storage Tank P to Changeover Cocks
24 Link changeover cock to Drain Tank C
25 Drain Tank C to Link Changeover Cock
26 Branch from Tank C (Weighted Cock)
27 Purifier L to Storage Tank P
28 Changeover Cock to Gearbox B
29 Purifier L to Sludge Tank S
30 Purifier L to Heater M
31 Heater M to Purifier L
32 Teeport Cock to Changeover Cock
33 Changeover Cock to Purifier
34 Drain from Filter M to Tank C
35 Take off from Tank P
36 Take off from Tank O
37 Take off from Tank N

FIG. 10–6—*Single engine lubricating oil circuit (Manned Engine Room: Independent Pumps)*

in the cooler become filled with cold, highly viscous lubricating oil of which it may be difficult to raise the temperature as the hot oil continues to circulate through only a portion of the tubes and continues to rise in temperature as the cooler is in effect reduced in its heat transfer surface area.

In an installation of any size it is usual to include purification of the lubricating oil by centrifuges and permanent connexions are made to the engine and also to the gearbox lubricating oil system for this process to be carried out. Figure 10–6 shows a typical single engine system using a scavenge pump with cross connexions and with facility for purification by centrifuging.

In twin engined or multi engined systems there are again, as with the jacket water, the alternatives of combining or separating the systems for each engine. With lubricating oil, however, there are stronger arguments in favour of keeping the systems separate. The lubricating oil has not only to be regulated in temperature but must be cleaned and kept clean. It is highly desirable, therefore, that if the oil in one engine becomes contaminated by some accident, that this contamination is not transferred to the lubricating oil system of another engine.

10.14. *Fuel systems*

For installations in which it is intended to burn only diesel oil its low viscosity enables the use of a fairly simple system, as it will flow under gravity at all times. An example of such a system shown in Figure 10–7 consists of a service tank placed at a height sufficient to give adequate flow through a filter to the fuel injection pumps. The service tank is filled from

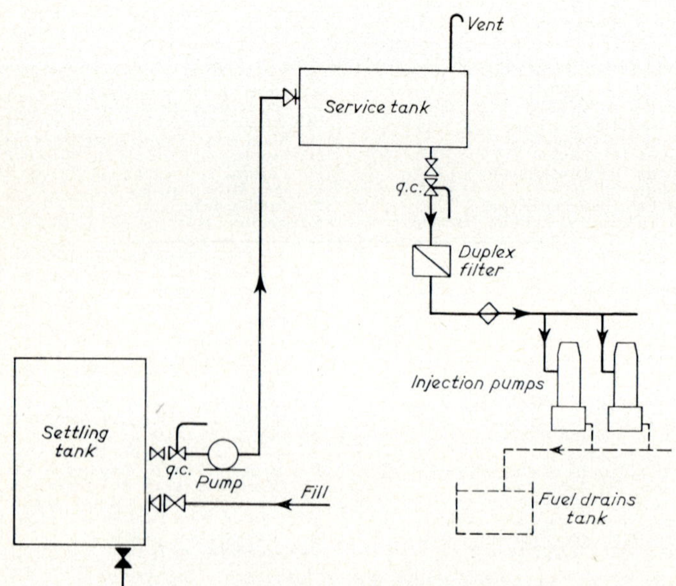

FIG. 10–7—*Diesel fuel system, gravity feed.*

a settling tank which in turn is filled from the bunkers. The settling tank is necessary in order to separate any water which may have found its way into the oil during its period in the bunkers. With oils of low viscosity and low specific gravity the settling process is quickly carried out. The filter between the service tank and the engine is a duplex type so that a changeover can be made and one section cleaned whilst the engine continues to run. In all but the smallest vessels remote operated valves would be included as shown in the figure so that fuel can be turned off in the accidental occurrence of dangerous leaks or fires or other hazards.

When heavy fuel oils are used a more complex system is required. It is usual to have some diesel oil available so that the engine can be run on a light fuel before shutting down for overhaul, as this makes for much easier handling of the fuel injection equipment. Heavy fuel systems, therefore, embody changeover valves or other connexions fulfilling this function. Because of its viscous nature heavy fuel oil cannot be relied upon to flow by gravity and it is necessary to arrange for pumping equipment to transfer it at all stages, and heating is usually required to reduce its viscosity to a reasonable level for pumping. Although settling tanks are provided, the high specific gravity of heavy fuel oils usually demands purification by centrifuges for water extraction. For centrifuging the oil has to be heated further so that its viscosity is suitably low.

Clearly, for safety reasons, it is advisable to heat the oil to a temperature no higher than the minimum required at any of the stages.

At admission to the engine fuel injection equipment the oil should be heated to give a viscosity of about 80 to 90 seconds Redwood No. 1, in order to obtain adequate spray characteristics for combustion. At the purifiers a viscosity of 100 to 150 seconds Redwood No. 1, will be required; elsewhere the temperature need be only sufficiently high to reduce the viscosity to a level at which the oil can be pumped. As well as lagging the pipes, both for safety reasons and also to retain the heat, it is necessary to provide trace heating of them particularly if it is required to start up from a completely cold condition. The trace heating may be by steam or by electricity.

A typical system for heavy and light fuels is shown in Figure 10–8(a). From the settling tanks fuel is pumped via centrifuging equipment to the service tanks. A single centrifuge is shown for each fuel system with standby cross connexions so that either can perform the duty of the other. In either system two centrifuges connected in series as purifier and clarifier can be incorporated if desired.

Heavy fuel is drawn from the heated service tank and pressurized by the feed pump. After passing through a further heating stage it is fed to the injection equipment through a micronic filter. To ensure even distribution of temperature at the injection pumps, over the range of flow rates required by the engine to meet various demands of load, it is usual to provide a recirculating system (Figure 10–8(b)), with the fuel passing continuously through the inlet gallery of each injection pump and returning via a pressurizing valve to the inlet side of the feed pump. A similar recirculating system may be used for the distillate fuel or alternatively it may

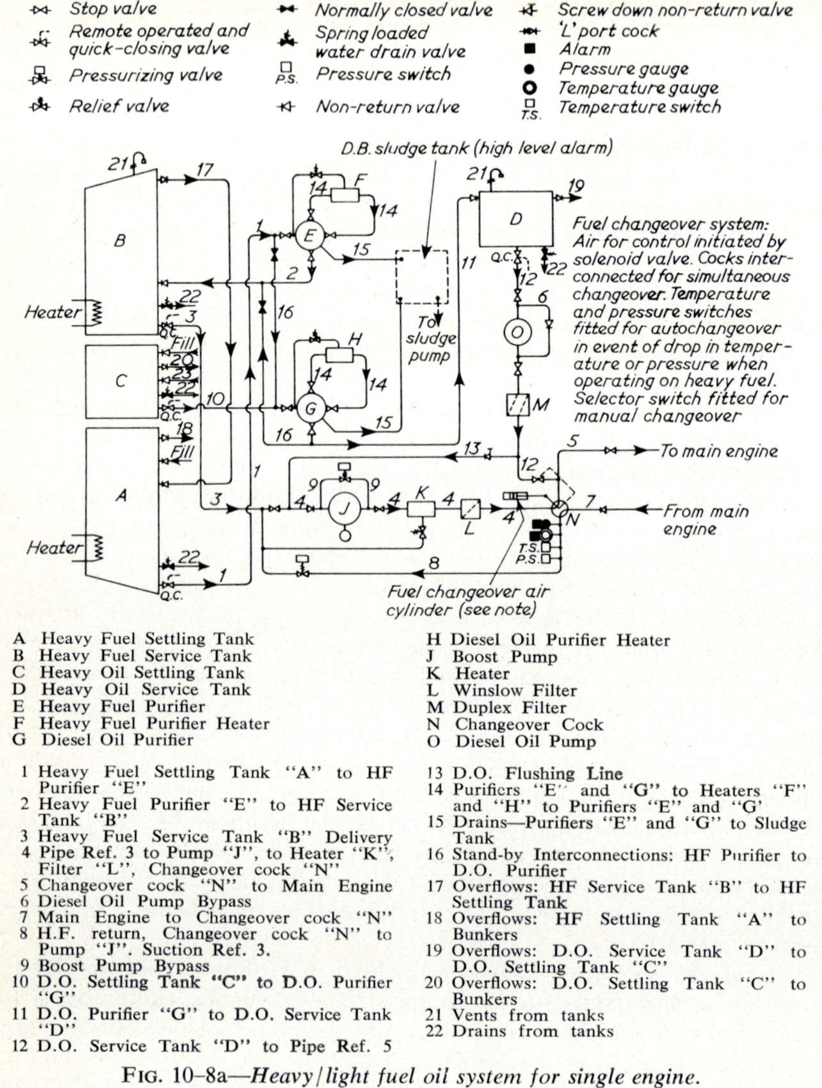

FIG. 10–8a—*Heavy/light fuel oil system for single engine.*

be converted to a non-circulating system by the operation of changeover valves arranged as shown in Figures 10–8(a) and 10–8(b).

10.15 *Waste heat recovery*

The diesel cycle has a thermal efficiency of about 40%, making it the most thermally efficient of all practical heat engine cycles. Nevertheless some 60% of the heat supplied in the fuel is rejected as low temperature energy. Some of this heat is carried away by the jacket water and lubricating oil systems but more than half of it is rejected to the exhaust gas. Some of this heat can be employed in various ways to increase the overall efficiency

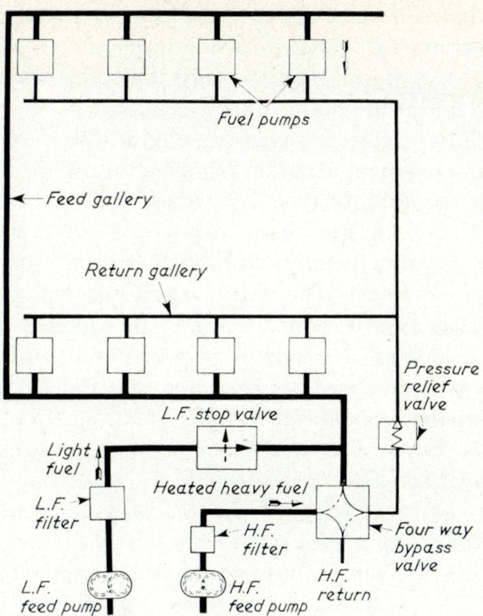

FIG. 10–8b.

of the machinery installation. However, to do this effectively, it is essential that the ship should be on a service which demands that the engines be operated at over 50% of their rated load for most of the time, otherwise it is difficult to justify economically the extra equipment and complication.

The exhaust gas and the jacket water are the two sources of waste heat most commonly used. The jacket water can be used for high vacuum flash evaporators for the production of fresh water, for accommodation heating and for boiler feed water heating. The exhaust gas heat can be used to raise steam which can find a use in fuel heating and steam tracing when heavy fuel is used. The steam can also be used for oil cargo temperature regulation, for various domestic services, for de-icing services and for electrical power generation.

There are usually available about 5·5 kg per brake horsepower hour of exhaust gas at a temperature of 300° to 400°C or even higher, depending on the engine. The engine builder is able to give information concerning his own products regarding the amount of heat available. Exhaust gas heated boilers are, necessarily, constructed to offer as little resistance to the flow of exhaust gas as possible. Clearly a high pressure drop would interfere with the operation of the engine, particularly if it is turbocharged. A pressure loss in the exhaust system greater than 300 mm of water may be regarded as excessive. A check should be kept on exhaust back pressure in service and cleaning of the boiler carried out whenever necessary.

10.16. *Air start systems*

Most medium speed engines are started by compressed air which is admitted to the cylinders by a distribution system to coincide with the power stroke in the cycle at each cylinder. Air is piped from the receivers to the starting air master valve which admits air to the distributor. On larger engines the distributor handles only pilot air, operating servo valves at each cylinder to admit the main supply of air.

In many designs of direct reversing engine the camshaft is positioned by compressed air or by hydraulic means controlled by compressed air. This air and that for the starting distributor and pilot valves is generally at a lower pressure than the main supply of starting air and the controlling mechanisms are usually combined to be operated by a single manoeuvring lever or wheel. The main supply of high pressure starting air is held by a master valve which cannot be opened until an interlock indicates that the camshaft is correctly in position according to the direction shown by the control mechanism. It is not unusual for other interlocks to be incorporated in the system so that the starting air master valve cannot be operated unless they are in the "safe" position, notably the shaft turning gear must be in the disengaged condition.

The number and capacity of air receivers depends on the type of engine. A direct reversing engine obviously needs the availability of a larger capacity of starting air than does a uni-directional engine. A typical air

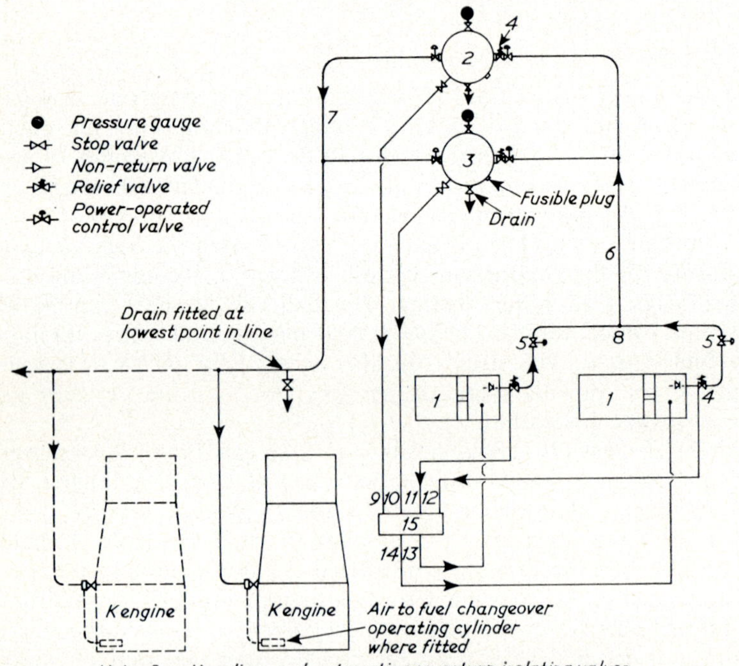

1 Compressor Set
2 Air Receiver
3 Air Receiver
4 Relief Valve
5 Stop Valve
6 Charging Line
7 Starting Line
8 Tee Piece
9 No. 2 Air Receiving Tank Control Line
10 No. 1 Air Receiving Tank Control Line
11 No. 1 Compressor Discharge Control Line
12 No. 2 Compressor Discharge Control Line
13 No. 1 Compressor Command Line
14 No. 2 Compressor Command Line

FIG. 10-9—*Air start system—automatic control.*

start system for a twin engine installation is shown in Figure 10–9. There are two compressors and two air receivers; pipe lines run from both the receivers to both the engines. Either compressor is capable of charging either receiver and either of the two engines can be started with air from either of the two receivers. These features are typical of those required for safety reasons by Lloyds Register and other classification societies.

The pipe which conducts air from the compressors to the receiver must be separate from that which conducts air from the receivers to the engines. Compressed air is apt to contain a good deal of water which will separate out in the pipes and if the run of these is not carefully designed it can result in water being carried into the engines by the starting air, causing extensive damage. If at all possible the piping should be arranged to drain towards the receivers as these can be drained of water easily. Draining of air receivers should be carried out regularly as a routine matter.

BIBLIOGRAPHY

BUDD, E. B. and WILKINSON, H. C., *Modular design of marine auxiliaries,* Trans. I. Mar. E., 1968, Vol. 80, No. 9.

BUCHANAN, G. I., *Merchant ship applications of medium speed geared diesel engines and associated auxiliary machinery,* Trans. I. Mar. E., 1970, Vol. 82, No. 1.

BUTLER, G. and ISON, H. C. K., *Corrosion and its prevention in water,* Leonard Hill, London 1966.

EVANS, U. R., *The corrosion and oxidation of metals,* Arnold, London, 1960.

CHAPTER ELEVEN

Propulsion Machinery

11.1. *Relation between engine and propeller power*

When a propeller is rotated behind a vessel it absorbs power in the form of torque and rotational speed, and converts it, more or less efficiently, into thrust and linear speed. The relationship between the power absorbed and the revolutions per minute approximates to a law of the form $P \propto N^3$, where P is the power and N is the rotational speed. Some authorities prefer to use $P \propto N^{2\cdot 8}$ as being a little more accurate.

The main engine provides this power, again in the form of torque and rotational speed, but for most applications of medium and high speed engines the torque is lower and the speed higher than that which is suitable for the propeller and a speed reduction gear must be included between engine output and propeller shaft. Some means of reversing the thrust of the propeller is needed which may be provided by direct reversing of the engine, or by a reversing gear train in the gearbox or by a variable pitch propeller. Each of these means is in present day use.

11.2. *Limits of engine power*

In Chapter 2 the limits to engine output were considered and it was seen that each design of engine has limits of b.m.e.p. and rev/min which cannot be safely exceeded. It is customary for engine manufacturers to ensure that these limits are not overstepped by fitting limit stops to the fuel pump and governor. A stop restricting the movement of the fuel pump rack (or corresponding component) in the direction of increasing fuel will determine the maximum amount of fuel injected into the cylinder per cycle and so restrict the b.m.e.p. to a maximum value. For a given engine design this corresponds to a limiting maximum torque. The setting of this limit stop will be carried out on the test berth at the full rated rev/min and usually corresponds to the maximum continuous rating of the engine. Changes in the volumetric efficiency of fuel injection equipment with rev/min may result in a variation in the limiting torque produced down the speed range but for the majority of engines it may be assumed that the maximum output characteristic is one of constant torque at all revolutions, see Figure 11–1. At the high end of the speed range a limit is set by restricting the governor speed setting mechanism to give the maximum rated rev/min when carrying the maximum continuous rated b.m.e.p. If the torque load on the engine is reduced the governor will generally respond in a manner giving a slight rise in rev/min as zero torque is approached. This is known as governor droop and is also shown in Figure 11–1. If engine speed is reduced at any constant torque load it eventually reaches a value at which the running becomes irregular and unstable. The idling speed is set just a little above this point. In the case of turbo-charged engines insufficient airflow may prevent high loads being carried at low rev/min. These conditions result in limits at the low speed end of the range which are again shown in Figure 11–1. These same limits can also be displayed on power and rev/min co-ordinates as shown in Figure 11–2.

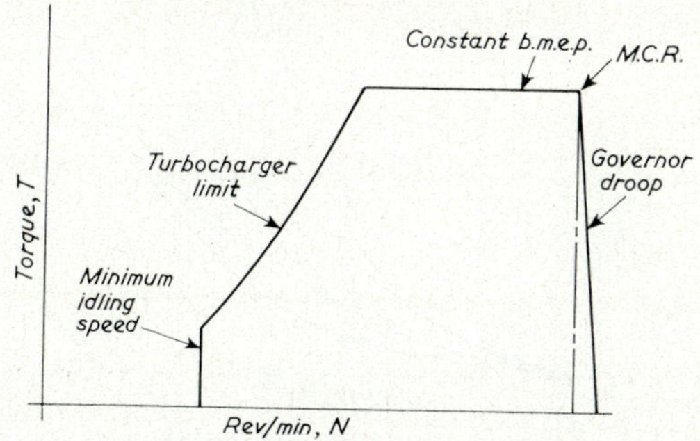

FIG. 11–1—*Engine output limits.*

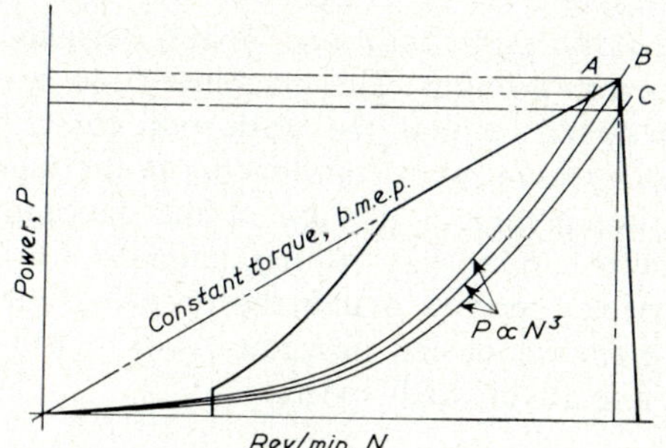

FIG. 11–2—*Engine output and propeller demand.*

11.3. *Power demand of a propeller*

As the ratio of engine rev/min to propeller rev/min will have a fixed value the propeller law curve may be plotted on the same co-ordinates. Making allowance for the power losses in the gearbox this will give a curve of $P \propto N^3$ form as shown at A, B and C in Figure 11–2. If the engine, propeller and gearbox are correctly matched the propeller curve will pass through the maximum continuous rating point of the engine, as shown by B. A propeller that is too small will reach full rev/min at less than full torque as at C and a propeller that is too large will absorb the full torque of the engine before full rev/min is attained as at A. In either of the cases A and C the full rated power will not be available and the intended speed of the vessel may not be achieved in consequence. This illustrates the importance of correctly matching the components forming the propulsion machinery. Some engine builders are prepared to allow small adjustments to torque and rev/min stops to assist in propeller matching, provided that the maximum continuous rated power is not exceeded. Whether this is possible or not depends on the criteria which are considered to limit the rating for that particular design of engine.

If the resistance to forward motion of a vessel having a fixed pitch propeller is increased, for example by using it to tow another vessel, the propeller law curve will become steeper although still of $P \propto N^3$ form as shown in Figure 11–3. A heavier tow will steepen it further, corresponding

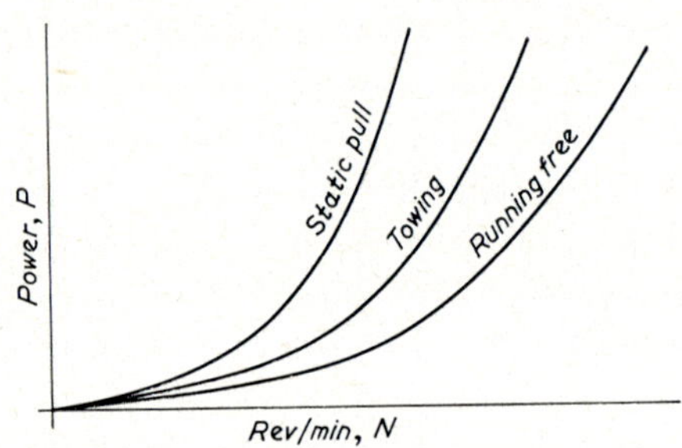

FIG. 11–3—*Effect of increasing the resistance of the vessel.*

to a lower forward speed of the vessel until the most sharply rising curve is produced by the static pull condition. Vessels designed to tow must take these circumstances into account. The machinery and propellers of docking tugs are usually designed so that the static pull curve passes through the maximum continuous rating point; this results in the highest possible value for the static pull but precludes the use of the maximum installed horse-power once the tow is under way. A tug intended for continuous towing will have machinery designed to match the propeller curve at the intended towing speed and a similar design procedure will be followed for a trawler which tows gear at a relatively high speed.

11.4. *Variable pitch propellers*

A variable pitch propeller has blades which can be moved to take up various pitch angles. Any one setting results in a particular propeller law curve for a vessel of given resistance and a whole family of propeller law curves can be produced for such a vessel by altering the blades to different settings. A similar set can be formed for any given increase in resistance of the vessel. It is therefore possible, within the design limits of the propeller, to choose settings which would match the maximum continuous rating of the engine under any condition from static pull to running free, and so utilize the maximum continuous rated power of the engine to obtain the highest tow rope pull and the highest speed attainable in the circumstances. Compared with a fixed pitch propeller the variable pitch propeller makes better use of the power available from the engine over a wide range of diverse conditions. On the other hand it has a disadvantage in that the efficiency is not so high as that of the fixed pitch propeller operating at optimum conditions but this is slight and obviously only of importance in a vessel that spends its time regularly at one speed and in one loaded condition.

A variable pitch propeller can also be used to improve the economy of operation of a vessel of constant resistance but which is called upon to operate at different forward speed during its service. If lines of constant fuel consumption of the kind shown in Figure 2–8, are added to the limits of Figure 11–2 their appearance is as in Figure 11–4. The family of curves

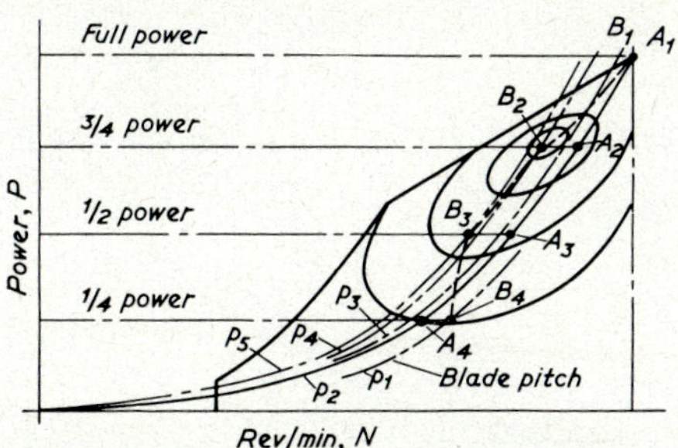

FIG. 11–4—*Optimizing fuel consumption by controllable pitch propeller.*

for a variable pitch propeller and vessel of constant resistance lies across these lines of constant fuel consumption and by choosing combinations of pitch angle and rev/min, the operating line for various speeds can be made to pass through the regions of lowest fuel consumption as is shown. It must, of course, be remembered that alteration of propeller pitch slightly affects propeller efficiency and therefore the relative location of load points and minimum fuel consumption points must be kept in mind in establishing the best relation between optimum engine thermal efficiency and optimum propeller efficiency. In Figure 11–4, the fuel consumption points at full load, three quarter load, half load and quarter load positions with a fixed pitch propeller are indicated at A1, A2, A3 and A4, whilst the improved fuel consumption points obtainable with variable pitch propeller are indicated at B1, B2, B3 and B4, on the dotted line passing through the isofuel consumption loops. Bearing in mind the change in propeller efficiency it may be that in practice optimum performance will be attained between these two settings. Such settings of pitch angle corresponding to rev/min may be selected automatically by appropriate control gear once they have been determined.

11.5. *Two engines geared to one propeller*

By gearing two or more medium speed engines together to drive a single propeller a high powered installation can be designed to occupy a small space and to have a low weight. Benefits in economy due to the use of slower turning higher efficiency propellers and the ability to use one engine only for low speed operation can be justifiably claimed for the arrangement as also can increased reliability and availability of the ship.

When two engines of equal power are geared to one propeller the relationship is as shown in Figure 11–5. Curve *A* represents the torque–rev/min characteristic of each single engine and curve *B* that of the two engines combined. Curve *C* represents the power demand of a fixed pitch

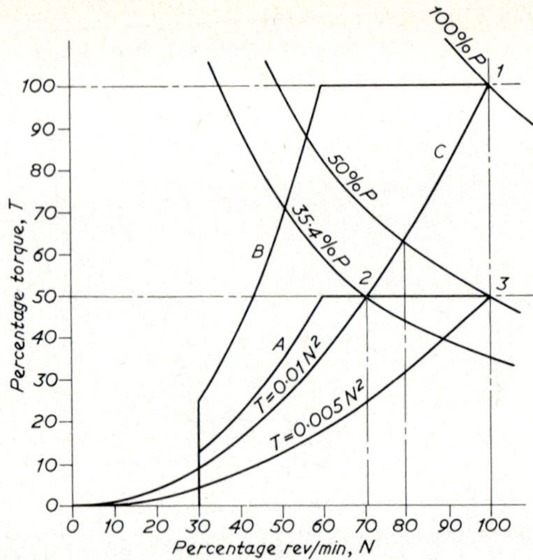

FIG. 11–5—*Torque—speed curves for two engines driving one propeller.*

propeller matched to absorb the full power (service rating) of the two engines together at full rev/min corresponding to point 1. The vessel may be propelled by one engine alone developing its full torque (equal to half the total torque) as at point 2. The revolutions will be reduced to 0·707 of full rev/min and the power available to $0{\cdot}707 \times 0{\cdot}5 = 0{\cdot}353$ of full power giving the ship 0·707 of its full speed.

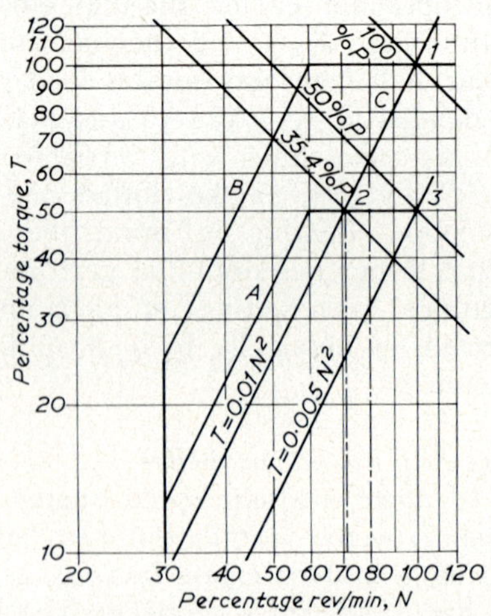

FIG. 11–6—*Torque—speed curves for two engines driving one propeller. log–log scales.*

If a controllable pitch propeller is fitted the pitch can be reduced when running on one engine permitting it to develop its full rev/min as well as its full torque as at point 3. The power is 0·5 of total full power and the vessel's speed is 0·8 of full speed.

When examining cases of this nature the curves can be plotted easily and quickly on "log–log paper" as shown in Figure 11–6 which has been drawn using percentage scales. Curves of constant power and power proportional to (rev/min)^n become groups of parallel straight lines which are easily located. The curves and points in Figure 11–6 correspond to those in Figure 11–5 and are lettered and numbered accordingly.

Vessels engaged on routes where reduced speed is required for a significant proportion of the running time can benefit economically from the installation of twin geared engines. As can be seen from Figure 11–7

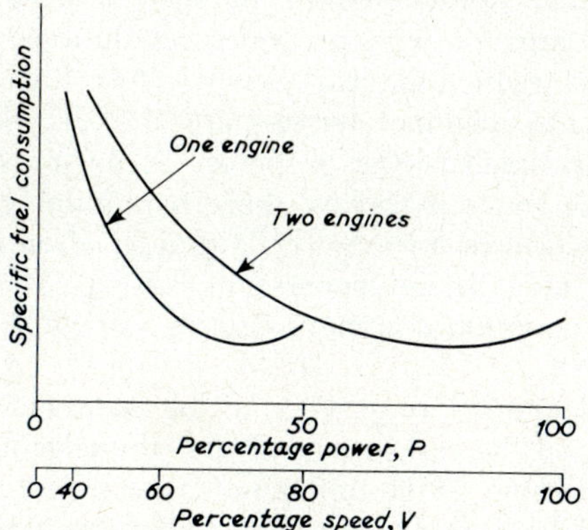

FIG. 11–7—*Fuel consumption, power and speed for two engines driving one controllable pitch propeller.*

which is drawn for twin engines and a controllable pitch propeller, the rise in specific fuel consumption at reduced power and speed can be countered by changing to single engine propulsion at speeds below 80% of full speed.

Installations may be designed with two engines of unequal power geared together or with three or four engines. The reduction in the speed of the vessel as a result of part of the total power not being used is easily calculated, as follows.

Let full power, speed, torque and rev/min be represented by P_f, V_f, T_f and N_f respectively,

and let the available power, speed, torque and rev/min be represented by Pa, Va, Ta and Na respectively.

The propeller will absorb the available torque and for a fixed pitch propeller the rev/min will be reduced.

Thus $$\frac{Ta}{T_f} = \left(\frac{Na}{N_f}\right)^2$$

$$\therefore \quad \frac{Pa}{P_f} = \frac{Ta\,Na}{T_f\,N_f} = \left(\frac{Ta}{T_f}\right)^{3/2}$$

and $$\frac{Va}{V_f} = \left(\frac{Pa}{P_f}\right)^{1/3} = \left(\frac{Ta}{T_f}\right)^{1/2}$$

For a controllable pitch propeller the full rev/min can be used.

Thus $$Na = N_f$$

$$\therefore \quad \frac{Pa}{P_f} = \frac{Ta\,N_f}{T_f\,N_f} = \frac{Ta}{T_f}$$

and $$\frac{Va}{V_f} = \left(\frac{Pa}{P_f}\right)^{1/3} = \left(\frac{Ta}{T_f}\right)^{1/3}$$

11.6. *Speed reduction gearboxes*

Medium and high speed engines, because of their crankshaft revolutions being high compared with propeller revolutions, are almost always installed with speed reduction gearboxes. A few high speed craft, naval applications and cross channel ferries, are the exceptions in which the engines are direct coupled to the propeller. In most cases the engines are derated, running at rev/min below their maximum rated speed with a corresponding reduction in horsepower, although often not a proportionate reduction as there may be an increase in b.m.e.p. above that which can be carried at the maximum continuous rating, as will be appreciated from a study of Figure 2–8.

Once a speed reduction gear is to be included in the propulsion machinery installation the revolutions of the propeller may be chosen from propulsion considerations alone unfettered by engine requirements, its size being restricted only by the aperture and hull form. The efficiency of such a propeller may be considerably higher than is usually the case with propellers for direct coupled low speed engines, the improvement outweighing the losses in the gearing.

The speed ratio of reduction gears is generally of the order of 3:1 to 4:1 for single engines of about 1,000 hp and below. For installations of higher power, particularly for twin engines (or multi-engines) driving a single propeller and ships of large displacement, ratios of 4:1 to 6:1 are used.

11.7. *Gears for two or more engines and single propeller*

Twin engines of equal power are the most popular form of multi-engined geared installations. The basic arrangement is seen in Figure 11–8(a), using either fluid couplings or mechanical clutch-couplings between the engines and the gears. If mechanical clutch-couplings are used in conjunction with a fixed pitch propeller and direct reversing engine the clutch surfaces will have to be of adequate size to withstand the demands of crash reversal conditions and also clutches must be situated so that the heat generated during these manoeuvres can be dissipated as described in article 11.11. By using quill shafts for the input pinions the clutches can be separated from the couplings and placed aft of the gears, as in Figure 11–8(b). This arrangement often meets the requirements more easily and uses less space.

The quill shaft drive also permits the accommodation aft of the gearbox of an electric generator which can be driven by one of the engines without turning the propeller. If there is a demand for large amounts of electric power whilst in port such an arrangement has much to commend it. Figure 11–8(c) shows separate auxiliary drives to generators to be used during passage only. These auxiliary gear drives can be speed increasing to permit the use of smaller higher speed electrical generators.

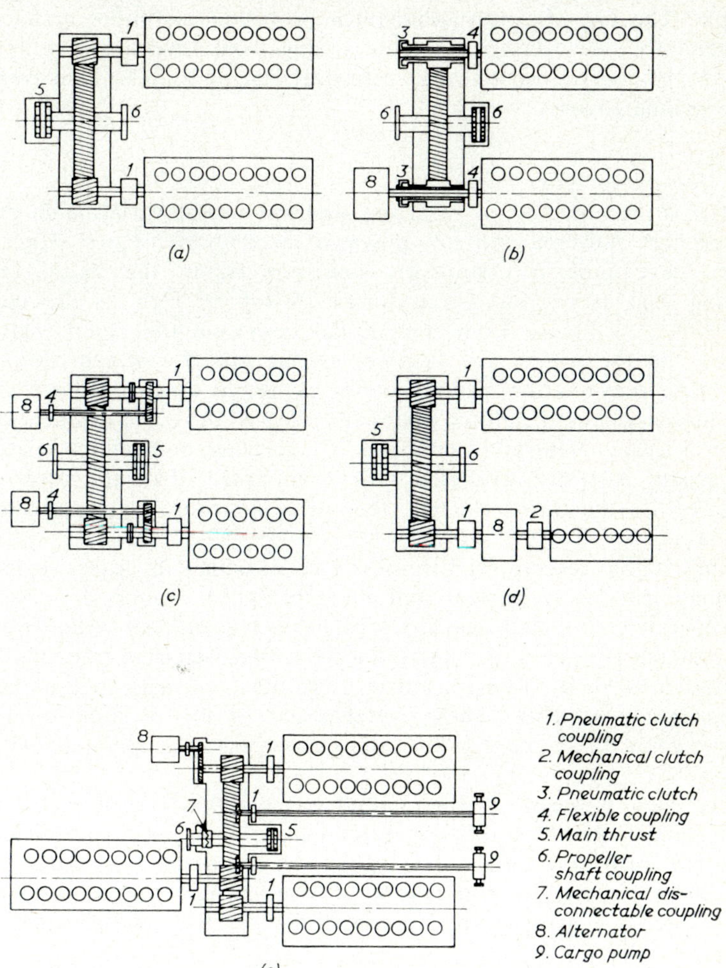

FIG. 11–8—*Arrangements of geared engines.*

The use of quill shafts for the input pinions with clutches or clutch couplings aft gives space to accommodate the thrust block forward of the main gear wheel; a position which many designers consider to be preferable from the point of view of the strength of the seating.

Figure 11–8(d) shows a design in which the two engines geared together are of widely differing power, sometimes termed a "father and son" arrangement. The smaller engine drives a generator (in port or under way) and can also assist in propelling the ship; the larger engine is used for propulsion but can also provide power to drive the generator as well during passage if it is desired not to run the smaller engine. The advantages of a controllable pitch propeller with this combination are obvious.

Three or four engines may be geared to one propeller and the variety of combinations of auxiliary drives is endless, Figure 11–8(e) shows one example. The use of controllable pitch propellers simplifies many of the arrangements, overcoming problems arising from reversing requirements. They are, of course essential if ac electrical machinery is to be driven from the propelling engines.

11.8. *Reversing gears*

It is in the method of reversing that the widest variation in practice between high powered and low powered installations is seen. High speed engines developing comparatively low powers, in the main, rely on reversing gear boxes for astern thrust. A typical form is the epicyclic train shown in Figure 17–6. For ahead operation the clutch plates are engaged and the whole gear revolves as one unit, for astern running the band brake prevents the rotation of the carrier in which the bevel pinions are mounted and as the clutch plates are released the drive is taken through the bevel gears. Neutral is obtained by releasing both clutch plates and band brake. A speed reduction gear is mounted abaft the reverse gear and embodies the thrust bearing, the whole assembly being built integral with the engine.

This simple reverse gear train and construction of gearboxes is quite inadequate for larger powers and medium speed engines are associated with separately mounted gearboxes of heavier scantlings. When reversing is carried out by gearing it takes the form of a separate train within the box, the ahead or the astern train of gears being engaged by the operation of appropriate clutches. These clutches are mostly of multi-drive plate design compactly arranged within the gearwheels and operated hydraulically.

For high powered medium speed engines the cost of incorporating reversing gear trains in the speed reduction boxes becomes higher than the cost of providing direct reversing of the engine and the latter method is the more popular in installations of large power, particularly for twin and multi-engine arrangements where flexibility for manoeuvring can be claimed in addition.

Controllable pitch propellers offer a means of obtaining astern thrust that is being increasingly adopted by all sizes of high speed and medium speed engines.

11.9. *Flexible couplings*

The separate mounting of engine and gearbox, the higher torques and the larger sizes of components make it essential to have some form of flexible coupling between the engine and the gearbox. This coupling must not only protect the gears from misalignment and impact loading but must be chosen to yield a shaft system free from severe torsional vibration critical speeds throughout the running range. This latter requirement is often a major feature of the design of the coupling. Many couplings are based on the principle of transmitting the torque from the driving member to the driven member through a flexible element which may be in the form of steel springs or of rubber blocks. In both these cases it is possible to manufacture transmission elements covering a wide variation of flexibility and so obtain a range of couplings of different stiffnesses. Selection of the appropriate stiffness value enables torsional vibration to be controlled by "tuning" the shaft system.

Some flexible couplings are designed so that they have non-linear torque deflexion characteristics. This gives them a "detuning" property which is of assistance in dealing with torsional vibration. There are other designs of coupling which go further than this and are arranged to damp torsional vibrations to which they are subjected.

11.10. *Fluid couplings and electro-magnetic couplings*

An alternative to directly mechanically coupling the engine and gearbox is to be found in the use of either fluid couplings or electro-magnetic couplings. In fluid couplings the torque is transmitted between the driving and driven members by means of an hydraulic fluid which circulates between the two. In the case of electro-magnetic couplings the torque is transmitted by magnetic flux. Both these couplings have the great advantage that they isolate the engine and its vibrations completely from the gearbox and the stern gear. With both of them, however, there is some loss of power due to the fact that there must be a slip between the driving and driven members in order to transmit the torque. In fact the more torque being transmitted the greater is the slip. Figure 11–9 shows the torque slip

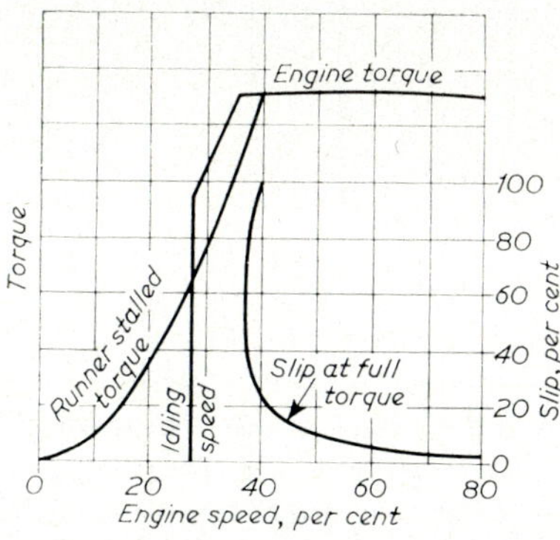

FIG. 11–9—*Torque slip characteristics of fluid coupling.*

characteristics of a fluid coupling. The size of the coupling is determined so that the amount of slip and the power loss is not excessive; about 3% of the engine power is usually considered tolerable. This power loss appears as heat in the hydraulic fluid or as heat in the electrical circuit of the electro-magnetic coupling. It is essential to provide adequate means of dissipation of this heat.

Electro-magnetic couplings have not been particularly popular as they are large in size, heavy and expensive. Fluid couplings have, however, been used quite frequently.

Both fluid couplings and electro-magnetic couplings provide a ready means by which the engine may be disconnected from the propeller. This can be a useful feature if the engine is sometimes used for purposes other than propulsion on occasion and is practically essential when two or more engines are geared together to drive one output shaft.

11.11. *Clutches*

Disconnexion may be desired when mechanical flexible couplings are used and if this is the case then some form of mechanical clutch must be incorporated. Mechanical clutches can take the form of disks, drums or cones and they may be operated manually, pneumatically or hydraulic-

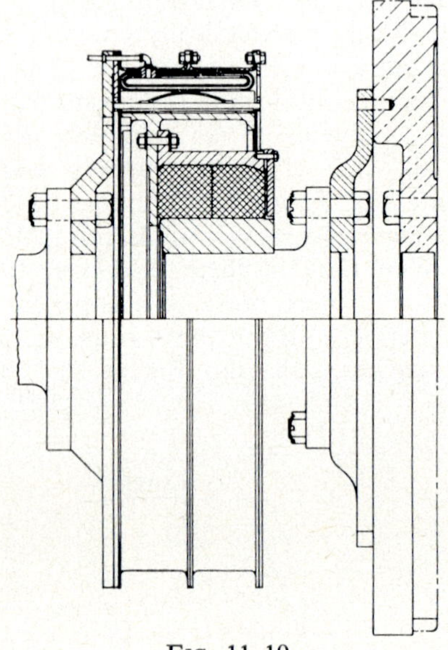

FIG. 11–10.

ally. On engines of over one hundred horsepower manual operation is usually impossible and pneumatic or oil operation is the rule. In order to save space clutches and couplings are frequently combined, the clutch being built around the flexible resilient centre forming the coupling. With this form of construction drum type or cone type couplings are the most easily accommodated. Examples of these are shown in Figures 11–10 and 11–11.

In determining the size of a clutch the first requirement is that it should be capable of carrying the maximum steady torque in the shaft system. However, in many cases the clutch is required to perform a far more arduous duty. This occurs when crash reversal of the machinery is carried out in order to stop the ship in the shortest time and distance possible when it is moving at full speed. During this manoeuvre the propeller is required to turn astern whilst the ship continues to move ahead.

With direct reversing engines and fixed pitch propellers the sequence of carrying out the manoeuvre is as follows. Fuel is cut off from the engines by closing the fuel pump racks and the engine and propeller rev/min fall together, the ship continuing with virtually no change in speed until there is zero torque in the shaft, which is achieved in some seven to ten seconds.

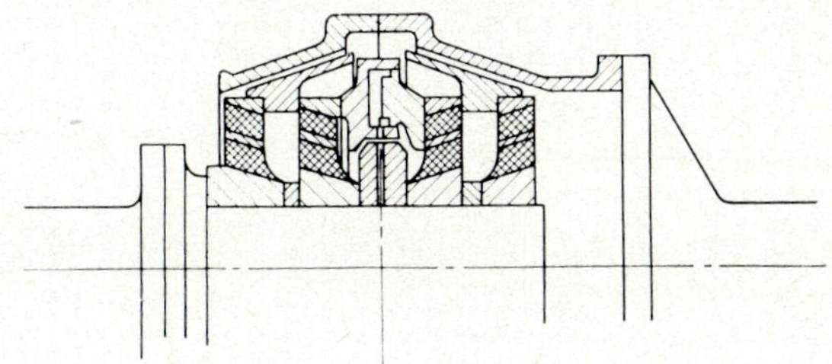

FIG. 11–11—*Cone type clutch-coupling.*

The clutches are then disengaged and the engines reversed. The engines must now be run at a speed sufficiently high above idling to ensure that they will not stall during subsequent re-engagement of the clutches. In each clutch one half is now turning astern driven by the engine and the other half is turning ahead driven by the trailing propeller, and, as re-engagement takes place energy is dissipated at the clutch friction surfaces in the form of heat. The clutches must be large enough and so designed that this operation does not cause them to burn out. If this re-engagement is performed properly the vessel will be moving ahead at a speed practically equal to its full speed when the manoeuvre commences and the propeller trailing rev/min and torque will be high and hence the rate of energy dissipation at the clutches will also be high. If time is allowed for the ship to slow down the trailing rev/min, torque and rate of energy dissipation will all be reduced but an unacceptably long stopping time or distance may result. A curve showing the variation of torque with propeller rev/min during the manoeuvre will have an appearance as shown in Figure 11–12. The vertical and horizontal axes of torque and rev/min respectively have been extended into the negative regions. Negative torque and negative rev/min represent torques and rotational speed in the astern direction. Starting at the condition of full torque and full ahead revolutions, point A, engine power is cut off and the rev/min falls until an equilibrium position is reached at which the propeller trails, the only torque in the shaft being that imparted by the trailing propeller to overcome friction in bearings and gears, point B. Between B and C the propeller continues to turn ahead but losing rev/min as torque in the astern direction of rotation is provided either by brakes or by clutches in the act of being engaged to the engine which, by now, has been reversed and is running astern.

The torque speed characteristic of the engine in the astern direction is shown by the shaded area in the bottom left hand quadrant of the graph. In the case of most engines this characteristic is precisely the same as that of the ahead characteristic and corresponds to the outline constructed in Figure 11–1. If the torque–rev/min curve of the propeller passes into this shaded area then the clutches can be engaged without waiting for the ship to slow down and the period of engagement is as indicated on the graph. If

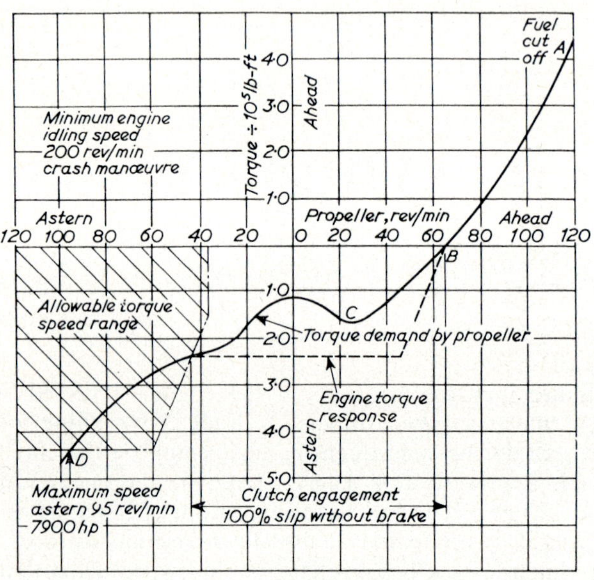

FIG. 11–12—*F.P.P. torque—rev/min characteristics for crash stop.*

the torque–rev/min curve does not pass through the shaded area then it will be necessary to allow the ship to slow down until a condition is reached in which this characteristic has lower torque values and enters the shaded area. In any event at the final astern rev/min the torque must not exceed that which the engine is capable of giving. (Refs. 1, 2 and 3)

Clutches therefore must be selected not only to be capable of transmitting the maximum torque of the engine but also of dissipating the energy which is converted into heat at their friction surfaces when undergoing a manoeuvre of this kind. Clutches that are combined with couplings and placed separately between the engine and gearbox are located in a position where they can be made amply large to cater for this condition.

When a reversing gear is used precisely similar actions take place as far as the clutches are concerned. Because of the demands of space within the gearbox, clutches operating astern gears are often restricted in size and are frequently not capable of coping with crash reversals at the full speed of the vessel. Some of the energy to be dissipated can be absorbed by a brake fitted on the propeller shaft. This is used to slow down and hold the propeller shaft stationary. The propeller can be held in this condition until the vessel slows down sufficiently for the clutches to be engaged without fear of burning out. Used in this way a brake can shorten the time taken for reversal of a ship when the clutches are not large enough to cope with the dissipation of energy for a crash reversal. However, if the clutches are sufficiently large to dissipate the energy then the addition and use of a shaft brake will not significantly improve the time for a crash

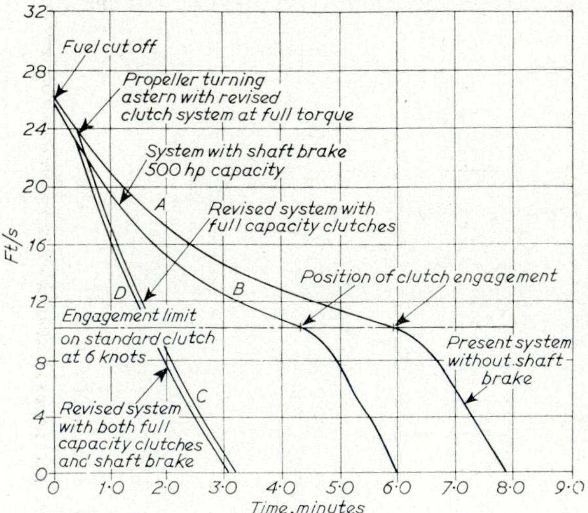

FIG. 11–13—*F.P.P. time taken to stop.*

reversal. Figure 11–13 illustrates four cases for a vessel having the following details:

Length B.P.	102 m (335 ft)	Engine speed—8·77 rev/sec (525 rev/min)
Breadth (MLD)	16·8 m (55 ft)	rating—1·400 MN/m² (203 p.s.i., b.m.e.p.)
Draught (MLD) loaded	7·8 m (25 ft 6 in)	No. engines—2
Displacement loaded	8380 tonnes (8260 tons)	No. cylinders—6
Block coefficient	0·64	Propeller dia.—4·19 m (13·75 ft)
Loaded speed	7·96 m/s (15·5 knots)	P/D—0·865
Power	3500 kW (4700 S.H.P.)	B.A.R.—0·55
Estimated loaded prop. rev/min	2·58 rev/sec (155 rev/min)	

At A is seen the speed time relationship for a vessel with clutches designed to take the full torque from the engine but not large enough to dissipate the energy necessary for early engagement. At B is shown the condition when a shaft brake is added to the system capable of being applied at the start of the manoeuvre. The time to bring the vessel to rest is reduced by these means from 7·9 minutes to 6 minutes but by using a clutch system of adequate size the time can be reduced to 3·2 minutes as shown by curve C. The use of a brake in conjunction with the large clutches results in only a negligibly small further improvement in stopping time, as can be seen from curve D. The corresponding ahead reach for case A is 22 ships lengths, for case B is 17 lengths, for case C is $8\frac{1}{2}$ lengths and for case D $8\frac{1}{4}$ lengths.

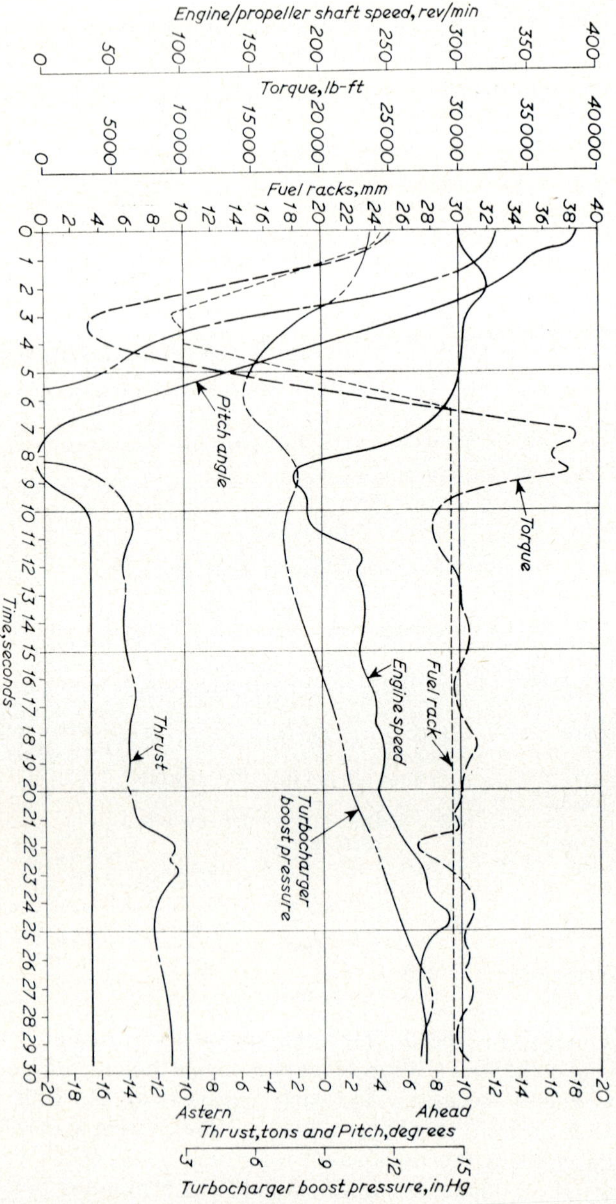

FIG. 11–14—*C.P.P. emergency stop by full reversal of pitch.*

11.12. *Astern thrust by controllable pitch propeller*

In the case of controllable pitch propellers there is no need when reversing for any disengaging and re-engaging of clutches. When the pitch is altered to the astern thrust position the forward thrust immediately dies away and astern thrust loading the backs of the blades builds up rapidly to a peak followed by an equally rapid reduction to a steady value remaining virtually constant until the ship is dead in the water. The sharp peak of thrust corresponds with peak torque reaction from the propeller. This cannot be allowed to reach a value much higher than the full torque of the engine, as if it did the engine would stall. In fact any excess above the full torque causes a momentary fall in engine speed. All these features are exhibited in Figure 11–14. Control of the rate at which the pitch of the blades is altered may be necessary to avoid such excessive torque. In this case, therefore, the clutches may be selected on the basis of the maximum engine torque plus a small margin.

REFERENCES

1. HENSHALL, S. H., *Development of a standardized power pack,* Trans. I. Mar. E. IMAS '69.
2. ADLEY, A. A. and LEA, K. E., *Selections, applications and installation of medium speed machinery systems.* Paper to branches of I. Mar. E.
3. GOODWIN, A. J. H., IRVINE, J. H. and FORREST, J., *The practical application of computers in marine engineering,* Trans. I. Mar. E., Vol. 80, 1968.

CHAPTER TWELVE

Introduction to Vibration and Noise

12.1.

As machinery has developed its power has increased in proportion to its mass and this has tended to intensify vibration and noise. Some vibration problems are concerned with avoiding high alternating stresses in the machinery, whilst many others are concerned with avoiding uncomfortable conditions in various parts of the ship. The level of noise or vibration that can be tolerated varies with the class of ship—obviously passenger ships can tolerate very little—and also with the location within the ship. Standards are continually and quite rightly being raised and the subject continues to grow in importance and in interest for marine engineers. A complete study of vibration associated with diesel engines is far beyond the scope of this book and all that is attempted in this chapter and the next is an introduction to some fundamental considerations necessary for an understanding of the more common manifestations. There are excellent standard works on the subject some of which are listed in the bibliographies at the ends of these two chapters.

12.2. *Natural vibration*

When a body which is held in place by elastic members is displaced and released it will start to vibrate. In moving it from its initial equilibrium position work is performed which is stored as strain energy in the supporting members. On releasing the body it will be accelerated towards the original position, the strain energy of the supports being converted into kinetic energy of the body, increasing its velocity. When it reaches the original position all the strain energy will have been converted into kinetic energy and the velocity of the body will be a maximum. The motion of the body will continue causing it to overshoot the original position and to give up its kinetic energy whilst strain energy is once again stored in the supports, but this time in the opposite sense. The body will come to rest when all the kinetic energy has been converted into strain energy, the motion will then be repeated in the opposite direction and this process will continue indefinitely. A vibration of this kind in which, after the initial displacement, no external forces act and the motion is sustained by internal forces and exchange of energy is termed a free or natural vibration.

12.3. *Damped vibration*

In practice the energy in the system is gradually dissipated, usually in the form of heat, by both external and internal resistances to the motion of the body. The vibration then dies away and the body finally comes to rest at its original equilibrium position. This type of vibration is termed damped vibration.

12.4. *Forced vibration*

The operating cycles and inertia forces of an engine give rise to rhythmic disturbances. If such a periodic disturbing force is applied to the body it will vibrate with the same frequency as this applied force. This form of vibration is termed "forced" vibration.

12.5. *Resonance*

If the frequency of the forced vibration coincides with the frequency of the natural vibration of the system a condition known as resonance will arise, in which quite a small periodic force can have a large and sometimes spectacular effect.

Vibration and resonance are not confined to mechanical systems, they can have acoustic, hydraulic, electrical, magnetic or other forms as well.

12.6. *Forms of mechanical vibration*

Mechanical vibrations can occur in three forms. Figure 12–1 shows a

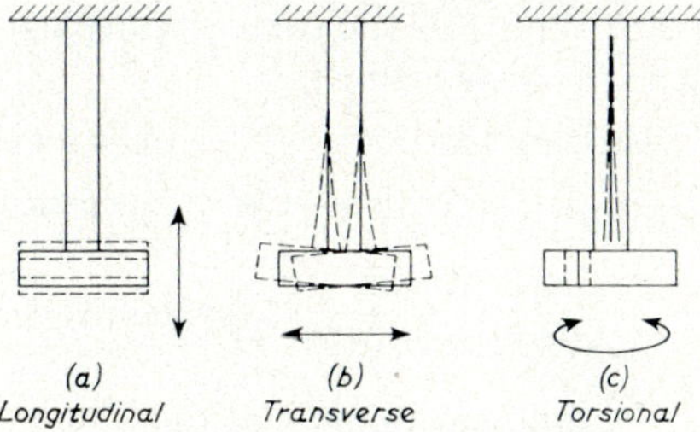

FIG. 12–1—*Forms of vibration*

simple system consisting of a flywheel suspended by a light shaft. The flywheel is assumed to have mass and to be rigid. The shaft is assumed to be elastic and of negligible mass and rigidly fixed at its upper end.

All particles of the flywheel may vibrate along straight paths parallel to the axis of the shaft. The shaft will be alternately compressed and extended. This is termed longitudinal vibration.

All particles of the flywheel may vibrate along straight paths perpendicular to the axis of the shaft. This will cause the material of the shaft to be alternately bent and straightened in opposite directions; this is termed transverse vibration.

All particles of the flywheel may vibrate along circular arcs whose centres lie on the axis of the shaft. In other words, the flywheel may rotate in alternate directions and the material of the shaft will be alternately twisted and untwisted in opposite directions. This is termed torsional vibration.

This chapter deals with the first two forms of vibration. Torsional vibration will be dealt with in the next chapter.

12.7. *Frequency of free vibration*

In the simple systems described the displacement of the flywheel from its equilibrium position will cause a force to be exerted on it due to the stiffness of the shaft which will be proportional to its displacement. Its acceleration will, therefore, also be proportional to its displacement. Text-books on theory of machines show that if a body oscillates about an equilibrium position in such a way that its acceleration towards the equilibrium position is directly proportional to its displacement from the equilibrium position it will have simple harmonic motion. The periodic time for a complete oscillation is given by:

$$t = 2\pi \sqrt{\frac{\text{Displacement}}{\text{Acceleration}}}$$

and the frequency of vibration is given by:

$$n = \frac{1}{2\pi} \sqrt{\frac{\text{Acceleration}}{\text{Displacement}}}$$

Let M = mass of flywheel

S = stiffness of shaft, *i.e.* the force required at the flywheel to produce unit displacement.

Then the restoring force $= S \times$ displacement

but the restoring force also $= M \times$ acceleration

$$\therefore \quad \frac{\text{acceleration}}{\text{displacement}} = \frac{S}{M} \text{ which is a constant.}$$

thus the motion is simple harmonic and the frequency, n, is given by:

$$n = \frac{1}{2\pi} \sqrt{\frac{S}{M}}$$

The static deflexion due to the weight of the flywheel is $\delta = Mg/S$.

Thus $$n = \frac{1}{2\pi} \sqrt{\frac{g}{\delta}} \qquad (12.1)$$

In British units when δ is in inches

$$n = \frac{3{\cdot}13}{\sqrt{\delta}} \text{ vibrations per second.}$$

$$\text{or } n = \frac{187{\cdot}8}{\sqrt{\delta}} \text{ vibrations per minute.}$$

In metric units when δ is in cm.

$$n = \frac{5}{\delta} \text{ vibrations per second.}$$

$$\text{or } n = \frac{300}{\delta} \text{ vibrations per minute.}$$

12.8. *Frequency of damped vibrations*

Figure 12–2 shows a body mounted on a spring and also included in

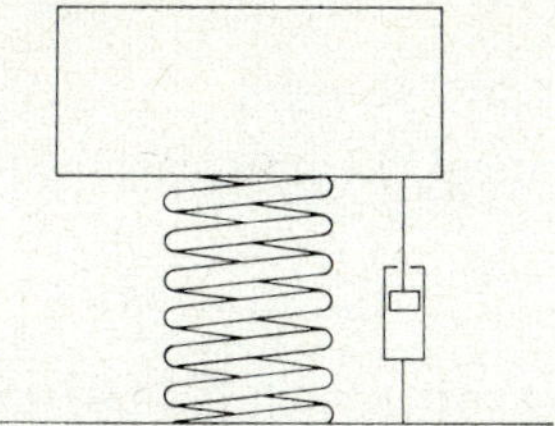

FIG. 12–2—*Mass-spring system with damping.*

the diagram is a dashpot which is intended to represent a damping resistance. The diagram is representative, for example, of a generator set mounted on flexible mountings and equivalent to the simple system we have just examined with the addition of the damping. The resistance to movement of the body by which it is damped is provided partly by the medium in which the vibration takes place and partly by internal friction or hysteresis of the material of the elastic mountings. For example, if these are rubber mountings the hysteresis damping can be quite high. Or it may be a purposely introduced damper of some kind. In considering the damping it is usual to regard it as a force proportional to the speed of movement of the body which is perfectly correct for viscous damping. The conditions of the motion of the body may then be analyzed in the following way:

Let M = mass of the body
S = stiffness of the spring
f = damping force per unit velocity
y = displacement at time t.

The acceleration is then given by: $\dfrac{d^2y}{dt^2}$

and the inertia force acting on the body $= M\dfrac{d^2y}{dt^2}$

the damping force $= f\dfrac{dy}{dt}$

and the elastic force $= S.y.$

As the algebraic sum of these forces must be zero

then $$M\frac{d^2y}{dt^2} + f\frac{dy}{dt} + S.y. = 0$$

or $$\frac{d^2y}{dt^2} + a\frac{dy}{dt} + b.y. = 0$$

where $$a = \frac{f}{M}, \quad b = \frac{S}{M} = \frac{g}{\delta} = (2\pi n)^2$$

n being the natural frequency.

The solution to this differential equation may be found in mathematical textbooks or standard works on vibration and given as:

$$Y = AC^{-a/2} \cos \sqrt{b - (a/2)^2}\, t.$$

Where A is the initial displacement.

From this can be obtained the frequency of the damped vibration

$$n_d = \frac{1}{2\pi} \sqrt{b - (a/2)^2} \qquad (12.2)$$

If there is no damping then $a = f/M = 0$

and
$$n_d = \frac{1}{2\pi} \sqrt{b} = \frac{1}{2\pi} \sqrt{g/\delta}$$

which agrees with equation (12.1).

The effect of damping on the frequency of the vibration is to reduce it to a lower value than the natural frequency. This effect, is however, very small unless exceptionally large damping forces are present.

12.9 *Amplitude of forced vibration*

Of great practical importance is the case when the body is subjected to a periodic disturbing force. For example the body may be a diesel generating set supported on flexible mountings and subjected to a small out of balance force originating within itself. Such a force would be harmonic in nature and can conveniently be represented by:

$$F = \cos \omega t$$

where
$$\omega = 2\pi n_f,$$

n_f being the frequency of the periodic force.

The equation of forces on the body is now:

$$M \frac{d^2y}{dt^2} + f \frac{dy}{dt} + S.y. = F \cos \omega t$$

or
$$\frac{d^2y}{dt^2} + a \frac{dy}{dt} + b.y. = c \cos \omega t$$

a and b have their previous values and $c = F/M$.

The complete solution of this differential equation is:

$$y = Ae^{-a/2} \cos \sqrt{b - (a/2)^2 t} + \frac{c}{\sqrt{(b - \omega^2)^2 + a^2\omega^2}} \cos (\omega t - \beta)$$

where $\tan \beta = \dfrac{a\omega}{b - \omega^2}$

The first term will be recognized from article 12.8 as representing the transient damped vibration which dies out rapidly leaving the steady state forced vibration maintained by the periodic force represented by the second term. Its amplitude is given by the maximum displacement of the steady state:

$$Y_{max} = \frac{c}{\sqrt{(b - \omega^2)^2 + a^2\omega^2}}$$

Dividing numerator and denominator by b

$$Y_{max} = \frac{c/b}{\sqrt{(1 - (\omega^2/b)^2) + (a^2\omega^2)/b^2}}$$

Now $c/b = F/S = \Delta$ the deflexion produced by a static load F.

So we can write

$$Y_{max} = D \,.\, \Delta \tag{12.3}$$

where D is called the dynamic magnifier because it is a factor by which the static deflexion produced by F must be multiplied to obtain the amplitude of the forced vibration.

$$D = \frac{1}{\sqrt{(1 - \omega^2/b)^2 + (a^2\omega^2)/b^2}}$$

Since $\quad b = g/\delta = (2\pi n)^2$ and $\omega = 2\pi n$

$$D = \frac{1}{\sqrt{[1 - (n_f/n)^2]^2 + (a/2\pi n)^2\,(n_f/n)^2}} \tag{12.4}$$

If the vibration were undamped a would be zero and

$$D = \frac{1}{\sqrt{[1 - (n_f/n)^2]^2}} \tag{12.5}$$

Equation 12.5 shows that for an undamped vibration at resonance when $n_f = n$, D is infinitely large.

Equation 12.4 shows that for the damped vibration D is always finite even when $n_f = n$.

Figure 12–3 shows a family of curves each having a different value for

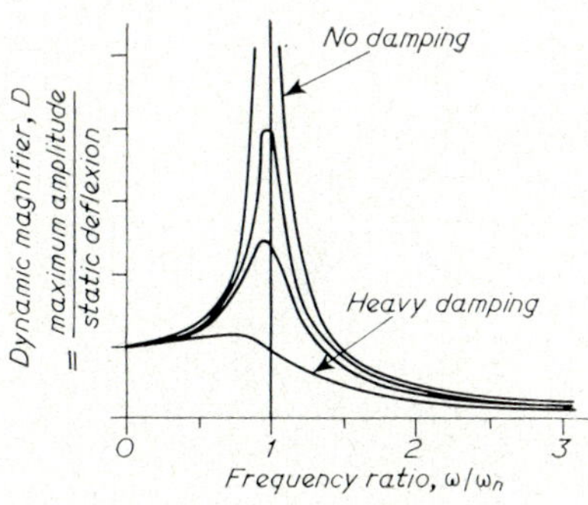

FIG. 12–3—*Response with different degrees of damping.*

the degree of damping. The upper curve is the undamped system in which the amplitude would reach infinite values; the other curves indicate how the amplitude of forced vibration is diminished by damping. It can also be noticed how, as damping increases, the maximum amplitude is attained at frequency values below the resonant frequency of the undamped system. This effect, however, is very small unless damping is very great.

12.10. *Transmission of vibration*

So far we have considered the vibration of the body itself. In many cases what is of most concern is the amount of vibration that is transmitted to the structure which supports the spring mounting; in other words it is a question of vibration isolation and interest lies in the "transmissibility" which is the ratio of the transmitted force to the impressed force. In an ideal case this would be made zero but in practice the aim is to make it as small as possible. Once again if the body mounted on the spring is regarded as an engine on spring mounts, the periodic force can be imagined as being generated internally in the engine. If this force varied very slowly the mass will move with it, compressing and extending the spring and the force at the lower end of the spring will vary completely in accordance with the compression of the spring; in other words the transmissibility will be unity.

Now suppose the force variation to be very rapid, that is an alternating force of very high frequency. In this case the heavy mass of the engine will not be able to follow this alternating force sufficiently rapidly and its own vibration will be small as has already been seen and illustrated in Figure 12–3. The compressing and extending of the spring will, therefore, be very small and the force variation at the lower end will also be small.

In between these two conditions lies the case at resonance where the applied force has the same frequency as the natural frequency of the engine on its mountings. Here the compressing and extending of the spring is very large and the transmitted forces correspondingly will also be very large.

Damping influences transmissibility in a rather different way from that in which it influences vibration. Figure 12–4 shows the family of curves obtained when transmissibility is plotted against frequency ratio for different values of damping. All the curves intersect at a point which corresponds to a frequency ratio of $\sqrt{2}$.

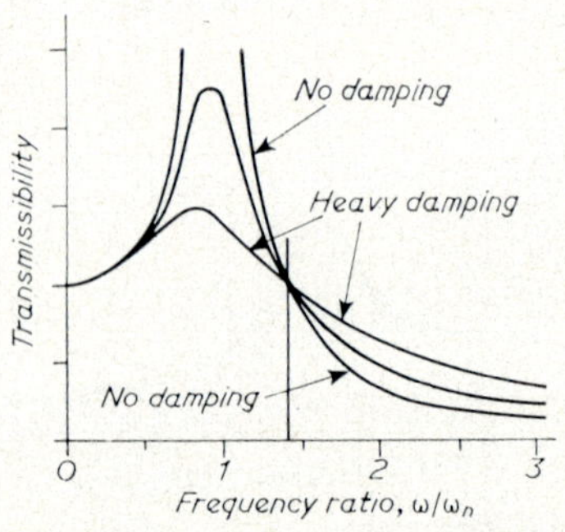

FIG. 12–4—*Transmissibility with different degrees of damping.*

First consider the curve without damping. At values for the frequency ratio above $\sqrt{2}$ the transmitted force is less than the disturbing force whilst at values below $\sqrt{2}$ the transmitted force is greater than the disturbing force and at resonance it can become infinite. For undamped or very lightly damped spring mountings, therefore, the natural frequency must be very low so that in operation the disturbing frequency ratio is very high, giving operation in the low transmissibility region.

Looking now at the curves showing different degrees of damping it can be seen that damping only reduces transmissibility when the disturbing frequency is lower than $\sqrt{2}$ times the natural frequency, which is the region in which spring mounting makes matters worse. On the other hand for all values of disturbing frequency to natural frequency ratio where the spring mounting helps, the presence of damping increases the transmissibility. For flexible mountings to be of use they must, therefore, be arranged to give the system a natural frequency that is lower than the forcing frequency. If they are very flexible then the natural frequency is best arranged so that it is lower than that corresponding to the starting rev/min of the engine. In some practical cases such a low frequency cannot be used for various reasons and if this is so and a stiffer mounting is used, it is essential to include some damping if trouble is to be avoided in running up to full speed of operation.

12.11. *Vibration of a beam*

If the supporting structure of a vibrating mass is in the form of a beam we have the case of transverse vibration. Textbooks on strength of materials show that the deflexion of a beam carrying a concentrated load is given by:

$$\delta = \frac{1}{K} \cdot \frac{WL^3}{EI}$$

where the value of the constant K depends on the conditions at the beam supports. The deflexion is therefore proportional to the load and the stiffness of the supports can again be represented by S the load per unit deflexion and the same equations apply that were developed for longitudinal vibration.

In most practical cases the vibrating system is more complicated and consists of a number of unequal masses spaced at different intervals along the beam. The weight of the beam itself will also have to be taken into consideration and it may not necessarily be uniformly distributed. This makes it more difficult to calculate and forecast the natural frequency and other properties of the system, but it does not alter in any way the effects of resonance and damping upon amplitude which operate in exactly the same way as for simple systems.

A beam does not always vibrate in a simple manner; as Figure 12–5 shows, a number of modes of vibration are possible. The points where the amplitude of vibration of the beam is zero are termed nodes, they do not necessarily coincide with the points of support of the beam. The points where vibration amplitude is greatest are termed antinodes.

12.12. *Hull vibration*

Hull vibration is a practical case. The hull is a beam supported by a fluid and carrying a complex system of loads. As shown in Figure 12–6 it can vibrate with two, three or more nodes. There will be a different frequency of vibration for each manner which will be higher as the number of nodes is increased. Excitation of these modes of vibration may arise from resonance with unbalanced forces or couples of an engine, the propeller or other periodic disturbance. The unbalanced forces of an engine are less

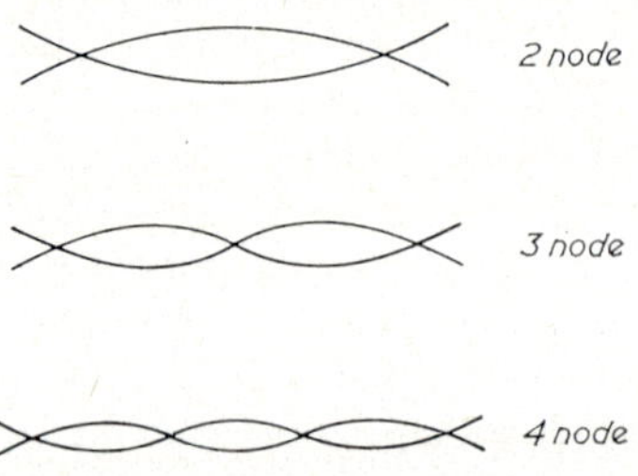

FIG. 12–5—*Modes of flexural vibration.*

likely to excite hull vibration if the engine is located at a node. On the other hand unbalanced couples are less likely to excite vibration if the engine is situated at an antinode, see Figure 12–7. Some cases of hull vibration have been cured by the addition of out of balance weights to the shafting, which have the effect of altering an out of balance force to a couple or vice versa.

12.13. *Noise*

Closely allied to the subject of vibration is that of noise. A vibrating surface causes the molecules of the gases forming the surrounding air to follow its motion. The elasticity of the air results in this motion being communicated to other molecules in the form of a series of radiating waves of

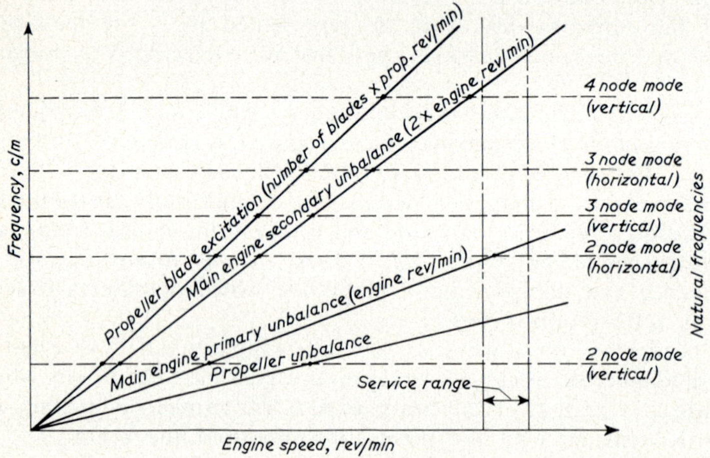

FIG. 12–6—*Hull vibration; typical sources of excitation.*

compression and rarification. When these pressure waves reach the human ear they stimulate it and give the sensation we call sound. If the vibrations are simple harmonic and of constant frequency we hear a pure tone. If they are a mixture of vibrations of varying amplitude and frequency we usually describe it as noise, noise being unwanted sound. The noise from a diesel engine consists of a mixture of regularly repeated transient vibrations, the repetition of some being of sufficiently high frequency to be recognizable as tones amongst the general noise.

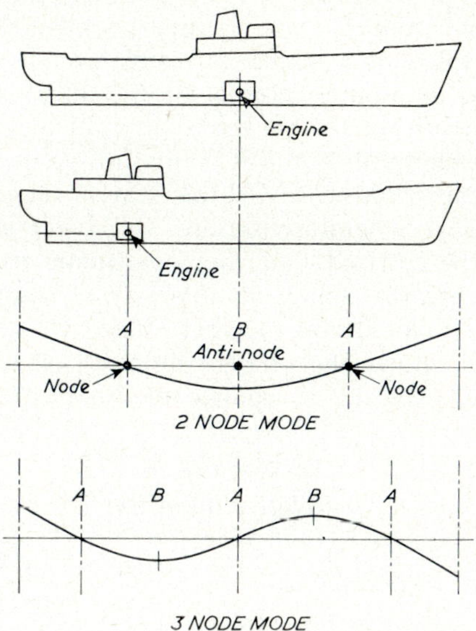

FIG. 12–7—*Engine position and hull excitation.*

12.14. *The decibel scale*

The intensity of sound is usually specified in terms of the magnitude of alternating pressure and occasionally as its energy equivalent. The ear copes with a very wide range of sound intensities. At the frequencies to which the ear is most sensitive a sound pressure of 0·001 microbars can be heard, while the ear is sometimes subjected to pressures one million times this value. Because of this very wide range it is convenient to use a logarithmic scale to describe values and the one in common use is the decibel scale.

Originally this scale was defined using as a unit the bell which is ten times larger than the currently used decibel (dB).

The value of one sound relative to another in bells is numerically equal to the logarithm of the ratio between the two sound energies. When the value in bells is derived from pressures it is necessary to multiply the logarithm of the pressure ratio by 2 because the sound energy is proportional to the square of the pressure, thus we have

$$\text{value in bells} = 2 \log \frac{P_1}{P_2} = \log \frac{E_1}{E_2}$$

where P_1 and P_2 are the sound pressure levels being compared and E_1 and E_2 are the corresponding sound energy levels.

It has been found more practical to use the unit of one tenth of a bell and hence the decibel. This involves multiplying the numerical value of the result in bells by a factor of ten, thus we have

$$\text{value in decibels, dB} = 20 \log \frac{P_1}{P_2} = 10 \log \frac{E_1}{E_2}$$

The scale thus provides for the definition of the magnitude of one sound pressure level compared with another. For the sake of providing an absolute scale the value of a sound pressure level is often just stated in decibels and (unless some other reference is specified) the value is taken to be relative to a pressure of 0·0002 microbar, which is close to the sound intensity just detectable by the ears of young people in the frequency range in which the ear is most sensitive. On this scale a range of sound intensity giving very faint to very loud sensations will cover an interval of some 120 dB. It should be realized that the logarithmic nature of the decibel scale implies a far greater change in intensity of sound pressure level at high decibel values than at low decibel values for the same change in decibels. Table 12–1 shows the relative intensity of sound energy corresponding to the decibel scale. A change of 10 dB from 10 to 20 dB is

TABLE 12–1

Decibel scale	*Relative intensity of energy*	
10	10	10 dB change ≡ 90
20	100	
30	1,000	
40	10,000	
50	100,000	
60	1,000,000	
70	10,000,000	
80	100,000,000	
90	1,000,000,000	
100	10,000,000,000	
110	100,000,000,000	
120	1,000,000,000,000	10 dB change ≡ 90×10^{10}

equivalent to a change of 90 units of energy but a change of 10 dB from 110 to 120 is equivalent to 90×10^{10} units of energy.

The total sound pressure level of two engines running together cannot be obtained in the decibel scale by adding together the separate values. It is necessary to add the two energy levels.

If each engine has a sound level of 90 dB,

then $\quad 20 \log P = 90 \text{ dB}$

i.e. $\quad \log P^2 = 9$

The energy level is proportional to $P^2 = 10^9$

The energy level of the two engines together will be proportional to $2\,P^2 = 2 \times 10^9$

i.e.

$$\log 2\,P^2 = \log 2 + \log 10^9$$
$$= 0{\cdot}3 + 9$$
$$= 9{\cdot}3 \text{ bells}$$

or

$$10 \log 2\,P^2 = 93 \text{ dB}$$

If one engine had a value of 90 dB and the other 80 dB then together they would create a sound pressure level of $10^9 + 10^8 = 1{\cdot}1 \times 10^9$ units which in the decibel scale

$$= 10 \log 1{\cdot}1 \times 10^9$$
$$= 90{\cdot}4 \text{ dB}$$

12.15. *Measurement of sound*

A sound level meter consists of a microphone, an amplifier, a series of weighting circuits and an output meter. The microphone picks up the sound waves in the air and translates them into an electronic signal which is passed to the amplifier.

The human ear responds differently to sounds of different frequencies and intensities and in order that the sound level meter may indicate values related to the impression of loudness registered by the ear the amplifier signal is passed through "weighting" circuits before its magnitude is displayed on the output meter. This type of meter is calibrated directly in decibels.

In comparing the noise level of different engines the position of the microphone relative to the engine is of the utmost importance. If the engines being compared differ considerably in size it is almost impossible to obtain a valid comparison from a single microphone position.

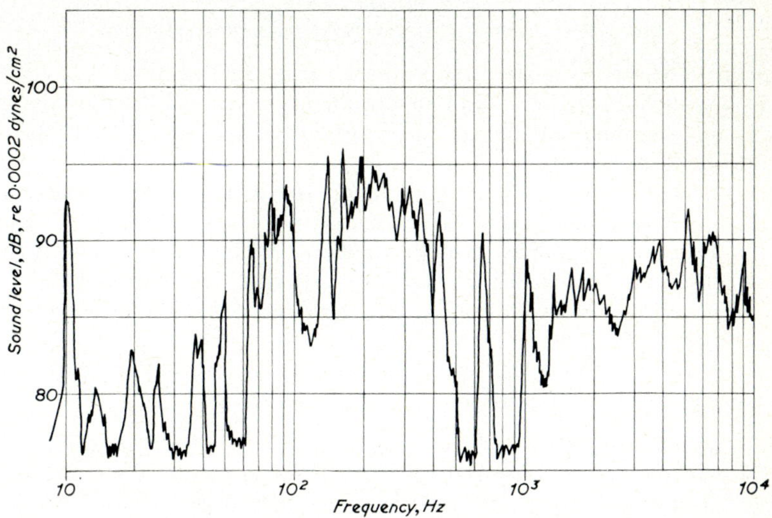

FIG. 12–8—*Typical narrow band frequency spectrum.*

12.16. *Frequency spectra*

The ear possesses the ability to distinguish between different frequencies as well as different intensities, and a complete analysis of the noise from an engine must include frequency measurements if it is to be adequately related to the impressions received by the ear. Observation of the frequency of a sound is also of assistance in identifying its source and hence in attempting to eliminate or reduce the noise.

The amplified output from a microphone may be passed through frequency selector or tuned filter circuits designed to pass narrow bands of frequency, Figure 12–8 (in some equipment a band width of $\frac{1}{3}$ octave is used, Figure 12–9) so that the sound energy in different parts of the acoustic

FIG. 12–9—*Typical $\frac{1}{3}$ octave band frequency spectrum.*

spectrum may be separately determined. The identification of sources of noise is greatly helped by running the engine at two, or preferably three, speeds and obtaining an acoustic spectrum for each. Sources of engine noise fall into three categories:

1. Frequency proportional to engine speed, *e.g.* firing impulses, fuel injection components, valves and gear.
2. Constant frequency independent of engine speed, *e.g.* torsional vibration, crankcase doors. With this class the amplitude of vibration may vary with engines rev/min and sometimes resonance may occur.
3. Frequency affected by engine rev/min but not proportionately *e.g.* turbocharger noise.

12.17. *Reverberation*

Sound is reflected from the surfaces according to their nature and shape. Hard metallic surfaces reflect a very high proportion of the sound. A sound that is released in an enclosed space bounded by such surfaces will be reflected again and again taking some time to die away. This is known as reverberation. If the sound is repeated continually with a frequency such that the repetitions occur before the reverberation has ceased, the sound level will build up to a saturation point. The sound levels caused by an engine operating in the engine room of a ship may thus reach a value several decibels higher than that resulting from the same engine in the relatively open atmosphere of a large test shop.

12.18. *Noise reduction*

The noise that is heard in an engine room is largely mechanical in origin but possibly includes air intake noise if the latter is situated in the engine room. The problem is at its most severe in manned machinery spaces where personnel can be subjected to high noise intensity for long periods. The hard surfaces which form the engine room boundaries reflect the sound pressure waves causing the energy to build up or reverberate and the noise level becomes very high. If these surfaces are lined with soft, sound absorbent materials the noise level in the engine room and also the noise transmitted through the decks and bulkheads is reduced. Such insulation can be expensive to carry out as large quantities are required and the material must be carefully selected as many sound absorbent materials are also excellent absorbers of oil and dirt.

Undoubtedly the greatest reduction in noise level is to be obtained by enclosing the engine. This method is only practicable with the engines of small boats. With larger engines it becomes impossible because of the necessity of accessibility for maintenance and because of heat dissipation restrictions when running. If it is impractical to enclose the engine then protection to the personnel can be afforded by an acoustically shielded area or a completely enclosed acoustically lined control room.

Noise can be carried by the structure of the ship as well as through the air and can cause vibrations in places remote from the engine room, these vibrations giving rise to noise. Noise of this type often emanates from a panel or a duct passing through a cabin or a public room. Sometimes the trouble can be cured by local stiffening of the panel concerned. In passenger ships, however, where a very low noise level is required in all parts of the ship a better solution is to adopt anti-vibration mountings for the engine or at the very least to use non-metallic contact mounts or chocks.

As combustion air has to be provided to the engine and its exhaust has to be evacuated, openings for the air and exhaust are necessary and these can be sources of sound which cannot be included in any form of enclosure. However, very effective silencers have been developed to deal with noises that arise from these two sources. The noises are caused by periodic interruptions in the flow gases. Naturally aspirated engines have characteristic low frequency sounds at air intake and exhaust whilst turbocharged engines tend to emit a high frequency whine.

Silencers may be of the absorption or reflection type. In the first type the silencing is accomplished by reducing the alternating energy by friction in porous materials with a large internal surface such as asbestos, glass wool or metallic wool. This material is usually arranged so that the gas or air stream is separated from it by a partition with perforated walls. This type of silencer is popular for air intakes. Because the absorption material can also accumulate oil and other impurities that are in the exhaust gas, reflection type filters are more usual on the exhaust side. These consist of a number of chambers with communicating tubes which may be arranged in series or in parallel or in a combination. The chambers and tubes are designed as an acoustic resonant system which will pass low frequency but block all oscillations above a certain limiting frequency. The cross sectional area of the passages through the silencer is arranged to be greater than that of the exhaust pipe to which it is to be attached.

The requirements of silencers both air and exhaust are obviously that there should be minimum resistance to flow, that they should occupy the smallest possible space and have the lowest possible weight consistent with adequate reduction of noise. Figures 12–10 and 12–11 indicate the principles of construction of absorption and reflection silencers.

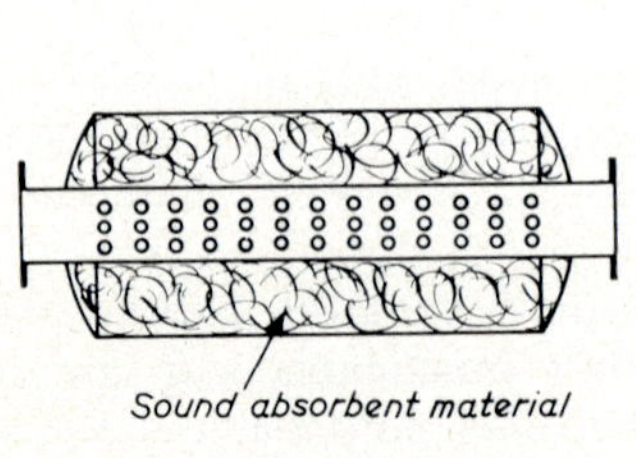

FIG. 12–10—*Absorbent silencer.*

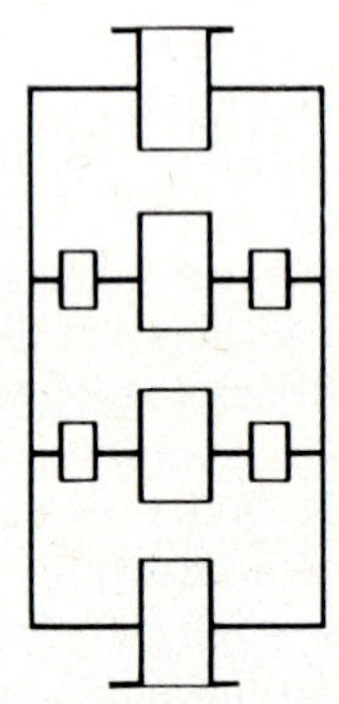

FIG. 12–11—*Reflection silencer.*

BIBLIOGRAPHY

AUSTEN, A. E. W. and PRIEDE, T., *Origins of diesel engine noise.* Symposium: Engine noise and noise suppression. Proc. I. Mech. E. 1958, 19.

BERTODO, R. and WORSFOLD, J. H., *Medium speed diesel engine noise.* Proc. I. Mech. E. 1968–69 Vol. 183 pt. 1 No. 6.

BISHOP, R. E. D. and JOHNSON, D. C., *The mechanics of vibration.* Cambridge University Press.

CREDE, C. E., *Vibration and shock isolation.* John Wiley and Sons Inc., New York. Chapman and Hall Ltd., London.

FIELDING, B. J. and SKORECKI, J., *Identification of mechanical sources of noise in a diesel engine: sound emitted from the valve mechanism.* Proc. I. Mech. E. 1966–7. Vol. 181, part 1, No. 19.

FIELDING, B. J. and SKORECKI, J., *Identification of mechanical sources of noise in a diesel engine: sound originating from piston slap.* Proc. I. Mech. E. 1969–70. Vol. 184, part 1, No. 46.

MORSE, P. M., *Vibration and sound.* McGraw-Hill Book Co. Inc.

VAN SANTON, G. W., *Mechanical vibration.* Philips Technical Library.

SKORECKI, J., *Vibration and noise of diesel engines.* ASME paper No. 63-OGP-2.

CHAPTER THIRTEEN

An introduction to torsional vibration

13.1. *The nature of torsional vibration*

The simple system of a flywheel suspended by a light elastic shaft rigidly fixed at its upper end which was illustrated in Figure 12–1 and mentioned at the beginning of the last chapter can vibrate torsionally as well as transversely and longitudinally.

Let q be the torsional stiffness of the shaft, which means the torque required to twist the shaft through unit angular displacement, that is one radian. $q = CJ/L$ where C is the modulus of rigidity of the material of the shaft and J is the polar second moment of area of the cross section of the shaft and L is the length of the shaft.

Let I be the mass moment of inertia of the flywheel. At any instant the restoring couple equals $q \times$ the angular displacement but the restoring couple also equals $I \times$ the angular acceleration.

$$\text{Thus } \frac{\text{angular acceleration}}{\text{angular displacement}} = \frac{q}{I} \text{ which is a constant.}$$

The motion is therefore simple harmonic and the frequency is given by:

$$n = \frac{1}{2\pi}\sqrt{\frac{q}{I}} = \frac{1}{2\pi}\sqrt{\frac{C \,.\, J}{L} \cdot \frac{1}{I}}$$

The important things to note are that if a flywheel with a larger moment of inertia is used the frequency will be lower and if a shaft with a greater stiffness is substituted the frequency will be higher. In practical cases the problem is seldom concerned with a system as simple as this one, but even the most complex systems are made up of elements of this simple kind and in analyzing them this simple system is fundamental. Even with the most complex system it is true to say that in general, increasing the mass moments of inertia lowers the natural frequency and increasing the stiffness of the shafts raises the natural frequency.

Most practical systems consist of a multiplicity of rotors on a complex shaft assembly, the whole rotating and transmitting power whilst vibrating torsionally. As with other problems in vibration the essential procedure is to avoid resonance of exciting forces with the natural frequency. Correct estimation of the natural frequency is therefore of great importance. The effects of the rotation and transmission of power usually do not alter the natural frequency and for analytical purposes it is satisfactory to assume that the whole system is at rest apart from the vibration and then having determined the natural frequency of the effects of vibration to add or superimpose the effects of power transmission.

13.2. *Two rotor system*

Let us now look at a simple two rotor system as shown in Figure 13–1 where two rotors are mounted at each end of a uniform shaft. The system will be supported in bearings which take no part in the vibration and are therefore not shown. If the rotors are turned in opposite directions twisting the shaft and then released they will commence to vibrate; moving in opposite directions and reaching in sequence their extreme positions and the mean equilibrium position at the same time in each case.

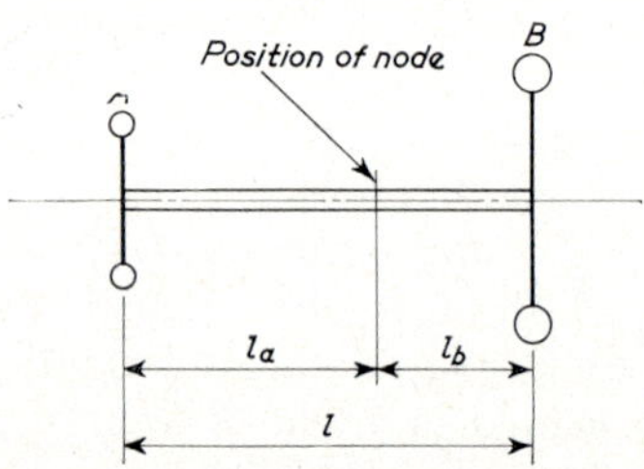

FIG. 13–1—*Two rotor system.*

The shaft will be alternately twisted and untwisted in opposite directions. At some point along the shaft the material will be undisturbed by the vibration. This is not to say that it will not be stressed because, of course, the whole of the shaft will be subject to a uniform torsional shear stress. The point at which the material is undisturbed by the vibration is termed the node. The shaft could be clamped at the node and each portion on either side of the node would vibrate with the same frequency but in opposite phase. Each portion of the system on either side of the node is then equivalent to the simple system which we examined at first. Using the subscripts a and b to distinguish the properties of the rotors and shafts on either side of the node we can write

$$n = \frac{1}{2\pi}\sqrt{\frac{q_a}{I_a}} = \frac{1}{2\pi}\sqrt{\frac{CJ}{L_a}\frac{1}{I_a}} \tag{13.1}$$

and

$$n = \frac{1}{2\pi}\sqrt{\frac{q_b}{I_b}} = \frac{1}{2\pi}\sqrt{\frac{CJ}{L_b}\frac{1}{I_b}} \tag{13.2}$$

As the frequencies of the two parts of the system are the same we can equate these expressions and derive an equation

$$\frac{1}{2\pi}\sqrt{\frac{CJ}{L_a}\cdot\frac{1}{I_a}} = \frac{1}{2\pi}\sqrt{\frac{CJ}{L_b}\cdot\frac{1}{I_b}}$$

$$\therefore \qquad L_aI_a = L_bI_b$$

or

$$\frac{L_a}{L_b} = \frac{I_b}{I_a} \tag{13.3}$$

From this we see that the node divides the length of the shaft in proportion to the moments of inertia of the two rotors. The frequency can be obtained by substituting the values for either rotor and portion of shaft into equations 13.1 or 13.2.

13.3. *Elastic line*

In Figure 13–2 a straight line has been drawn at an angle through the node. As the shaft is subject to a uniform torsional stress this line represents the twist along the shaft or the torsional strain. Where the line intersects the

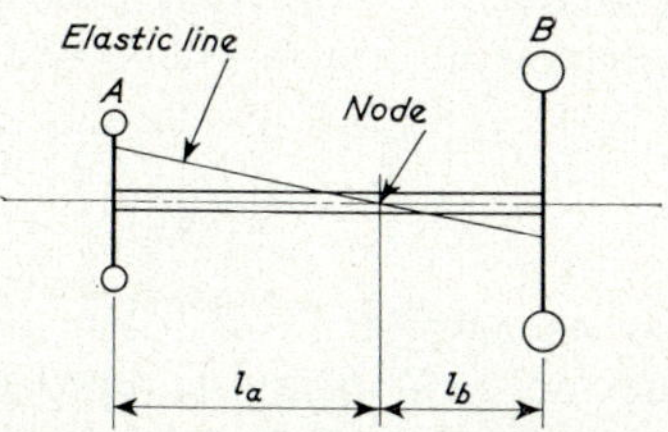

FIG. 13–2—*Elastic line for two rotor system.*

planes of the rotors A and B it will show the relative amplitude of angular deflexion of the rotors. This line is known as the elastic line. The amplitude of the vibration may vary according to a number of factors but the relative displacement of the rotors will always be the same and in proportion to the vectors shown by the elastic line.

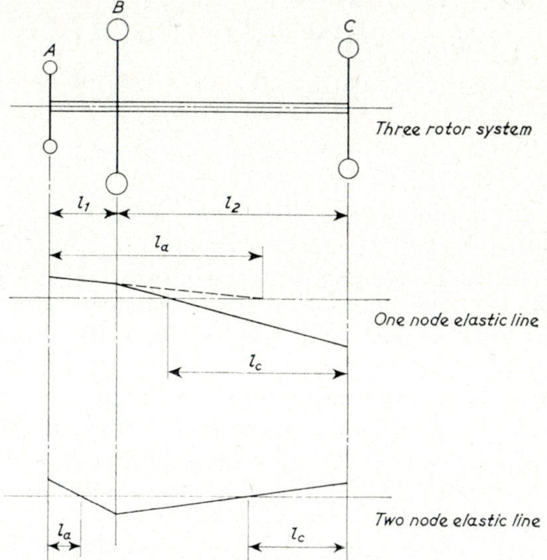

FIG. 13–3—*Three rotor system.*

13.4. *Three mass system*

Figure 13–3 shows a slightly more complex system in which there are three rotors mounted on a uniform shaft. It is possible for this system to vibrate in two ways as indicated by elastic lines drawn on the figure. The rotors A and B may swing in one direction whilst the rotor C swings in the opposite direction, causing a vibration having a single node which lies between B and C. The other form of vibration may involve the rotors A and C turning in the same direction, whilst rotor B turns in the opposite direction and then there will be two nodes, one between A and B and one between B and C. As before, all the rotors will reach their extreme positions at the same instant and will pass through the mean equilibrium position at the same instant with both forms of vibration.

The system can be analyzed by breaking it down into the now familiar small elements. The frequency of rotor A will correspond to

$$n_a = \frac{1}{2\pi}\sqrt{\frac{q_a}{I_a}} = \frac{1}{2\pi}\sqrt{\frac{CJ}{L_a}\cdot\frac{1}{I_a}} \qquad (13.4)$$

The frequency of rotor B will also be given in the same way but the stiffness of the shaft will correspond to it being clamped at both the nodes; this stiffness is given by

$$q_b = \frac{CJ}{L_1 - L_a} + \frac{CJ}{L_2 - L_c}$$

$$= CJ\left(\frac{1}{L_1 - L_a} + \frac{1}{L_2 - L_c}\right) \qquad (13.5)$$

and the frequency

$$n_b = \frac{1}{2\pi}\sqrt{\frac{q_b}{I_b}} = \frac{1}{2\pi}\sqrt{\frac{CJ}{I_b}\left(\frac{1}{L_1 - L_a} + \frac{1}{L_2 - L_c}\right)} \qquad (13.6)$$

The frequency for rotor C similarly to that of A is given by

$$n_c = \frac{1}{2\pi}\sqrt{\frac{q_c}{I_c}} = \frac{1}{2\pi}\sqrt{\frac{CJ}{L_c}\cdot\frac{1}{I_c}} \qquad (13.7)$$

The frequency of all three rotors must be the same so that from these expressions we can derive an equation in either L_a or L_c. This equation will be a quadratic equation and will give two roots, one of which will be the solution for the one node vibration. If it is L_a that we have chosen to find then the distance will not be a true node but will be the projection of the elastic line between rotors A and B to a point on the shaft about which rotor A would oscillate with the appropriate frequency. Substitution will give the length L_c which gives the position of the true node. The shape of the elastic line for each case should be noted in Figure 13–3.

13.5. *Multi rotor system*

Practical cases to be investigated will usually concern an engine driving some machinery or a propeller. To ease the labour of calculating the natural frequency of a complex system of this kind it is usual to reduce it to an equivalent system as shown in Figure 13–4. An equivalent rotor inertia takes the place of the mass inertia effects of the rotating and reciprocating parts at each cylinder and similar equivalent rotors are introduced to represent the flywheel and the driven electrical machine or the propeller. The shafting in between the oscillating masses is shown in the simplified system as of uniform diameter. This is done to ease the labour of calculation; usually the diameter of the main journals of the crankshaft is chosen and the stiffness of the shafting between each of the oscillating masses expressed as equivalent lengths of this diameter. The determination of the equivalent rotors for the running gear at each cylinder and for the equivalent lengths for the crankshaft elements that separate them have been the subject of many investigations during the time that torsional vibration has been studied. These are far too lengthy to explain in this chapter and the interested reader is referred to the Handbook, (ref. 1), in which reliable formulae are given for carrying out these processes. In practice most engine builders acquire a wealth of empirical knowledge about their own engines

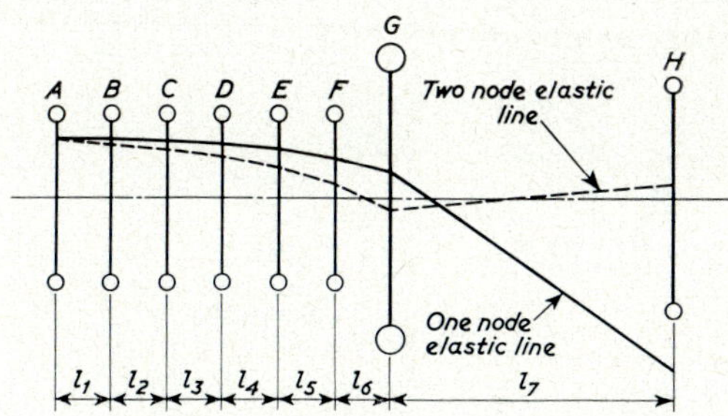

FIG. 13–4—*Multi-rotor system.*

from which they derive their own standard values for the equivalent rotor inertias and the equivalent lengths.

We have seen that a two rotor system can vibrate with one node between the rotors and that a three rotor system has two modes of vibration, a one node and a two node. In general a system with a large number of rotors will have as many modes of vibration as the number of rotors minus one. Fortunately vibrations with large numbers of nodes are not of great practical importance and it is usually sufficient in systems such as that shown in Figure 13–4 to investigate only the one node, the two node and occasionally the three node vibrations. Although it is possible to break down this multirotor system into simple elements as was done in the case of the three rotor system, the equations which would result from equating the frequency expressions would be too involved for practical solution by manual methods, although perfectly feasible for computer methods. A different approach is therefore adopted. Using the notation given in Figure 13–4, θ with the appropriate subscript would be the amplitude in radians of the torsional oscillation of a given rotor. Let $\omega = 2\pi n$, where n is the vibration frequency per second. The maximum angular acceleration of a rotor is $\omega^2\theta$ when the rotor is at the extremity of its swing and the torque which the rotor exerts on the shaft is given by

$$T = I\omega^2\theta$$

When vibrating freely the only torques which act on the system are the inertia torques and at any instant the algebraic sum of these must be zero, hence when all the rotors are in their extreme positions

$$\Sigma T = 0$$

$$\therefore \qquad \Sigma\,(I\omega^2\theta) = 0 \qquad (13.8)$$

In the simplified system the shaft is assumed to have negligible inertia so that there is no change of torque along the shaft. Consequently the difference between the amplitudes of vibration of adjacent rotors will be equal to the angle of twist of the shaft due to the torque transmitted from one rotor to the next.

Starting at rotor A the maximum inertia torque is T_a and this is also the torque transmitted on the length l_1 of the shaft. The angle and twist of l_1 is given by

$$\theta_a - \theta_b = \frac{T_a}{q_1} \qquad (13.9)$$

$$\therefore \qquad \theta_b = \theta_a - \frac{T_a}{q_1}$$

$$\text{i.e.} \qquad \theta_b = \theta_a - \frac{I_a\omega^2\theta_a}{q_1} \qquad (13.10)$$

Similar expressions will follow for the amplitude of vibration of any of the other rotors.

These results are applied to the system in the following way. The amplitude of vibration θ_a is assumed arbitrarily to have a convenient magnitude, say 1 radian. A value is then arbitrarily assumed for ω. This value will be assumed based on experience or if none is available by reducing the system even further to a three mass system. The amplitudes of the other rotors may then be calculated using the equations above. Finally the sum of the products $I\theta$ for all the rotors is determined. According to equation 13.8 this sum must be zero for the correct value of ω. If the sum is not zero then another value of ω must be tried. An experienced technician can usually arrive at the correct value for ω in two or three trials. The amplitudes θ_a, θB etc., can be plotted at the rotors and their extremities joined by straight lines, the result being the elastic line of the shaft. The point of intersection of the elastic line with the axis will be the position of the node.

13.6. *Holzer tabulation*

The process is not so tedious as it sounds as it is eased by setting it out in what is known as a Holzer table. An example is given on pages 241 and 242 of one node and two node calculations for a simple system of an engine driving a propeller.

The first and sixth columns of the table are filled in from the known dimensions of the system and the second column is filled in from the assumed value of ω; the amplitude of the left hand rotor is assumed to be 1 radian in column three, and the first entries in columns four and five follow at once. The first entry in column seven is obtained by dividing the

Holzer Tabulations

Six-cylinder Engine, Reduction Gear and Propeller

One node tabulation. Frequency = 477 vib/min. $\omega^2 = 2480$ rad²/sec²

Column No.		1	2	3	4	5	6	7
Rotor		I	$I\omega^2$ ($\times 10^6$)	θ	$\delta T = I\omega^2\theta$ ($\times 10^6$)	$T = \Sigma I\omega^2\theta$ ($\times 10^6$)	q ($\times 10^6$)	$\delta\theta = \frac{\Sigma I\omega^2\theta}{q}$
A	Damper	361·3	0·899	1·0000	0·899	0·899	1239	0·0007
B	No. 1 Cyl.	316·2	0·787	0·9993	0·787	1·686	874	0·0019
C	No. 2 Cyl.	316·2	0·787	0·9973	0·786	2·472	874	0·0028
D	No. 3 Cyl.	316·2	0·787	0·9945	0·783	3·255	823·8	0·0040
E	No. 4 Cyl.	316·2	0·787	0·9906	0·780	4·035	874	0·0047
F	No. 5 Cyl.	316·2	0·787	0·9859	0·776	4·811	874	0·0055
G	No. 6 Cyl.	316·2	0·787	0·9804	0·772	5·583	787·6	0·0070
H	Flywheel	3957·0	9·853	0·9734	9·587	15·170	5·8	2·6164
I	Coupling	455·0	1·133	−1·643	−1·86	13·31	153·6	0·0860
J	Gears	176·9	0·440	−1·729	−0·76	12·55	18·46	0·6800
K	Propeller	2092·0	5·209	−2·409	−12·55	0		

TABLE 13.1 (continued)

Holzer Tabulations

Six-cylinder Engine, Reduction Gear and Propeller

Two node tabulation. Frequency = 1997 vib/min. $\omega^2 = 43730$ rad^2/sec^2

Column No.		1	2	3	4	5	6	7
Rotor		I	$I\omega^2$ ($\times 10^6$)	θ	$\delta T = I\omega^2\theta$ ($\times 10^6$)	$T = \Sigma I\omega^2\theta$ ($\times 10^6$)	q ($\times 10^6$)	$\delta\theta = \frac{\Sigma I\omega^2\theta}{q}$
A	Damper	361·3	15·8	1·0000	15·8	15·8	1239	0·0128
B	No. 1 Cyl.	316·2	13·83	0·9872	13·65	29·45	874·0	0·0336
C	No. 2 Cyl.	316·2	13·83	0·9536	13·18	42·63	874·0	0·0488
D	No. 3 Cyl.	316·2	13·83	0·9048	12·51	55·14	823·8	0·0670
E	No. 4 Cyl.	316·2	13·83	0·8378	11·59	66·73	874·0	0·0763
F	No. 5 Cyl.	316·2	13·83	0·7615	10·52	77·25	874·0	0·0884
G	No. 6 Cyl.	316·2	13·83	0·6731	9·31	86·56	787·6	0·1099
H	Flywheel	3957·0	173·0	0·5632	97·45	184·0	5·8	31·72
I	Coupling	455·0	19·90	−31·160	−620·0	−436·0	153·6	−2·84
J	Gears	176·9	7·735	−28·32	−219·0	−655·0	18·46	−35·481
K	Propeller	2092·0	91·47	7·161	−655·0			

first entry in column five by that in column six. This gives the twist of the length of the shaft between rotors A and B so that the amplitude of the rotor B is found by subtracting this from 1. The difference is entered on the second line of column three. The entry in the second line of column four then follows and is added to the first entry in order to give $\Sigma I\omega^2\theta$ for the rotors A and B. This sum is entered in the second line of column five and is divided by the second entry in column six to give the second entry in column seven. The process is repeated to complete the table. If the final figure in column five is not zero then the process has to be repeated with a different value for ω. For odd degrees of vibrations, that is the first and third nodes, a positive value of $\Sigma I\omega^2\theta$ shows that the assumed value of ω is too low.

13.7. *Gears*

Most medium speed engines are geared to the propeller and in constructing the equivalent simple vibrating system the effect of the gearing has to be taken into account. Both the equivalent lengths and the equivalent inertias will be affected by a factor equal to the square of the gear ratio. Again the interested reader is referred to the standard works for a complete explanation. The gears themselves will also have inertias which have to be taken into account and will appear as additional rotors in the system.

13.8. *Branched systems*

When two engines are geared together on to one propeller shaft the equivalent vibrating system appears as shown in Figure 13–5. This is known

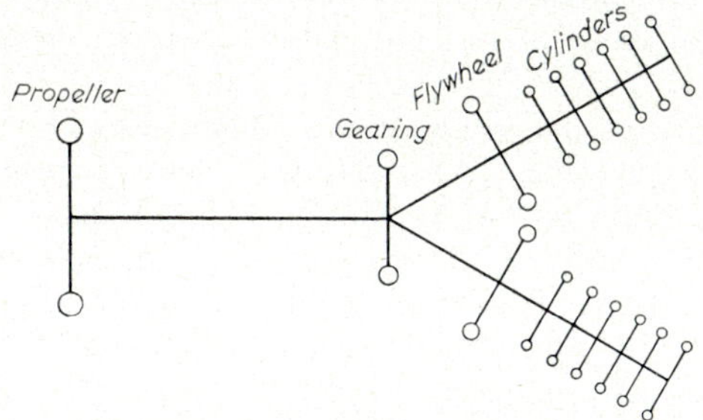

FIG. 13–5—*Example of a branched system twin geared engines.*

as a branched system. The two engines may vibrate in phase in opposition to the propeller in a one node vibration, the node being somewhere along the flexible propeller shaft. Or if flexible couplings are used between the engine and the gearing, nodes may appear in these two branches with the propeller and gearing oscillating together. A third possibility which has to be taken into account is that the two engines will vibrate in opposition with a node at the gearbox and the propeller shaft vibrating on its own with possibly a second node in the system. The natural frequency of each of these is determined by methods similar to those outlined for a simple shaft and though necessarily more lengthy are still dependent on the same principles.

13.9. *Critical speeds*

The disturbances which excite torsional vibration arise from the combination of gas loads on the pistons and inertia loads of the running gear to form the turning moment applied at each crankshaft throw. A four stroke single acting engine will exert a major peak of torque with a frequency equal to half the number of cylinders multiplied by the rev/min. If the engine is run at a speed where this frequency coincides with the natural frequency of the system the resonance will produce vibrations of high amplitude. This particular engine speed is known as the major critical speed.

The turning moment diagram for each cylinder is a complex curve repeated every two revolutions. Like any other repeating curve it can be represented by a series of component sine waves of diverse amplitude and phasing. The fundamental wave will correspond to the major torque peaks as just described; the other component waves will correspond to periodic disturbing torques occurring two, three, four etc., times per two revolutions. Each of these harmonics has the potential to excite the free natural vibration of the system if the engine is run at certain speeds. These speeds are termed critical speeds and are distinguished from each other by the number of vibrations that occur during each engine revolution. This number is termed the order of the vibration. When there are four vibrations per revolution for example there is said to be a fourth order vibration and the engine speed which produces resonance of this vibration is termed the fourth order critical speed. The fourth order critical speed would be the major or fundamental critical speed of an eight cylinder four stroke cycle single acting engine.

With four stroke cycle single acting engines any order of vibration in steps of one half is possible. The magnitude of the exciting torque of each order of vibration depends on the number of cylinders of the engine, the firing order, and the shape of the elastic line of the system when vibrating. Clearly for a six cylinder engine an eighth order vibration is not likely to be very great in torque amplitude but a four-and-a-half order vibration is possibly quite high. As a general rule if the cylinders are all on one side of a node then the orders which are important are those which are obtained by multiplying half the number of cylinders by integral numbers. This, however, is not an infallible rule and certainly should not be trusted if the elastic line shows the node to be amongst the cylinders. The exciting torques depend greatly on the relationship between the phasing of the firing of the cylinders and phasing of the vibration as denoted by the elastic line. Again the interested reader is referred to the BICERA Handbook, ref. 1. The table on page 245 shows the more important orders for engines with the more common arrangements of cylinders. Each order of vibration may have a resonant condition with each mode of vibration, that is to say that all the orders appear in the one node, two node, or three node vibrations.

TABLE 13.2

Important Orders

No. of Cranks	Four-stroke Cycle	Two-stroke Cycle
3	1½, 3, 4½, 6, 7½, 9, 10½, 12, 13½, 15.	3, 6, 9, 12, 15.
4	2, 4, 6, 8, 10, 12, 14, 16.	4, 8, 12, 16.
5	2½, 5, 7½, 10, 12½, 15.	5, 10, 15.
6	3, 6, 9, 12, 15, 18	6, 12, 18.
7	3½, 7, 10½, 14.	7, 14.
8	4, 8, 12, 16.	8, 16.
9	4½, 9, 13½, 18.	9, 18.

Figure 13–6 shows a spectrum of vibrations excited by different orders for a six cylinder engine covering both one and two node modes of vibration.

13.10. *Effects of torsional vibration*

The amplitude reached at each critical speed will depend not only on the exciting torque but also on the damping as with any other kind of vibration. Major critical speeds are those where the energy input is high compared with the damping and the only satisfactory system is one which avoids running on major critical speeds altogether. However, it may not be possible to avoid minor critical speeds or the flanks of some of the major

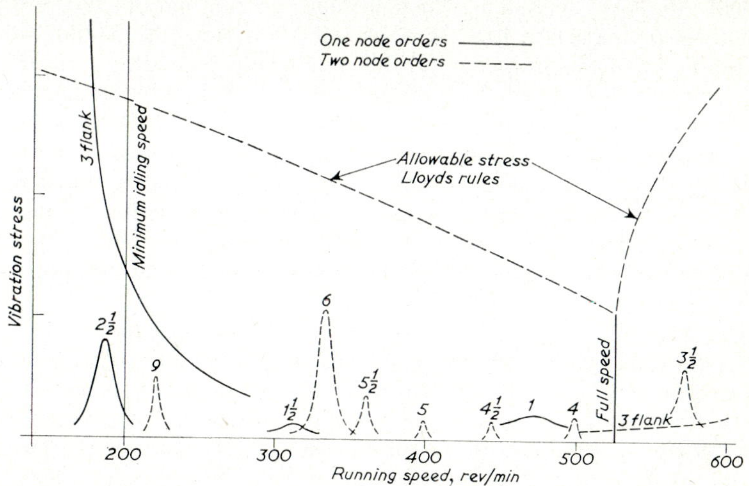

FIG. 13–6—*Typical spectrum of torsional vibration six cylinder four-stroke engine.*

critical speeds and the damping in the system must be high enough to ensure that the energy input does not cause amplitudes which would stress the shafts unduly or cause other damage.

The adverse effects of running with too high a torsional vibration may take several forms. Undoubtedly the most severe is failure of a shaft. Torsional fatigue failures are characterized by progressing in a helical direction along the shaft at 45° to the axis. A typical line of failure is sketched in Figure 13–7. The crankshaft is vulnerable to this kind of failure for a number of reasons. It is frequently the case that the node of one of the modes of vibration lies within the crankshaft near to the flywheel end.

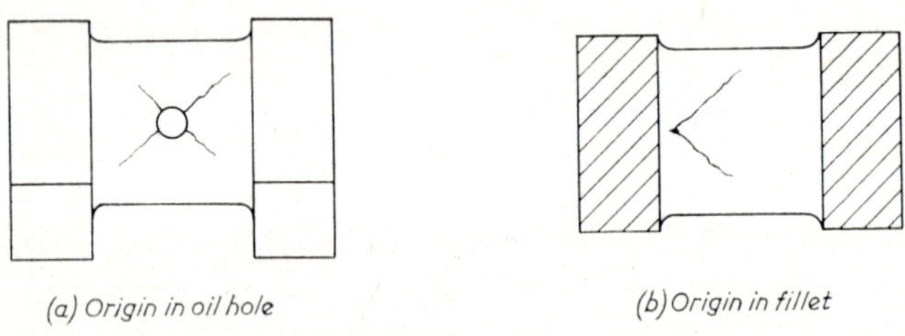

FIG. 13–7—*Typical torsional fatigue fractures.*

As well as the stresses due to torsional vibration and the torsional transmission of power the crankshaft is also stressed in bending when carrying out its function of converting reciprocating motion into rotary motion. These stresses are all combined. The crankshaft is also a complex shape in which it is difficult to avoid localities in which there is some concentration of stress. Positions of comparatively high stress occur in the fillet radii of journals and pins joining the webs at the inner positions in the plane of the crank throw. See Figure 7–8.

Shaft failure is not the only way in which trouble can arise from torsional vibration. A system which includes gearing may have a vibration problem when loads are low. In a propulsion system this could occur at low speeds when the propeller torque is not very high. If a critical speed occurs at such a low speed the vibratory torque will possibly be greater than the transmitted torque even though neither is very high. Because of back lash in gearing this will cause the gear teeth to separate and make contact alternately with high impact forces which can be very detrimental. The condition is termed gear hammer and must be avoided either by the introduction of sufficient damping to reduce the vibratory torques or by adjustment of the natural frequency of the system to alter the position of the critical speed.

The torsional vibration of a simple system of circular shafting and rotors would not be transferred in any way to the bearings supporting it but complex shapes of crankshafts distort in a way that throws loads upon the crankshaft bearings. Similarly gear trains suffering torsional vibration transfer their reactions to the bearings of the shafts. Some of the energy of the vibration is thereby transmitted to the bearings and thence to the frames and structure of the ship so that a critical speed of any magnitude is frequently manifest by vibration and noise which whilst it may be objectionable in itself is generally much more likely to be advertising a dangerous condition of the machinery.

13.11. *Avoiding torsional vibration*

In combating torsional vibration problems use is made of adjustments to the inertia, the stiffness and the damping in the system. For engines that run at a constant speed, for example, driving a generating set, adjustment of the natural frequency so that critical speeds do not occur near to the running speed is usually sufficient. Problems are more difficult in the variable speed engines used for propulsion. To adjust the inertia of any of the rotors in the system is not often easy and usually can be performed only in the increasing direction perhaps by adding another mass in the shape of a flywheel at the forward end of an engine. Most systems include a flexible coupling between the engine and the gearbox in order to avoid extreme fluctuations of load on the gear teeth. By suitable choosing of such couplings their stiffness or equivalent length can be arranged to overcome situations in which the natural frequency of the system would give rise to critical speeds at inconvenient intervals. The Holset coupling shown in Figure 11–10 is an example of this type of coupling. The buffers may be provided in different grades of rubber to give different values of stiffnesses. Other types use springs which are capable of being changed to give different rates.

Some spring couplings have two sets of springs or buffers and when operating at low torques only the first set of springs is compressed giving a low stiffness, whilst when a certain torque is exceeded the second set of springs comes into play and the stiffness is increased. This has the effect of altering the system according to whether the torque is above or below the point at which the second set of springs comes into operation, and the system will accordingly have two frequencies. By a device of this kind it is often possible to dispose critical speeds so as to give a wide range of operation free from vibration.

Some couplings are arranged so that they "detune" in the manner just described but continuously throughout the torque range. A mechanical coupling in which the driving and driven halves are connected by leaf springs can have the slots to receive these springs contoured so that the effective length of the spring is changed as torque is increased. With this device the frequency is changed not only as the transmitted torque is increased but also as the vibratory torque is increased. Couplings using this principle are known as detuning couplings.

Other designs of coupling make use of the same principle but in addition the springs and the pockets in which they work are filled with oil which is kept under pressure from the engine lubricating oil supply and any relative movement between the driven and driving halves of the coupling causes oil to be squeezed between fine clearances from one side of the springs to the other. This introduces damping as well as detuning the energy from a high frequency vibration becoming absorbed in the work done in transferring the oil back and forth.

13.12. *Dampers*

However, damping of this kind is not particularly effective at the coupling, because the coupling is usually situated not very far from the node where amplitudes are low. A similar design can, however, be arranged as a damper and when used in this way is best situated near the end of a system, for example the free end of an engine where amplitudes are great and damping is more effective. It operates as a damper by the fact that the driven half although not operating any machine can be made of sufficient mass to act as a flywheel and when the system is vibrating it will tend to turn with uniform motion rather than follow the vibrations; the driving portion will, of course, vibrate because it is rigidly connected to the shaft and the movement of the springs will transfer the oil through the fine clearances between the passages so absorbing the energy of the vibration and damping it.

There is a disadvantage in using lubricating oil as a damping medium in that the energy put into it heats it up and it therefore lowers its viscosity and makes it less effective as a damper. Dampers in which a viscous fluid is sheared, however, are correct in principle and the Holset viscous fluid damper is an elegant application. It is illustrated in Figure 13–8. A comparatively light outer casing is rigidly connected to the end of the vibrating shaft. Inside, carried quite freely, is a heavy mass and the only connexion between the outer casing and the inner mass is the viscous fluid with which the damper is filled. This fluid is a silicone fluid of higher viscosity than lubricating oil and with the property of altering its viscosity very little with temperature. When the shaft rotates the inner mass is carried round with the outer casing by the viscous shear of the fluid and will attain the speed of

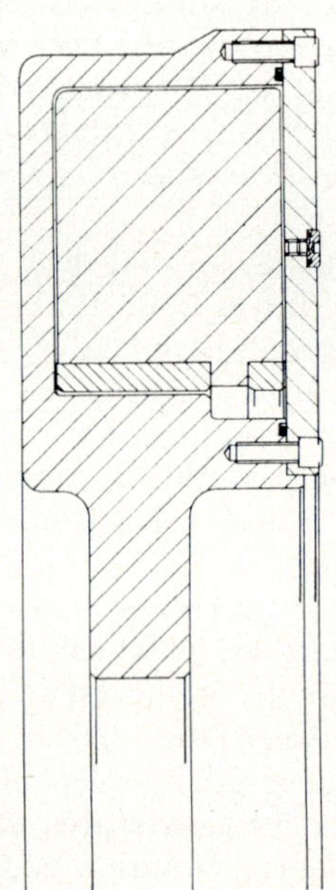

FIG. 13–8—*Viscous fluid damper.*

rotation of the shaft. When the shaft vibrates, however, the inertia of the inner mass does not allow it to follow the vibrations but it continues to rotate with more or less uniform motion. The outer casing does follow the vibration and there is consequent shearing of the viscous fluid which absorbs the energy of the vibration and so gives a large measure of damping. The energy will, of course, heat up the fluid but if this damper is situated in the open air its shape and its rotation are such that it loses this heat to the air in adequate quantities to maintain a state of equilibrium. If the damper is placed inside the engine then it may be necessary to arrange for cooling either by air or oil to be applied to it. These dampers are self contained sealed units which require no attention throughout their life.

This form of damper also acts as a detuner but in a different way from the detuning obtained by altering the stiffness of a coupling. Here, when the whole of the damper, the inner mass and the outer casing, rotates together it will have a certain inertia. At the onset of vibration, the increased shearing forces in the viscous fluid permit the casing to move separately from the mass, and the inertia of the casing alone will become the determining inertia in the natural frequency of the vibrating shaft. The vibrating system will therefore have a different frequency from that of the system when it is not vibrating. This detuning has the effect of restricting the amplitudes to which a vibration can build up. The vibration will start to build up at a critical speed corresponding to that of a system in which the whole inertia of the mass operates but as its amplitude rises this inertia will change to a lighter one and the critical speed will rise so that the system is no longer in resonance. If the engine speed is then increased, the mass will revert to the first condition and the critical speed will fall below that at which the engine is operating.

There are many types of torsional vibration damper and it is beyond the scope of this book even to list them. Many are described in the BICERA Handbook (ref. 1) and by Ker Wilson (ref. 2). They mostly consist of a seismic mass which may be freely supported as in the case of the viscous fluid damper just described, or it may be elastically connected to the shaft system by springs, by rubber or by low stiffness shafts. When the mass is connected to the shaft in this way it is usually because the damper concerned is designed to act chiefly as a detuner and to have the effect of altering the resonant conditions. Often it is effective in this only for the one particular critical speed for which it is designed. Frequently the design is also arranged to absorb and dissipate energy caused by the relative motion of seismic mass and casing, that is as well as detuning it also damps. The damping energy may be absorbed:

(a) by the shearing of viscous fluid
(b) by pumping oil through fine passages
(c) by causing friction surfaces to rub together, or
(d) by operating pendulum mechanisms which oppose the motion of the vibration.

Whatever the form of damper used it must be located in a position where the energy which it absorbs by damping the vibration can be dissipated, otherwise it may overheat and cease to function correctly. Most dampers are mounted outside the engine crankcase and dissipate the energy to the ambient air. Dampers which are mounted inside the crankcase may require special cooling arrangements.

13.13. *Torsiographs*

When it is necessary to measure torsional vibration the instrument employed is termed a torsiograph. The early instruments to be developed were simple mechanical types, a typical one being the Geiger.

It consists of an outer casing of drum form mounted so that it can rotate about a horizontal axis and made to follow the motion of the crankshaft either by being rigidly connected to the end of the shaft or belt driven from a convenient point. Inside the drum casing is mounted a seismic mass free to rotate on the same axis as the casing, and connected to it by a light spring of very low stiffness. When rotation of the crankshaft causes the instrument to rotate the motion of the seismic mass approximates very closely to constant angular velocity whilst the casing follows the motion of the crankshaft itself. If torsional vibration is present there will be relative movement between the two parts. This relative movement is detected and amplified by a system of levers which cause a pencil trace to be made on a paper strip. Two other series of marks are made on the same paper strip to show the revolutions of the crankshaft and time intervals. The trace is analyzed by manual inspection. Sometimes optical magnification methods are used to enlarge the trace and ease the task of analysis.

Mechanical instruments of this kind are very robust and are consequently still popular for use in service on ship board.

Robust electronic instruments have recently been developed and these have a number of advantages. They fall into two categories, the first detecting the motion of the shaft during vibration and the other detecting strain in the shaft.

Whereas the mechanical torsiograph is a self-contained instrument, the electronic version is a collection of equipment but the various functions that are carried out are virtually identical.

The electrical torsiograph detecting motion of the shaft is usually mounted directly on to the shaft at the free end. As for the mechanical version, the sensing element consists of seismic mass mounted freely on a light casing or spindle which is rigidly attached to the crankshaft. The mass may be driven by a very light spring connexion or simply by friction in its bearings. Relative movement between the two parts is made to provide an electronic signal. This signal is transmitted to an amplifier and then caused to make a trace which may be on a oscilloscope or it may be permanently recorded by photographic methods.

Amplitude measuring torsiographs, whether mechanical or electrical, must be mounted at a position on the shaft where the amplitude is significant and in this respect the free end of the shaft is usually satisfactory. The alternative electrical method uses strain gauges to obtain a direct measurement of the stress in the shaft and requires this equipment to be placed at a point on the shaft where stress is high, ideally near to a node. The strain gauges are arranged to give rise to an electronic signal which can be amplified and displayed.

The electronic methods have the virtue that wave analyzing equipment or frequency selector or filter circuits can be introduced, enabling the analysis of complex wave forms to be carried out electronically, the resulting traces displaying separately the vibrations of each critical speed examined. This saves a great deal of tedious manual analysis.

REFERENCES

1. NESTORIDES, E. J., *A handbook on torsional vibration.* Cambridge University Press, 1958.
2. KER WILSON, W., *The practical solution of torsional vibration problems.* Chapman and Hall.

CHAPTER FOURTEEN

Governing and Speed Regulation

14.1. *The control of power for propulsion*

The purpose of any prime mover is to produce power which can be controlled and directed according to the will of the human operator. When controlling a propulsion plant the operator is concerned with regulating the supply of thrust from the propeller to the ship. Ideally this is required to be capable of continuous variation from zero to a maximum in either the ahead or astern direction. This ideal can be closely approached within the limitations of the installed machinery and may be achieved either by varying propeller rev/min from zero to a maximum in each direction or by varying the pitch of the propeller. The former arrangement involves stopping and reversing the engine or disconnecting the propeller from the engine and reversing it by means of gearing. Every engine, therefore, needs to have a means for starting and stopping and for controlling the rate at which power is supplied. For direct reversing engines the direction of rotation must also be controlled.

Starting an engine involves a logical series of operations to provide the services essential to running, followed by an operation setting the engine in motion. Stopping is primarily a matter of cutting off the fuel supply but may also involve devices designed to arrest motion, to take care of essential services or to reset mechanisms ready for the next start.

In manned engine rooms the controls for starting and stopping are usually local to the equipment that is to be operated and the logical sequence of operation is determined by the engine room staff. Starting and stopping actions are required only from time to time most frequently in the case of ferries and least frequently in ocean-going ships where the services may be called into play only at comparatively long intervals of time. On the other hand the rate of supply of power has to be controlled continuously whilst the ship is under way and is almost invariably arranged to be automatic.

14.2. *Speed regulation*

Controlling the rate of supply of power takes the form of requiring that a shaft shall be turned at a predetermined mean speed of rotation, the mean torque to the shaft being supplied at the value required to achieve this condition.

Every power control problem must start with a consideration of the characteristics of both the load and the power supply. Some engine and load systems are stable and once the engine has been set to work at a certain rate it can be expected to continue at that rate unless some abnormal circumstances intervene. In the case of many prime movers the power remains substantially constant over a speed range because either the working fluid or the fuel is supplied at a constant rate as a result of a simple control setting. For example, steam flowing through a control valve orifice to a turbine or air and petrol mixture flowing through a set throttle opening to a spark ignition engine. Arrangements of this sort result in a torque characteristic which falls with increasing rev/min (torque being proportional to power divided by rev/min) giving stable operation in conjunction with many loads.

A diesel engine, however, has a fuel supply which is substantially constant per cycle for each setting of the fuel rack; thus power is proportional to rev/min and torque remains constant as rev/min is varied. Considering the whole range from idling speed to full speed the characteristic is modified by volumetric efficiency effects, to result in a slightly humped curve but, as has already been shown, for most practical purposes the torque can be regarded as constant and certainly this is so over any small part of the speed range.

The simplest form of control is to set the fuel pump rack to give the desired constant amount of fuel injection per cycle and hence a correspondingly constant torque; each setting of the rack thus resulting in a corresponding torque level as shown in Figure 14–1. Clearly, stable operation

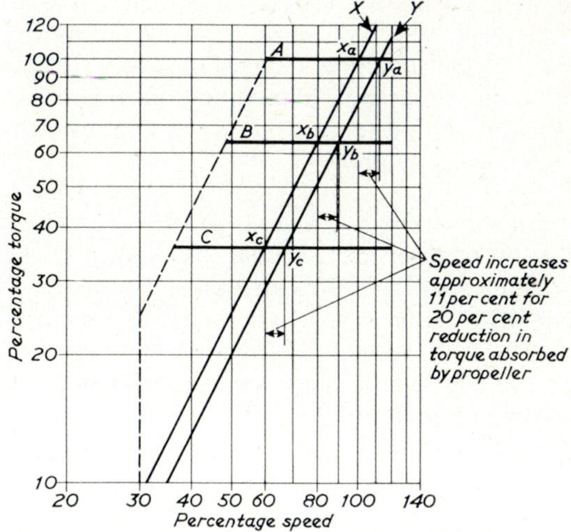

FIG. 14–1—*Speed regulation by fuel pump rack setting.*

will depend on the load characteristic; if the engine is directly coupled to a fixed pitch propeller having a torque characteristic which is unique and increases with rev/min then the result, as shown, is that for each fixed rack setting there will be a single operating point. Curves A, B and C represent torque output corresponding to three rack settings and curve X represents the torque absorbed by the propeller relative to engine rev/min; the operating points occur at x_a, x_b and x_c; with the vessel steadily under way, therefore, the rev/min can be set simply by setting the fuel pump rack.

A load which does not have a unique torque to rev/min relationship requires more complex control. If a controllable pitch propeller is used then the various torque curves resulting from different pitch angles will produce different rotational speeds for the same fixed setting of the fuel pump racks. Even with a fixed pitch propeller variations in the resistance to the propeller caused by rough weather can result in the speed of the engine varying rapidly and in an uncontrolled manner. For instance, a change in the propeller torque characteristic from curve x to curve y will shift the operating points to positions y_a, y_b and y_c respectively, corresponding to a substantial change in rev/min.

To safeguard against dangerous overspeed it is invariably the case that some form of governor is provided to limit the maximum speed of the engine and in the case of medium and high speed engines it is the usual practice to provide speed regulation governors which operate throughout the full range of rev/min from idling to full rotational speed.

14.3. *Feedback*

The use of a governor introduces the control technique of "feedback" to give proportional control. An instrument sensing the quantity to be controlled, (in this case the rotational speed,) feeds back this information to a device capable of comparing it with the value demanded (the set speed) and which in turn passes an instruction to an "actuator" to adjust the controlling mechanism (the fuel pump rack) according to the magnitude of the error. This control technique appears in several forms and is applied to the control of many situations. Sometimes the functions of sensing, comparing and actuating are all merged in one mechanism; this is the case with the simple mechanical governor.

14.4. *Mechanical governors*

Mechanical governors are familiar mechanisms, the need for speed regulation having existed from the early days of prime movers. The Hartnell type of spring loaded centrifugal governor is the form most widely used in marine work. The ball weights respond to the speed of rotation, their centrifugal force being balanced by the spring load as indicated in Figure 14–2. It is shown in textbooks on Theory of Machines that by taking moments about the fulcrum of one of the fly-weights the following equation can be derived:

$$p = 8\pi^2 . M \left(\frac{l_1}{l_2}\right)^2 . \frac{N_1{}^2 r_1 - N_2{}^2 r_2}{r_1 - r_2} \qquad (14.1)$$

where p = rate of the spring
M = mass of one weight
l_1 = distance from fulcrum to c.g. of weight
l_2 = distance from fulcrum to point of action on the sleeve
N_1 = equilibrium speed when c.g. of weight is at radius r_1
N_2 = equilibrium speed when c.g. of weight is at radius r_2.

14.5. *Sensitivity and stability*

The rate of the spring must be chosen so that the equilibrium position at a higher speed is such that the radius to the centre of gravity of the weights is greater than for the equilibrium position at a lower speed otherwise the governor will be unstable; *i.e.* for stability, N_1 must be greater than N_2 when r_1 is greater than r_2.

The mechanical limits of movement of the mechanism will result in a minimum radius and a maximum radius being possible for the centre of gravity of the weights. If there is a large difference between the equilibrium speed at the maximum radius and that at the minimum radius then the governing will be coarse. Conversely, if only a small change in speed is required to cause the governor sleeve to move over a large part of its maximum travel the governing will be fine or sensitive.

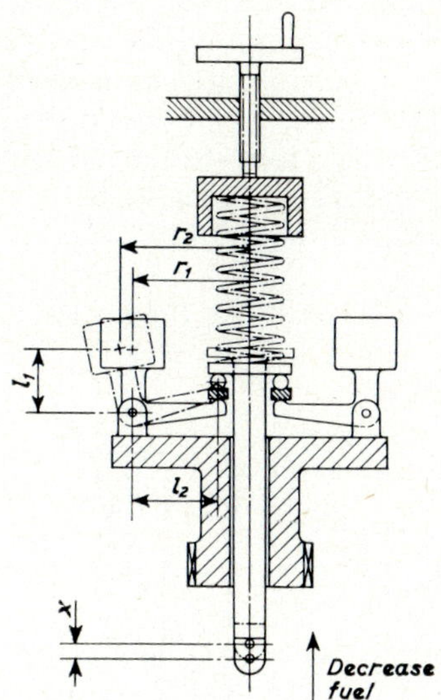

FIG. 14–2—*Hartnell type governor.*

It is possible to design a governor so that the equilibrium speed at the minimum radius is the same as that at the maximum radius, in other words there is only one speed which is the equilibrium speed for all positions of the weights, *i.e.* $N_1 = N_2$. With such a governor, if the rotational speed of the weights is gradually increased from a low value until the equilibrium speed is reached then the weights will suddenly fly out; similarly on reducing speed from a high value, when the equilibrium speed is reached the weights will again fly in, giving an "on-off" governing condition. In older textbooks a governor of this form is termed "isochronous". However, it is in reality an on-off switch incapable of exercising stable control. In modern usage the term isochronous is applied to sophisticated governing systems embodying reset action by means of which the speed control setting is automatically adjusted following a change of load to restore operation at the original set speed.

It may be noted that it is also quite possible to design a governor in which the equilibrium speed at the minimum radius is higher than at the maximum radius. With such a mechanism when raising the rotational speed from a low level a value will be reached at which the weights will suddenly fly out. This will occur immediately the equilibrium speed at the minimum radius is exceeded. On reducing speed from a high level the rev/min will have to fall to a lower value, namely that of the equilibrium speed at the maximum radius, before the weights will fly in. These conditions of totally unstable governing are of no use for continuous control purposes; but may be used for tripping devices (*e.g.* overspeed trip).

14.6. *Droop*

The sensitivity of the governor has so far been considered in relation to the positions of minimum and maximum radius of the weights. In practice the governor will be connected to an engine and the positions of zero fuel and maximum fuel will, in general, lie within these positions of minimum and maximum radius. The sensitivity of the engine and governor combination will therefore be different from that of the governor as a mechanism alone. When considering the engine and governor combination the difference between the no load speed and the full load speed is known as the governor droop.

The effect of droop is illustrated in Figure 14–3; where curves A, B and

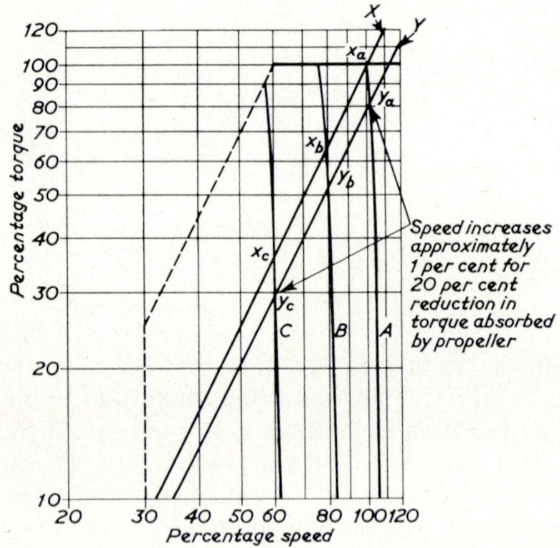

FIG. 14–3—*Speed regulation by governor setting.*

C represent the droop at three different speed settings of the governed engine. If the engine drives a propeller with the torque characteristic shown at X then the operating speeds at each setting will be as indicated by the points x_a, x_b and x_c. If the torque absorbed by the propeller is changed (*e.g.* by a change of pitch, assuming it to be a controllable pitch propeller) to the characteristic shown at Y, then the corresponding operating points will be y_a, y_b and y_c which are only slightly different in speed from x_a, x_b and x_c respectively. This may be compared with the large change in speed which occurs if the engine is ungoverned and controlled by fixed rack settings as shown in Figure 14–1.

A small governor droop means a small difference between the full load and no load steady rev/min, but it also means a rapid swing from full fuel to no fuel if a small rise in speed occurs. When a change in load occurs an interval of time is bound to elapse before the engine torque alters to match. This delay arises from several causes: friction has to be overcome so that a small change in speed occurs before the governor takes any action; inertia of the components requires a finite time for the fuel pump rack to reach its new position and when it is in the new position no effect can begin to be felt until combustion takes place in the first cylinder to fire after the change has been made; the effect will not be complete until all the cylinders have completed a cycle. During this period the speed is continuing to change and a sensitive governor will continue to adjust the fuel pump rack position so that over correction occurs. This will result in too large a change in speed and will be followed by a correction in the other direction. Equilibrium may be reached only after several oscillations.

A large governor droop giving a slower response to the change in speed will not have such a strong tendency to overshoot and equilibrium will be reached without a large number of rapid alterations and in many cases it will be reached in less time than with a sensitive governor.

14.7. *Governor effort and work output*

When a governor experiences a change of speed there is an imbalance of forces on the sleeve between the centrifugal effects from the weights and the force from the spring. This resultant force is termed the governor effort and it decreases to zero as the sleeve and weights move to their new equilibrium position. More precisely, and in order to compare different governors, the effort is defined as the mean force exerted during a one per cent change in speed. When this is multiplied by the corresponding travel of the sleeve the result is termed the work output or work capacity.

The delay in operation, or lag, is a function of the speed of response and the work output of the governor, the inertia of the fuel pump control gear and the friction. It may become so large that the governor operation is completely out of phase with changes in engine speed and a state of continuous oscillation is set up, a condition known as "hunting". The lag is minimized by lighter gear, reduced friction, less sensitive or more powerful governor.

The effort exerted by the spindle of a simple governor experiencing a change of speed is sufficient to actuate the fuel controls of a small engine directly but not the heavier gear of a larger engine. A more powerful governor of the same simple kind will involve heavier weights and a much stronger spring. As the speed setting of the governor is varied by applying a force to this spring the more powerful the design of the governor the greater will this force be and there is, of course, a limit to the manual effort that can be exerted. Furthermore, the speed of response of the heavier parts is much slower.

14.8. *Hydraulic servo governors*

In the governing process the elements of all control functions can be recognized. The rotating weights sense the speed and compare it with a demand which is represented by the spring load. The result of the comparison is a signal which is the resulting effort. This signal is imparted to the fuel control racks so that they increase or decrease the fuel supply as required. For a large engine requiring a powerful governor with quick response the centrifugal ball head may be used as a speed sensing mechanism only and its output signal multiplied to a value which will actuate the fuel control racks by means of a servo system, usually hydraulic. Some of these servo systems are shown in Figure 14–4.

Because the centrifugal force of the weights is proportional to the square of the rotational speed the response is not linear if a spring of constant rate is used. This deviation from linearity may be important in some servo systems and can be overcome by using a trumpet shaped spring having a variable rate to match the square law and so result in proportional control signals from the centrifugal speed sensing unit.

The simple servo system shown in Figure 14–4(a) is one in which the

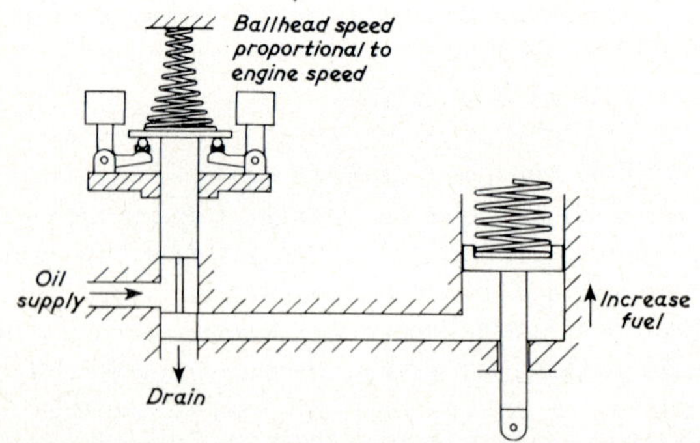

FIG. 14–4a—*Simple hydraulic servo governor (unstable).*

pilot valve is positioned proportionally to speed but this merely controls the rate at which oil is supplied to the fuel rack positioning piston, the final position of which is bound to be either full fuel or no fuel. Such a governor will therefore have zero droop and infinite sensitivity.

For the sake of stability it is necessary to build in feed back from the fuel rack positioning piston. This may be done as shown in Figure 14–4(b) by a linkage connecting the control piston to the fuel rack positioning piston. Alternatively it can be arranged as in Figure 14–4(c) by a linkage which reacts on the speeder spring to change the force which the centrifugal weights have to balance.

14.9. *Compensation*

If it is desired to control the speed of an engine within very close limits the governor can be designed to have a reset action. This is termed compensation and results in a governor which has a speed droop which is transient but which over a long period of time is isochronous in the modern sense. As shown in Figure 14–5 a floating lever connecting the speeder rod pilot valve and the receiving piston is urged by a centring spring to an equilibrium position. So long as the receiving piston is in this equilibrium position centring the valve requires that the fly weights be always in the same position. For a fixed speeder spring setting this means that the centrifugal head must be running always at the same speed. The receiving piston is displaced from its equilibrium position by a flow of oil initiated by movement of a transmitting piston which moves with the fuel rack servo. If there were no leakage from this compensating hydraulic system the receiving piston would move as though rigidly connected to the servo and would result in permanent speed droop. However, the adjustable leak in the form of a needle valve which is provided between the compensating hydraulic system and the oil reservoir allows the centering spring to return the receiving piston slowly to its initial position after a disturbance by

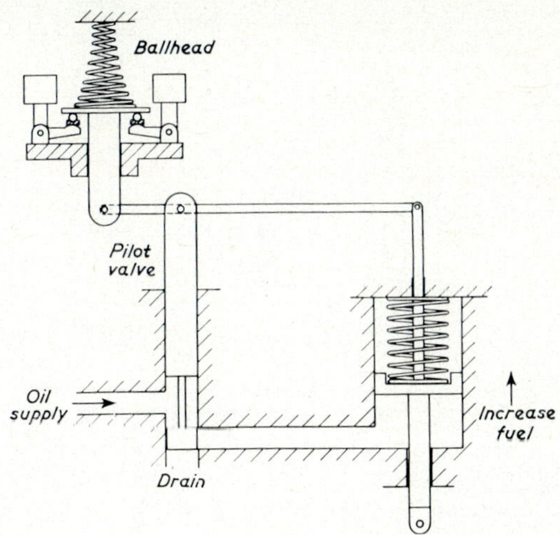

FIG. 14–4b—*Hydraulic servo governor with feedback between pistons.*

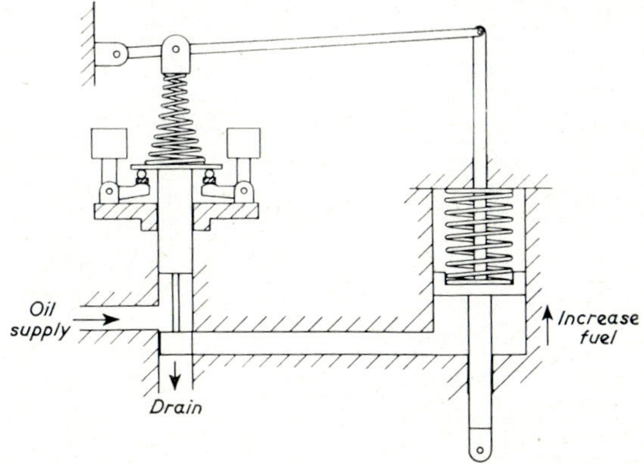

FIG. 14–4c—*Hydraulic servo governor with feedback to speeder spring.*

forcing oil out or drawing oil in through the needle valve as required. As this occurs the speed setting of the governor slowly returns to its original value although the servo and therefore fuel rack are permitted to remain at a different load position.

14.10. *Hydraulic speed sensing*

The centrifugal ball head is by no means the only speed sensing device that may be employed. Figure 14–6 shows a governor with a hydraulic speed sensing unit. It consists of a positive displacement gear pump which is driven by the engine, the oil displaced by it being caused to flow through an orifice in the piston A. The pressure drop across this orifice is compared with the demand pressure drop set by the speeder spring C; the resulting position taken up by the valve B is used to control the pressure acting on piston D which actuates the fuel linkage.

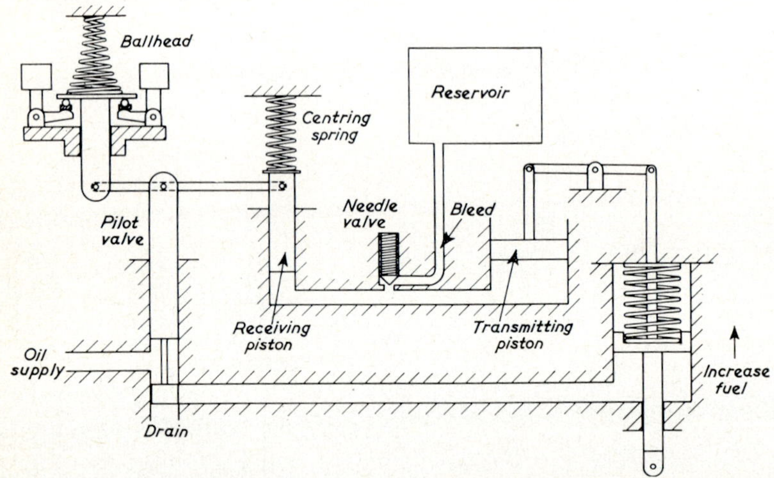

FIG 14–5—*Compensation or reset action to provide isochronous governing.*

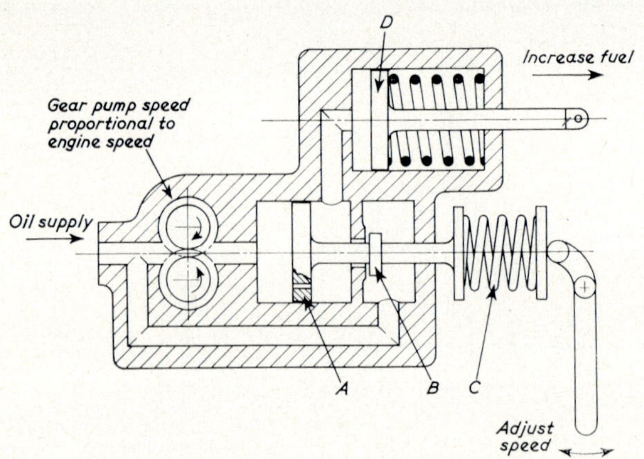

FIG. 14–6—*Hydraulic speed sensing.*

14.11. *Electronic governor*

A recent development is the electronic governor. The speed is sensed in the form of electrical impulses generated by an electronic transducer in proximity to a toothed wheel. These pulses are compared in frequency with the demand speed which is fed into the controller also in the form of pulses. Any deviation produces a signal which is transmitted to the actuator electrically and causes the actuator to use hydraulic power to move the fuel control linkage. In electronic governors the speed sensing unit, the control unit and the actuating unit are all separate and can be located at positions on the engine or in the control room best suited to the function of each individual component.

14.12. *Load increase on highly turbocharged engine*

Highly turbocharged engines are limited in the rate at which they can accept increase of load. Increased fuel supplied to the cylinder can only be burnt if the air is supplied accordingly and the turbocharger cannot provide increased flow of air until the exhaust energy has been increased by burning some fuel. A rapid increase in the rate at which fuel is supplied will result only in the production of exhaust smoke and will not increase the speed or load carrying capacity of the engine. The rate at which the fuel supply is increased can be moderated to avoid this happening by utilizing a signal produced by the air manifold boost pressure or the turbocharger speed. This moderating of the rate of increase of fuelling can be carried out mechanically or hydraulically but it is probably in its neatest and most elegant form as a small additional circuit in the controller of an electronic governor.

14.13. *Load sharing*

When two engines are geared together they are required to share the load between them equally. Because they are mechanically connected together the two engines will obviously run at the same rotational speed and if the governor of one engine is set to demand a slightly higher speed than the governor of the other engine, then that one engine, in attempting to run faster, will carry a greater share of the load. In Figure 14–7 is shown the shape of the governor droop curve from the full load rev/min up to the no load rev/min. If the speed setting of two engines is slightly different the governor droop curves for the two engines will follow lines similar to those shown in the figure. As both engines are running at the same speed the vertical line represents this common speed and it will intersect the two droop curves at different levels of load. The system is quite stable and the engines can be made to carry the same load by adjusting the speeder settings so that the droop curves coincide. Small differences caused by the manufacturing tolerances of the governor and the engine may produce droop curves of slightly differing shape for the two engines, in which case coincidence can only be achieved at one load and one speed. For precise load sharing at all loads adjustment to the speeder settings will be necessary whenever there is a load change. It is obvious from Figure 14–7 that sensitive governors giving steep droop curves will result in a larger disparity in the loads taken by the engines for the same difference in speed setting. Operation is eased by providing governors with coarse droop but fine adjustment for speed setting control.

Sometimes a large droop is unacceptable, for instance when the load is an alternator and the frequency is to be maintained within close limits. Load sharing must then be performed by a more sophisticated system of governing in which a signal relating to the fuel rack position of each engine is fed back to the governors and compared with the position of the rack on the other engines sharing the load. The relative positions of the racks are then adjusted continuously by the governor system.

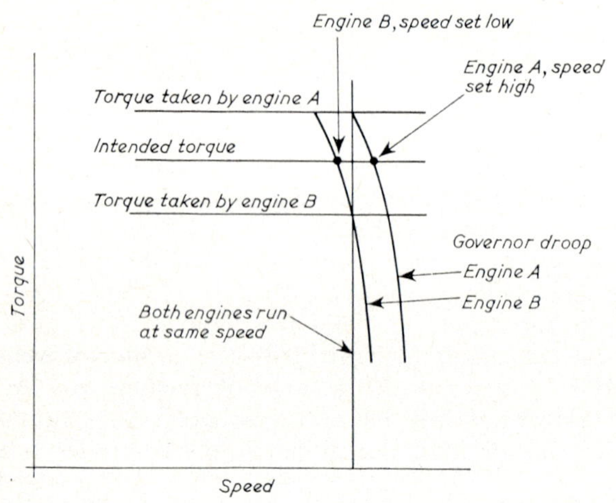

FIG. 14–7—*Load sharing between two engines geared together.*

14.14. *Load regulation*

Speed sensing systems may be used to modify the load characteristic as opposed to modifying the power supply characteristic. Speed regulation is a matter of maintaining the equilibrium between the torque output of the driving machine and the load. In the classical application of governors to prime movers the output torque of the engine is adjusted to match the requirements of the load. In suitable circumstances it is equally feasible for the torque demand of the load to be adjusted to match the torque available from the engine. One example of this is the use of a speed sensing system or governor in controlling the pitch of a controllable pitch propeller. An increase in speed can be made to produce an increase in pitch so that the load on the system is immediately increased also. The load torque will thus exceed the torque supplied and the speed will fall until the set value is once more resumed, and vice versa.

14.15. *Electric power generation*

The principal means of generating electricity on board motor ships is by auxiliary diesel generator sets, comprised of medium or high speed engines coupled directly to d.c. or a.c. generators. The engines used for this purpose are basically of the same design as those for propulsion, but are required to run in only one direction and at constant speed. Both the mean speed and cyclic variations in speed have to be controlled within limits much closer than those which can be tolerated for propulsion engines.

14.16. *Regulation of mean speed*

The design of electrical machinery is such that the e.m.f. (voltage) generated depends on rotational speed and field strength. The maintaining of the voltage between close limits is important and any deviations from the required value are immediately corrected by varying the excitation of the field electrically by means of the automatic voltage regulator (a.v.r.). It is nevertheless important for the sake of continued reliable operation that the designed rotational speed is maintained closely and this demands good governing. As electrical loads are applied or removed by the operation of switch gear the demands made on the engine and its speed regulation system are sudden and sometimes severe. The need for prompt and adequate response of the governor to load changes is obvious but it is not always realized that equally prompt governing is required when running with a steady electrical load. Consider the steady load to be represented by a constant resistance across the generator terminals. If a small increase in rev/min takes place the a.v.r. reacts to maintain the voltage constant and hence the power constant. As the same power is now being absorbed at an increased rev/min the absorbed torque must fall leaving an excess of supplied torque which will produce further increases in rev/min until the fuel rack is returned towards the closed position by the governor. This situation arises because the action of the a.v.r. system is generally very much quicker than the action of the engine governor system. It should be noted that in the absence of an a.v.r. the situation is stable as any small increase in rev/min would produce a corresponding increase in voltage, so that the power absorbed by a constant resistance would increase proportionately to the square of the speed. Torque absorbed would increase in proportion to speed giving an excess over the torque available so tending to restore the original speed.

14.17. *Cyclic variation of speed*

As well as close control of the mean speed by the governor the design of the generator set must avoid undue cyclic fluctuations in speed. The fluctuating torque output from every diesel engine results in a cyclic fluctuation in its speed. In the days when engines had few cylinders and ran at low speeds the corresponding cyclic variation in voltage could reach values which were unacceptable because of lamp flicker which was apparent to the human eye. In consequence limits were set to what was termed "Cyclic irregularity" defined as

$$\frac{\text{maximum speed} - \text{minimum speed}}{\text{mean speed}}$$

The values of cyclic irregularity were usually calculated from the turning moment diagram and the inertia of flywheel and generator rotor and assumed the shaft system including the crankshaft to be a rigid body.

The sensitivity of the human eye to flicker diminishes as the frequency is increased due to the persistence of vision, a phenomenon which, as is well known, has permitted the development of cinematography and a.c. lighting. Most modern auxiliary generator sets run at a high enough speed and have sufficient number of cylinders for the cyclic fluctuations to be at a much greater frequency than is troublesome and questions of cyclic irregularity now rarely arise. A modern approach to the calculation of cyclic fluctuation of the rotor would consider the flexibility of the shafting as well as the torque diagram from each cylinder and the inertias of the running gear, flywheel and rotor.

14.18. *A.c. generation and synchronous speeds*

The foregoing applies equally well to d.c. and a.c. generators but in the case of a.c. power generation the accurate maintenance of rotational speed is of even greater importance as not only the voltage but the frequency depends upon it. Frequency, in fact, depends solely on speed for a given design of alternator and is standardized usually at 60 or 50 cycles per second. 60 cycles is the more usual in marine application. During operation of a generator relative motion takes place between conductors and a magnetic field and one complete cycle of alternating current occurs as each pair of poles is passed by a conductor. An alternator with one pair of poles will, therefore, generate 60 cycles per second only if run at 60 revolutions per second, *i.e.* 3,600 rev/min; one with two pairs of poles must be run at 1,800 rev/min, and so on as listed in Table 14.1. Modern navigation equipment often requires precise control of frequency and hence very close governing of auxiliary generator sets.

TABLE 14.1

Synchronous Speeds

No. of pairs of poles	50 *Hz* *rev/min*	60 *Hz* *rev/min*
1	3000	3600
2	1500	1800
3	1000	1200
4	750	900
5	600	720
6	500	600
7	428	514
8	375	450
9	333	400
10	300	360
11	273	327
12	250	300

14.19. *Parallel operation*

Electrical power is required continuously and it is essential to have standby generating equipment available so that regular maintenance can be carried out or unexpected faults rectified without interrupting the supply.

The duties of some ships involve wide ranging demands of power and sometimes high loads are imposed for only comparatively short periods of time, *e.g.*, when handling cargo. To install a single set large enough to meet the maximum power requirements but to run it for most of the time at low loads would obviously be quite uneconomical. It is usual, therefore, for reasons of both reliability and economy to provide two or three auxiliary generator sets and in special cases there may be even more.

When it is required to start up a set and bring it on to share or take over a load already being carried by other generators, certain conditions and procedures must be observed. With d.c. machines it is necessary to ensure that the voltage of the incoming machine is exactly the same as that of the bus bars. This requires control of the speed at a steady value. With a.c. machines not only must the voltage and frequency be exactly the same as that of the bus bars, but the phase must be synchronized. Synchronizing gear is provided at the switchboard to indicate when the switches of the incoming machine may be closed and to achieve these conditions remote control of the generator set speed is often provided. Such controls usually take the form of motorized equipment fitted to the speed operating mechanism of the engine governors and operated by controls situated at the switchboard.

BIBLIOGRAPHY

Burman, P. G. and De Luca, F. *Fuel injection and controls,* Technical Press Limited, London.

Fuller, R. A. *Governing of diesel engines.* D.E.U.A., March 1958.

Welbourne, M. A., Roberts, D. K. and Fuller, R. A. *The governing of compression ignition oil engines.* Proc. I. Mech. E., 1959. Vol. 173. No. 22.

CHAPTER FIFTEEN

Surveillance Systems and Sequential Controls

15.1. *Control of temperature*

Control of the power output is a primary requisite but if satisfactory operation and optimum performance are to be achieved the various systems serving the engine must be controlled also. For example cooling water temperature should be maintained within close limits. By careful design and use of correct controlling components stable systems can be devised which require only periodic attention. The more complex the systems the more sophisticated are the controls required if human intervention at frequent intervals is to be avoided.

One of the simplest and yet most common control systems is that devoted to maintaining fluid at a constant temperature, particularly the temperature of the cooling water at outlet from the engine. From the smallest high speed engine to the largest of the medium speed engines the self operated temperature control valve is most frequently used for this purpose. This type of valve is commonly known as a thermostat.

An example is shown in Figure 15–1. It obtains its control action from

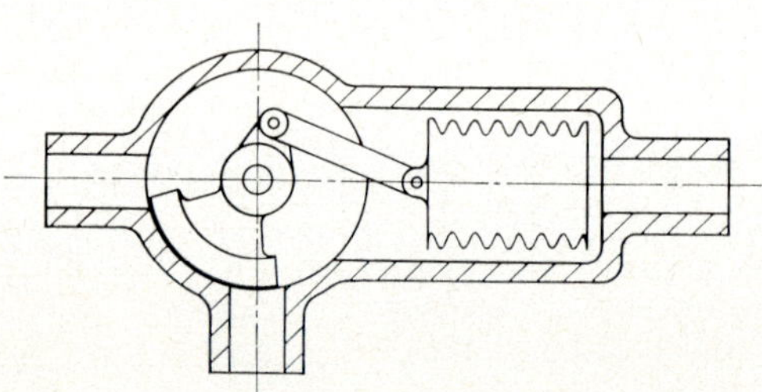

FIG. 15–1—*Principle of operation of thermostat*

the high volumetric expansion which occurs over the melting range of a specially prepared wax mixture. The wax is contained within a sealed capsule and as the temperature of the wax changes it expands and a diaphragm transmits motion to a plunger. This motion is transmitted via a linkage to a shutter which rotates within the valve body. The valve is basically a three-way cock and may be used as either a by-pass valve with the wax filled capsule sensing the temperature of the incoming fluid, or a mixing valve with the capsule sensing the temperature of the outgoing fluid. It operates only at the temperatures for which it is designed and cannot be adjusted. Also by its nature it must control the flow of the fluid, the temperature of which is being measured by the thermostatic element. In other words it can only be used for control of the primary fluid.

In performing its function it carries out a number of operations. It senses the temperature of the cooling water, it compares this with the demand temperature, which in this case is the melting range of the wax in the sensitive capsule, it signals the deviation by means of mechanical positioning of the plunger and by the same plunger it actuates the shutter which controls the flow of the fluid. These processes are fundamental and common to any control system maintaining a steady condition.

15.2. *Secondary fluid control*

If control of the secondary fluid is required, as outlined in Article 10.4, then these functions have to be carried out by separate components. There must be a temperature sensing element in the primary fluid, and this temperature must be compared with the temperature demanded, either by a controller or by the sensing element itself. A signal must then be sent to an actuator which operates the valve directing the flow of the secondary fluid. Such a system can, of course, be used in primary circuit control as well as secondary circuit control. The medium by which the signals are sent from a sensing element to the actuator may be electric, hydraulic or pneumatic and similarly the power supply of the actuator may be electric, hydraulic or pneumatic.

For temperature measurement there is a wide variety of sensing elements which includes liquid filled thermometers (*e.g.* mercury in steel), gas filled thermometers, resistance thermometers, thermistors and thermocouples.

Temperature sensing elements of all these types are normally mounted in a pocket in the pipe lines so that they can be replaced or repaired without necessitating a shut down. Temperature sensing elements should be sited close to the cooler or the engine outlet which it is desired to control in order to avoid transmission lag.

15.3. *Feed back*

For some purposes the signal sent to the actuator can be a simple on-off signal but in most cases a proportional control is necessary. In other words feedback is used to open the controlling valve more or less according to the amount and sense of the deviation from the required temperature. Redirection of the coolant by the actuated valve results in a correction to the temperature which the sensing element notices and ultimately the actuated valve takes up an equilibrium position.

15.4. *Sensitivity and Stability*

In sophisticated control systems controllers are used in which the ratio between the deviation and the output correction signal may be varied. This enables a choice to be made regarding the degree of sensitivity or coarseness of the control. Suppose a temperature controlling valve to be arranged as shown in Figure 10–2. (Either primary or secondary fluid control can be assumed). The valve opens or closes the passage to a by-pass and it is desired to control the temperature in the primary circuit at the sensing point to 70°C. Suppose the controller is set so that the by-pass passage is fully open at 65°C and fully closed at 75°C. If the engine is running light this control circuit will be endeavouring to keep the water hot by by-passing most of it; the valve will therefore be almost fully open and the temperature will be only a little above 65°C. Application of load to the engine resulting in more heat being added to the cooling water will raise its temperature causing the valve to close the bypass and to reach an equilibrium point approaching 75°C. It will do this fairly rapidly, perhaps passing through one oscillation, as shown in Figure 15–2(a). This is a relatively coarse rate of control and it may be desired to control the temperature within much closer limits. If the ratio between the signal and the response is altered temperatures much nearer to the desired level of 70°C can be produced and carrying out the same operation of putting load on the engine will produce a result as shown in Figure 15–2(b). A very sensitive and close control may produce a result as shown in Figure 15–2(c). Here the time taken for the system to settle down is very much longer and it will pass through a number of oscillations. This phenomenon is common to all control systems; a coarse control is stable and a finely sensitive control sacrifices stability. Both stability and sensitivity are desirable and the design of the control system must be such as to achieve an adequate compromise between sensitivity of control and stability.

15.5. *Offset and reset*

In all the three cases illustrated in Figure 15–2 it can be seen that there is a change in temperature level necessary in order to cause a response in the setting of the valve. This difference is known as off-set.

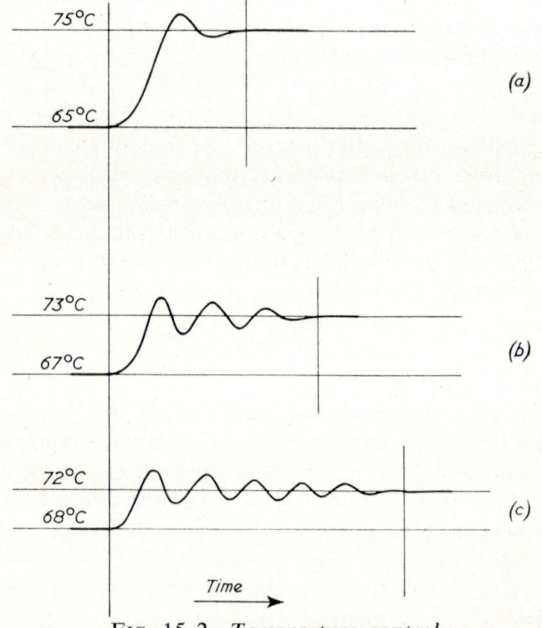

FIG. 15–2—*Temperature control.*

In the great majority of control situations for diesel engines, proportional control with moderate off-set is adequate. However, many control systems have a facility for adjusting the set point. If, after a change, such as the one just described, has taken place, the set point is readjusted then the controlled temperature can be made exactly the same as that which it was before the change of load on the engine. Sophisticated control systems can have built into them integral or reset action which will alter the set point in this way; thus they can combine the stability of coarse control with reset action to give both fine and close control. The reset action can be made at a fast rate or a slow rate and once again compromise is necessary because a fast rate leads to a less stable system.

So far the control system has been considered as applying to temperature regulation. Obviously the same principles apply exactly to the regulation of pressure and viscosity. Pressure sensing devices take the form of Bourdon tubes, strain gauged tubes or electronic transducers using capacity, inductive or piezo-electric effects. Viscosity sensing devices are based on the measurement of pressure difference across a capillary through which the fluid flows.

In the foregoing it will have been observed that the control processes are identical for the regulation of temperature or pressure and for governing, as described in Chapter 14. The history of development is responsible for differences in terminology.

15.6. *Remote operation*

With many small service craft, particularly docking tugs which are continually manoeuvring, it is advantageous to control the thrust of the propeller directly from the bridge. Remote control of reversing and speed was applied early to this type of vessel generally by means of mechanical linkages as the distances between control station and engine were not great.

The controls required were uncomplicated for a single uni-directional engine with oil operated reversing gear and could easily be combined on one lever on the bridge. The engine would be started, initially at least, in the engine room and then control transferred to the bridge: with the bridge control lever in mid position the gear would be in neutral and the engine running at idling speed. The first movement forward of the bridge control lever would cause the clutch in the ahead gear train to engage, sometimes accompanied by a small increase in fuelling rate (rev/min) to avoid stalling. Further forward movement would increase rev/min until full speed was reached. Similarly movement astern from the neutral mid position first engaged the clutch in the astern train and then increased the speed of the engine.

Rather more complex controls were needed in the case of direct coupled reversing engines. Processes of stopping the engine, repositioning the camshafts and restarting had to be designed for servo operation in order to reduce the effort to manageable proportions necessary for combination into a single lever bridge control.

In both cases skill and care was demanded of the master in going from ahead to astern in order to avoid stalling the engine, wrong direction running of the engine or burning out of clutches as has already been explained in Chapter 10. As the master ideally needs to be free from concern about the engine in order to devote his attention completely to the ship a requirement quickly grew for a system which automatically carried out engine manoeuvres on demand without danger of failing in any of these ways. The satisfaction of these requirements would also have the effect of making remote control attractive for larger vessels.

Systems have now been devised which carry out the reversing of the propeller in a number of steps following a logical sequence with delays where necessary and interlocks to prevent malfunction. These systems involve the transmission of small energy signals from the control station to the machinery with large energy actuation only at the points where mechanical operation of the machinery takes place. The switching of delays and interlocks in sequence are termed "logic systems" and they may be constructed to use hydraulic, pneumatic or electric or a combination of these means. The logic of a system is identical whether it is carried out by pneumatic, electronic or other means and the operation of a control system of this kind is often represented by a "logic" diagram.

The commencement of a logical sequence may be initiated manually or by a demand signal generated automatically. The former is invariably the case for manoeuvring; the latter is often used for such needs as starting up air compressors when the pressure in the receivers falls below a set level or to start up a generating set and bring it on to load in parallel with others when the demand for electrical power exceeds that which can be safely supplied by the sets already running.

15.7. *Logic systems*

Control logic systems are usually designed around a series of basic

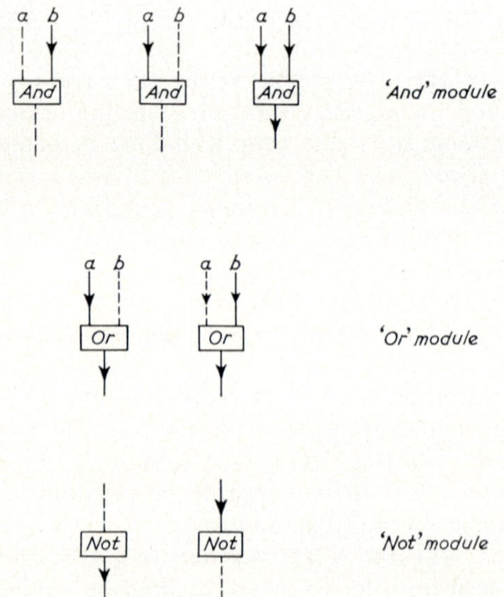

FIG. 15-3—*Basic "and", "or" and "not" modules.*

standard modules. Both systems and modules vary a little according to the manufacturing source. The examples given in the following paragraphs are typical and are intended to illustrate the principles of control system logic.

In Figure 15–3 are shown three typical modules. With the AND module input must be present at both (a) and (b) for the module to give an output. With either input signal missing there is no output.

With the OR module input at (a) OR input at (b) gives an output. The OR function used is not exclusive, thus if both inputs are present, there will still be an output.

The NOT module is sometimes known as an "inverter", due to its property of reversing a signal. Thus with no input present, an output is present; with an input present, there is no output present.

From AND, OR and NOT modules more complex modules can be built up.

A *SELECTOR MODULE* is shown in Figure 15–4 and is designed

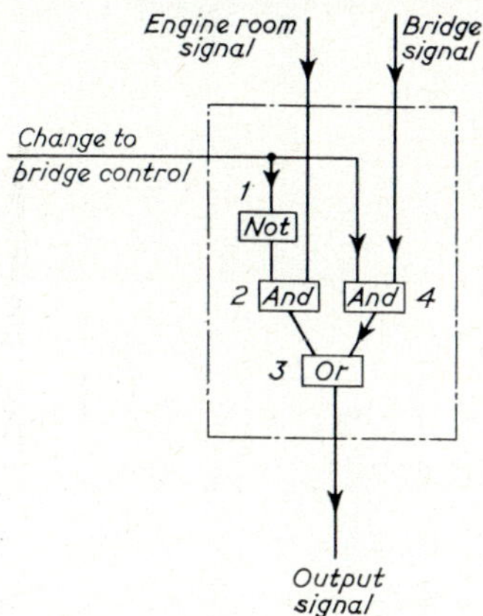

FIG. 15–4—*Selector module.*

to select and pass either bridge or engine room command signals. It will not allow both signals to pass. Several of these modules may be linked together and all operated by a common "change to bridge control" signal.

When a "change to bridge control" signal is applied to the module, it provides an input to AND module (4), but removes the input from AND module (2) by the action of NOT module (1). Therefore only a bridge signal may pass, via AND module (4) and OR module (3). Removal of the "change to bridge control" signal reverses the effect, allowing only an engine room signal to pass.

A *Delayed Build Up Module* (or simply DELAY) shown in Figure 15–5 is used mainly within other modules, but alone is used as a start signal delay in reversing engine systems, in order to allow the engine time to commence manoeuvring before the start signal reaches the engine.

When a signal is applied, there is a fixed delay before the output signal appears. Removal of the input signal immediately removes the output signal.

The *Delayed Cut-Off Module* is an example of the use of the delay module within another module. This module is used as a starting air preservation system, and is included in the start line of all the engines. Should the start signal be held on for too long, this module will automatically cancel the signal.

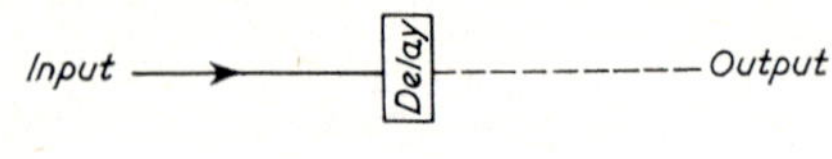

Delayed build-up module

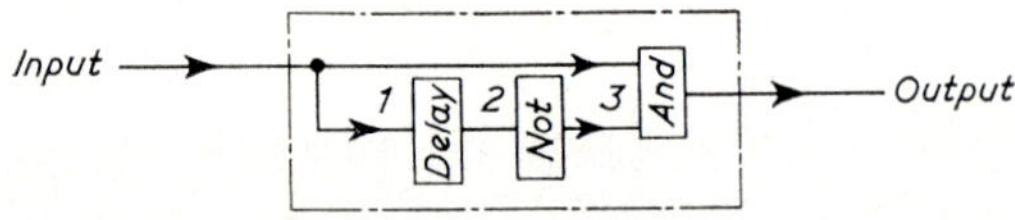

Delayed cut-off module

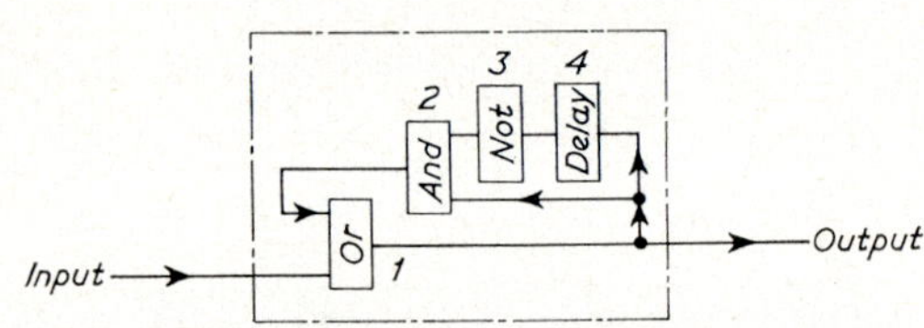

Extended hold-on module

FIG. 15–5—*Types of delay and hold-on modules.*

When an input signal is applied, there is an immediate output signal from AND module (3). An input is also applied to the DELAY module (1). When the delay expires, an output is passed to NOT module (2) which inverts, removing the signal to AND module (3). Thus the output from the module is removed. Removal of the input signal resets the module.

A further example is the *Extended Hold On Module* which is used to hold a signal on for an extended period, for instance to keep clutches disengaged during engine reversal.

When an input signal is applied it passes OR module (1) to give an immediate output. A feed also passes to AND module (2) which is already receiving its other input from NOT module (3). Thus AND module (2) supplies another input to OR module (1) which holds on the entire module giving an output signal. A feed also passes to DELAY module (4). When the DELAY expires, it gives an input to NOT module (3) which inverts, releases the holding action, removes the output signal, and resets the module.

A more complex module is the *Memory Module* shown in Figure

15–6. This may be used with reversing engines to ensure that the engines are supplied with either an ahead or astern signal, at all times whilst the command unit is moved to, or through the stop position.

An ahead signal passes OR module (1) to NOT module (2), which inverts giving no output. The ahead signal also passes OR module (3), giving an ahead output signal, and also a signal via OR module (4) to NOT module (5), which is thus held inverted. If the Ahead signal is removed,

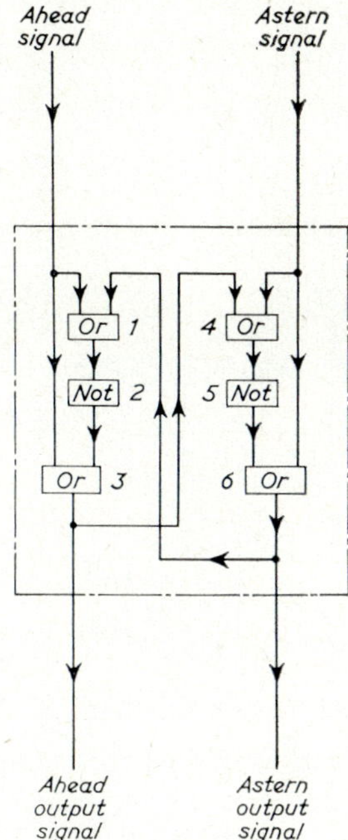

FIG. 15–6—*Memory module.*

the NOT module (2) inverts maintaining an ahead output signal via OR module (3). If an astern signal is now applied, the module is set in the opposite direction, and will hold an astern output signal as before. It is assumed that ahead and astern signals cannot be generated simultaneously.

Another module of more special application is the *Reversal Interlock Module*, Figure 15–7.

This module is to prevent accidental manoeuvring, as an opposite direction signal can only pass the module after deliberately applying a 'release interlock' signal.

Application of the release signal inverts NOT module (7), removing the inputs to AND modules (4) and (5), which now have no output to NOT modules (1) and (2). NOT modules (1) and (2) both invert providing inputs to AND modules (3) and (6). Thus either an ahead or an astern signal will pass to the module output.

If an ahead signal is applied it also passes from AND module (3) through OR module (12) to NOT module (10), and through OR module

(8) to AND module (4), and also through OR module (13) to NOT module (11). If the release signal is now removed NOT module (7) inverts giving an output to AND modules (4) and (5). AND module (4) thus gives an output to NOT module (2) which inverts, removing the input signal to AND module (6). Thus any astern signal is blocked.

If an astern signal is now applied it will be held out by the action of AND module (6) as mentioned above, until the release interlock signal is applied to invert NOT modules (1) and (2). As the astern signal passes, it

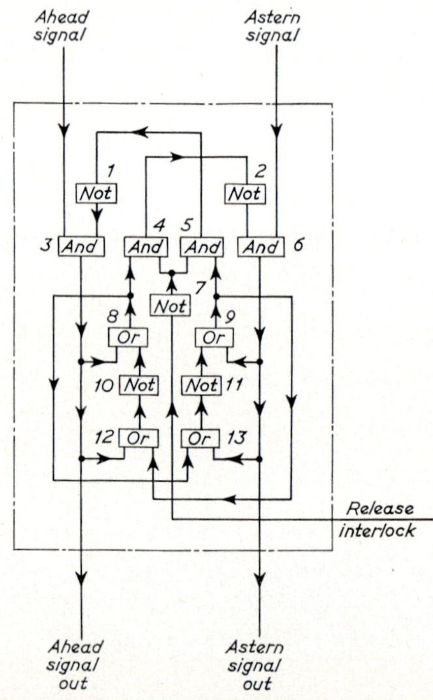

FIG. 15–7—*Reversal interlock module.*

will trip the interlock to the reverse direction. When the release interlock signal is removed, the interlock will be set for astern signals.

When the interlock is set in one particular direction (say ahead), ahead signals can be applied and removed at will, without affecting the interlock. When the ahead signal is removed, NOT module (10) inverts, giving an output via OR module (8) to hold the interlock as previously set.

Quite complex modules are built from the three basic modules, such as the *Automatic Starting And Clutch Control Module For Reversing Engines* shown in Figure 15–8 which sequences the timing and application of start and clutch control signals automatically.

The operation of the module is as follows:

In the stop position, the NOT modules (1) and (2) give outputs to the AND modules (3), (4), (5) and (6). If a run ahead signal is passed to

AND module (3) the output passes to AND modules (9) and (10), and module (10) is already receiving the other input signal from NOT module (14). The output from AND module (10) passes OR module (15) to give a START signal output.

When the speed of the engines rises above 40 rev/min ahead, a corresponding signal passes through OR module (8) to AND module (9), and to NOT module (1), which inverts, removing the input signal to AND module (4). Thus there is no input to AND module (10), and therefore the START signal output is removed.

If a run ASTERN signal is now applied to AND module (6), and the

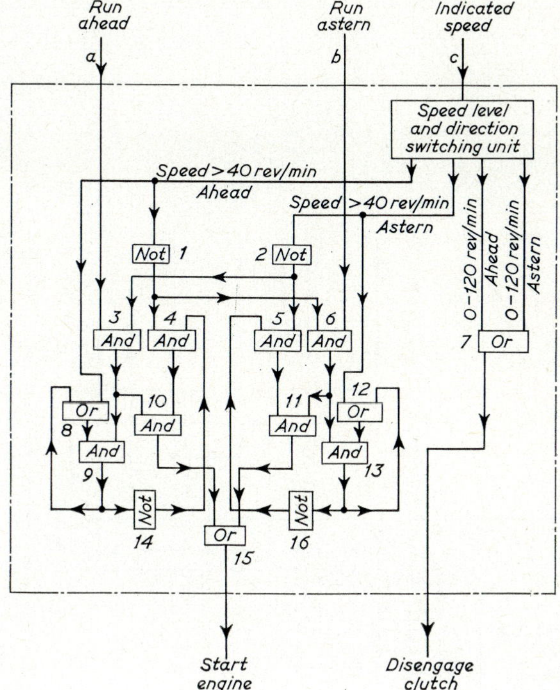

FIG. 15–8—*Automatic starting and clutch control module for reversing engines (one per engine).*

ahead signal removed, there is no immediate effect in the system. When the speed of the engines falls below 40 rev/min ahead, the corresponding signal is removed from NOT module (1) and OR module (8). NOT module (1) inverts passing a signal to AND module (6), which already has its other input present. AND module (6) passes a signal to AND module (11). AND module (11) already has its other input present, derived from AND module (5) and NOT modules (2) and (16), and therefore gives an output via OR module (15) as a START signal.

The engine will thus stop, and then start rotating astern. When the speed rises over 40 rev/min astern, the system sets itself in a similar manner to that described for going ahead.

In the event of the engine speed falling below 40 rev/min ahead or astern unintentionally, whilst the control lever is still in a run position (as may happen on loss of fuel, lubricating oil failure, etc.), the system is interlocked to prevent repeated starting attempts. This is done by maintaining a signal on NOT module (14) all the time the lever is in the ahead position (similarly for astern). This maintains a block on AND modules (4) and (10), and thus on the START output.

As an example of a complete control system Figure 15–9 shows the application to a single direct reversing engine in which bridge control is fully automatic, but engine room control is not.

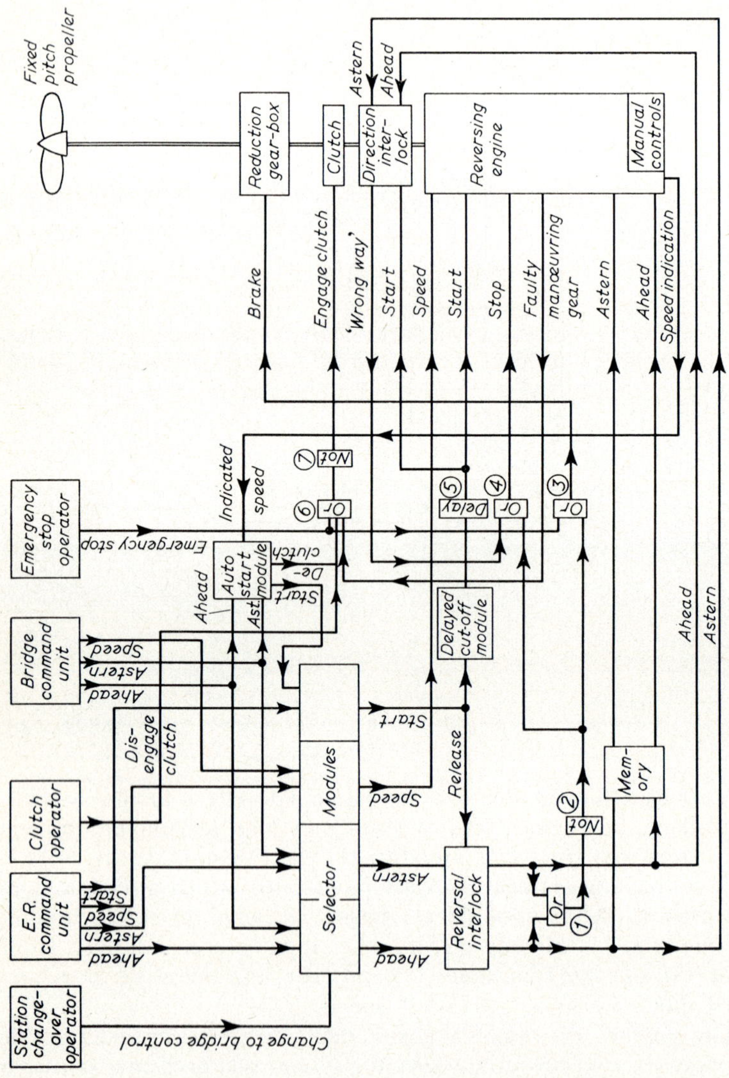

FIG. 15–9—*Modular logic diagram of a single reversing engine control system.*

When on Engine Room control, to start (assuming the previous running has been in the ahead direction), the engine room command unit is moved to the start ahead position. An ahead signal passes the selector module, the reversal interlock, and memory to the engine, maintaining the manoeuvring gear in the ahead position. An ahead signal also passes to the direction interlock with no effect, and via OR module (1) to NOT module (2), which inverts, releasing the brake and the stop signal to the engine.

The start signal passes the selector module to the reversal interlock, with no effect, and also via the delayed cut off module, and delay module (5) to the direction interlock and the engine. Thus starting air is admitted to the engine. The start signal also admits limited fuel to assist starting.

When the engine reaches 40 rev/min ahead under the influence of starting air, the engine room command unit is moved out of the start position to the desired speed setting. The start signal is removed from the reversal interlock which is thus set, and from the engine, cutting off main air and allowing the fuel racks to come under governor control. Engine speed is adjusted by movement of the command unit. When the engine speed rises above 180 rev/min, the clutch is engaged automatically by the auto start module.

If it is desired to run astern the command unit is brought to the stop position. The ahead and astern signals are removed from the reversal interlock, and thus from OR module (1), and NOT module (2) which inverts passing a signal to the brake, and to the engine stop line. The memory unit maintains an ahead signal on the engine to prevent the manoeuvring gear disengaging at speed.

When the engine speed falls below 180 rev/min the clutch is disengaged by the auto start module and when the engine speed falls to approximately 40 rev/min, the command unit is moved to the start astern position. The astern signal passes to the reversal interlock where it is blocked by the interlock setting until the start signal reaches the interlock, when it is released, allowing the astern signal to pass via the memory to the engine. The engine begins to manoeuvre, and in doing so sets up a block within the engine start system.

The start signal also passes the delayed cut off module to delay module (5), which is set to allow the engine time to manoeuvre.

The engine completes manoeuvring, and the delay module (5) expires, passing the start signal to the engine, and direction interlock.

However, the engine will still be in the final stages of ahead rotation, and thus the start signal applied to the direction interlock initiates a wrong way signal which passes via OR module (4) to the engine stop line thus cutting off fuel. The starting air which has been applied to the engine will act as a brake during the final stages of ahead rotation.

Immediately the engine stops and begins rotating astern, the direction interlock drops the wrong way signal releasing the stop signal to the engine. Fuel is then again admitted to the engine.

When the engine speed reaches 40 rev/min astern, the engine room command unit moves out of the start position to the desired speed setting. The start signal is removed, the reversal interlock resets, and the engine is allowed to come under governor control. When the engine speed rises above 180 rev/min, the clutch is engaged automatically by the auto start module. Engine speed may be adjusted by movement of the command unit.

When operating on Bridge Control the processes are fully automatic and it is only necessary to move the bridge command unit to the desired speed setting. The ahead or astern signals generated control the engine as described under engine room control, but the start signal is generated and controlled automatically by the auto start module.

Clutch control is automatic as previously described, and engine speed is again adjusted by movement of the bridge command unit.

A manual clutch operator may be provided to override the automatic control in an emergency and an emergency stop operator provided which passes a signal via OR module (6), to NOT module (7) to disengage the clutch and via OR module (3) to apply the shaft brake.

15.8. *Alarms and automatic shut down*

It is a short step from continuous indication of values of pressure, temperature or flow to signalling when a pre-set value for any of these parameters has been exceeded. Commonly an electronic signal is transmitted to a panel where it is caused to sound an audible warning and at the same time give a visual indication of the parameter concerned. Frequently a means is provided whereby the audible warning can be cancelled, the cancellation acting as an acknowledgement that the warning has been noticed. The visual indication usually remains until corrective steps have been taken. The most frequent use of this kind of alarm is in connexion with high temperature of cooling water and the low pressure of lubricating oil. In addition it may be extended to cover high temperature of lubricating oil, loss of pressure or loss of flow in cooling systems, exhaust gas temperature and fuel level in the service tanks.

The excessive deviation from normal levels of certain parameters may be arranged to initiate an automatic sequence to shut down the engine. When considering the advisability of installing such equipment the possibility of extensive damage to the engine consequent upon continuing to run it has to be weighed against the possibly more disastrous consequences to the ship of a sudden unexpected loss of power. The failure of the lubricating oil supply is so serious a matter for the engine and of itself likely to result in a sudden loss of power for the ship that it is general practice to arrange for automatic shut down if the lubricating oil pressure falls below a predetermined level. Often the circuits are arranged so that a warning is given when the lubricating oil pressure falls below one pre-set value followed by automatic stopping of the engine if the pressure continues to fall and reaches another lower pre-set value.

Sudden stoppage of one auxiliary engine may not be as serious as sudden stoppage of a main engine as the electricity supply to essential services is often shared by at least two generators. Automatic shut down sequences for auxiliary generators may therefore be arranged to follow from high coolant temperature or high lubricating oil temperature as well as from low lubricating oil pressure.

It is usual on engines of 300 horsepower and over to provide an overspeed trip arranged to shut down the engine if its speed should rise more than 15% above the full rated speed. If such an event occurs control of the supply of power has already been lost and no greater danger to the ship arises from shutting down the engine.

An overspeed trip is particularly necessary on auxiliary generator engines where the electrical load can be suddenly removed and on propulsion engines which may be disconnected from the propeller by operating a clutch or relieved of load by turning the blades of a controllable pitch propeller to zero pitch. It should be separate from the speed regulation governor and it usually operates by cutting off the fuel supply. A typical action is shown in Figure 15–10. It is a form of unstable governor as described in Article

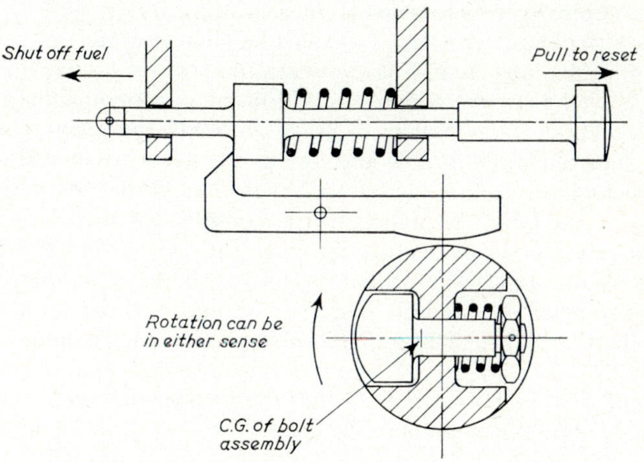

FIG. 15–10—*Overspeed trip—principle of operation.*

14.5. The centre of gravity of the bolt is eccentric to the shaft and its centrifugal force is counteracted by the spring until the designed speed of operation is exceeded. At this speed the bolt flies out striking the trigger and releasing the mechanism cutting off the fuel supply. To restore the fuel supply it is necessary to re-set the mechanism by hand thus prompting an investigation into the cause of the incident.

In the automatic operation of machinery having multiple engines geared together it is essential that any automatic shut down sequences should start by disconnecting the engine concerned from the gearbox by disengaging its clutch.

15.9. *Unmanned machinery spaces*

Obviously the combination of remote control, automatic operation, governing and the automatic maintaining of operating conditions enables machinery to be left unattended for considerable periods. The alarm systems indicating any faults that may arise, can be extended to give warnings in the bridge area or the "off duty" rooms of the engineers so that safety is ensured at all times.

Periodic attention is still necessary, of course, to carry out routine maintenance, daily and at other regular periodic times but this may be restricted to eight hour daytime watches during which all regular inspections are carried out. The machinery may then be left unattended for up to sixteen hour periods, but whatever the period the automatic systems, alarms and protection circuits and devices have to be substantially the same. Protection of the machinery spaces against accidents, particularly fire, is obviously necessary and Classification Societies have strict rules on the equipment that is necessary. In this present work protection of the engine itself is the only consideration.

When designing an installation for unattended operation there is a considerable temptation to include the largest possible number of alarms and it is well worth while giving careful consideration to the avoidance of unnecessary complexity which arises from a multiplicity of these devices. For example, a key parameter such as exhaust gas temperature can be used as a warning of high inlet manifold temperature or of faulty turbocharger operation as well as of the conditions within the cylinders, without either of these two other localities having to be monitored separately. The warning of high exhaust temperature would bring engineers to investigate and a routine checking procedure determined beforehand would quickly reveal the locality of the fault. An unnecessarily complicated installation will also result if an attempt is made to guard against two or more things going wrong at the same time. The probability of two faults developing at exactly the same instant is very remote and even should it occur an alarm raised by one of them is sufficient to initiate an inspection which should ultimately diagnose both.

It should also be borne in mind that the instruments are not necessarily more reliable than the machinery itself.

15.10. *Data logging*

The regular logging of data is a classical technique in the supervision of machinery conditions over a long period. It is an invaluable aid to maintenance and particularly useful in the prevention of breakdowns. Electronic data loggers have been developed which will record a large number of parameters automatically and most accurately. The use of such instruments is an adjunct to unattended machinery spaces as they will faithfully record the happenings at all times including the period when the installation is unmanned.

As in the case of alarms the monitoring of operating parameters can easily be overdone in the enthusiasm of applying data logging. A few critical points monitored regularly at reasonable periods in conjunction with the protective circuits are all that is required, otherwise an enormous amount of printed data is produced which can only be analyzed by very tedious searching.

Data logging equipment consists essentially of a means of scanning rapidly and continually a large number of parameters by means of analogue signals generated by various forms of transducer. These signals are converted into digital form for printing out. The equipment can embody a means for comparing the signals with pre-set limits and raising alarms if these limits are exceeded. A central data logger can be so used to take over the duties of all alarm systems. It is also possible and convenient to arrange for the print out to occur regularly at fairly long intervals, say every two hours but with an immediate print out if an alarm condition arises.

BIBLIOGRAPHY

CHESNUT, H. and MAYER, R. W., *Servomechanisms and regulating system design.* Chapman and Hall Ltd., London.

DICK, J. E. *The automation of diesel engine installations.* D.E.U.A. Publication No. 331. March 1970.

CHAPTER SIXTEEN

Some medium speed engines

16.1. *Allen Diesel Engines*

Allen Diesel engines are manufactured by W. H. Allen Sons and Company Ltd., Queens Engineering Works, Bedford, England, who have kindly provided the material for this section.

Production of Allen Diesel engines centres on two basic types which have similar construction features. They are suitable for both propulsion and auxiliary power uses. The larger is designated BCS37-E for in-line and VBSCS37-E for Vee form engines. A 12-cylinder engine is shown in Figure 16–1, it has a bore of 325 mm and a stroke of 370 mm and runs at

FIG. 16–1—*Allen 12-cylinder engine VBSCS 37-E.*

speeds up to 600 rev/min. The 16-cylinder engine, turbocharged and intercooled, is rated at 4000 bhp. The smaller type has a 9½ in bore and 12 in stroke; it is termed S12-D in-line and VS12-F Vee form, runs at speeds up to 750 rev/min and is available in pressure charged and also pressure charged and intercooled forms with 3, 4, 5, 6, 8, 12 and 16 cylinders. A 12-cylinder Vee is shown in Figure 16–2. The power range extends to 2800 bhp and a fuel consumption of better than 0·34 lb/bhp/h is achieved over a wide range of loads. (See Figure 16–3.)

A special "low profile" version of the 3- and 4-cylinder engines, as shown in Figure 16–4, is available for applications where headroom is at a

FIG. 16–2—*Allen 12-cylinder engine VS 12-F.*

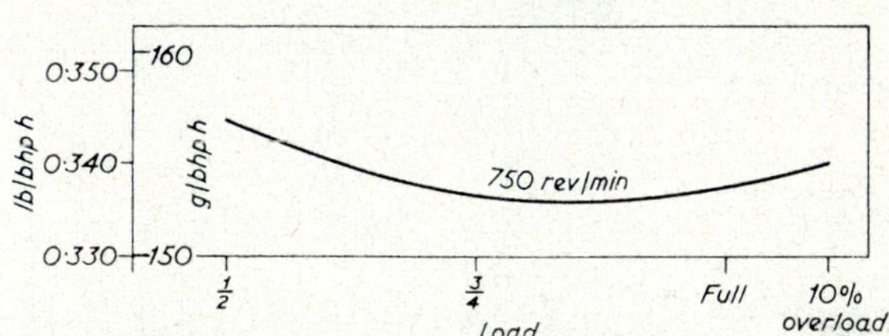

FIG. 16–3—*Fuel consumption, Allen S 12—D Engine.*

FIG. 16–4—*Allen "low profile" engine.*

premium with an overall height for piston withdrawal of 8 ft 1½ in. This design can also be supplied with electric or air motor starting equipment.

A detailed description of the in-line S12-D engine follows and a section appears in Figure 16–5. In all engine sizes the cast iron 'A' frame is of monobloc construction well ribbed to form a rigid structure. All working parts within the crankcase are accessible through doors provided at the front and back for each cylinder. The camshaft housing forms an integral part of the cylinder block casting. "Wet" type liners are fitted.

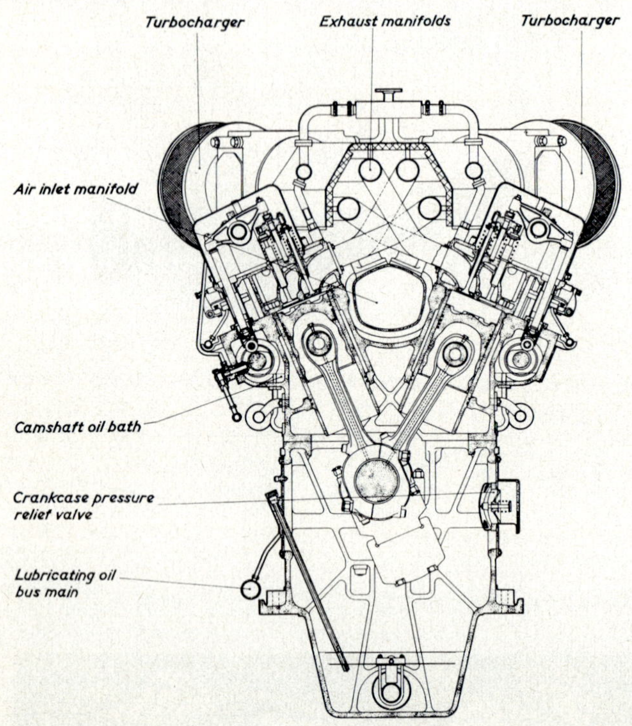

FIG. 16–5 (a) and (b)—*Section through Allen S 12—D engine.*

The deep section bedplate on which the engine frame is mounted embodies thin wall steel backed lined main bearings. An additional bearing is incorporated to carry the combined loads of the flywheel and part weight of the driven machinery. Axial location of the crankshaft is provided by renewable thrust rings. For marine auxiliary duty the bedplate design forms a wet sump to contain all the lubricating oil.

The crankshaft is a one-piece alloy steel forging incorporating balance weights where necessary. A solid half coupling forged integral with the crankshaft carries the flywheel to which the driven unit is coupled. Provision is made for the extension of the free end of the crankshaft to meet the requirements of special applications such as generator, bilge pump or compressor. Oil from each main bearing, fed through passages in the crankshaft lubricates the connecting rod bearings. Four cylinder, pressure charged engines having cranks at 180° are fitted with a secondary balancing gear as described in Chapter 6, article 6.6; it is shown in Fig. 16–6.

The main coupling bolts pass through the crankshaft half-coupling flange, flywheel and driven machine half coupling flange and two additional bolts are incorporated to retain the flywheel on the crankshaft when the driven machine is disconnected.

The H section connecting rods are of forged steel. A hollow boring in each rod conveys lubricating oil to the small end bush and gudgeon pin.

FIG. 16–6—*Secondary balancing gear.*

Each connecting rod has a detachable bottom cap. Prefinished thin wall steel backed bearings are fitted at the bottom ends and renewable bushes at the top ends.

The piston design provides large bearing areas and each piston is fitted with three pressure rings and one scraper ring above the gudgeon pin. The case hardened steel gudgeon pin is fully floating and is retained by circlips. The pistons are made of aluminium alloy with an integrally cast alloy iron carrier for the top piston ring.

The wet liners are made of close-grained cast iron. Synthetic compound sealing rings are fitted between the liner and frame to make a watertight joint on the lower land; a copper gasket is fitted on the upper land.

The camshaft is carried high in the engine frame and runs in its own oil bath. The forced feed supply for each camshaft bearing delivered through an oilway bored the full length of the camshaft, also maintains the camshaft oil bath level. A roller chain at the driving end of the engine transmits the drive from the crankshaft to the camshaft. An extension from

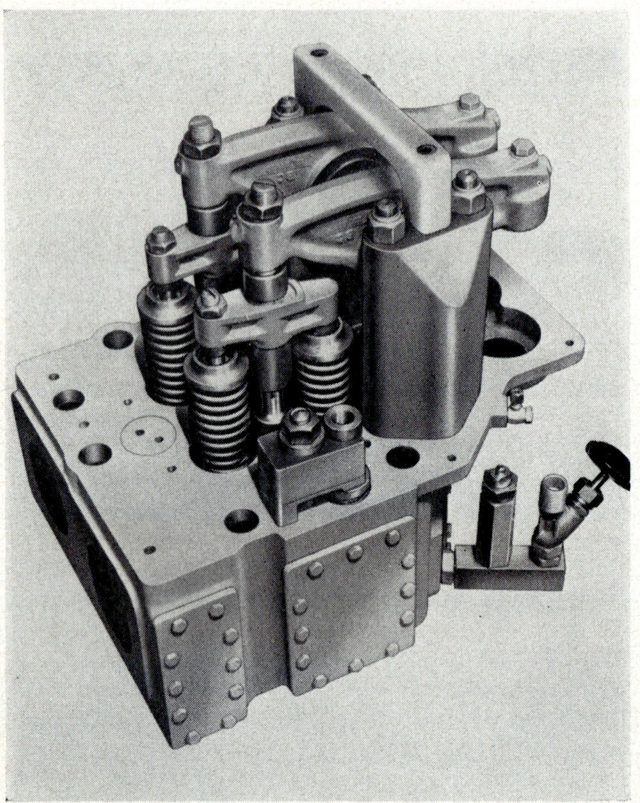

FIG. 16-7—*Cylinder head and valve rocker gear.*

the camshaft at the flywheel end drives both the governor and tachometer through a flexible coupling. One set of cam blocks (for the inlet and exhaust valves, starting air valve and fuel pump) is provided for each cylinder. The cam followers are of the guided roller type and operate the valve rockers by short push rods.

The separately detachable cylinder heads are made of alloy cast iron having a high resistance to thermal stress. The valve gear is totally enclosed and lubricated from the engine lubricating oil system. The centrally placed fuel injector is situated outside the valve covers so that fuel oil contamination of the lubricating oil is avoided. The valve gear is illustrated in Figure

16–7. In addition this feature enables the injectors to be withdrawn easily for servicing without disturbing the valve covers. Each valve has two springs. The valves seat direct in the cylinder heads on renewable seats of nickel iron alloy.

A separate camshaft-actuated fuel pump is employed for each cylinder. The fuel pump racks are linked through a control shaft to the engine governor which regulates the quantity of fuel injected into the cylinders according to the power output requirements.

In the Vee form engine the air and exhaust manifolds are both located in the centre to give maximum access to push rods, fuel pumps and cylinder heads.

The turboblower is mounted at the free end of the engine. The final exhaust flange can be positioned at an angle to suit a particular installation. An air intake filter silencer is normally fitted to each unit as standard, but provision may be made for ducting the air from a remotely located filter to suit the installation.

The intercooler comprises a matrix of finned tubes through which sea or raw water may be passed, in series with the lubricating oil cooler and engine heat exchanger.

One or two engine driven cooling water pumps may be fitted, as required. The materials used in their construction are stainless steel for the impeller shaft, gunmetal for the impeller and cast iron or gunmetal for the volute casing. Non-self-priming pumps are provided as standard but self-priming types may be incorporated when circumstances necessitate them.

A hydraulic governor is fitted. It incorporates a centrifugal speed sensing device controlling a suitably damped oil operated servo-cylinder through a pilot valve. Speed droop and load limit controls are provided as standard. Motor operated speeder gear can be fitted when remote control of the engine speed is required.

Whilst a variety of cooling systems may be adopted for marine auxiliary generating sets, the most commonly used is the simple closed circuit system. A tubular type heat exchanger is normally provided with each set to dissipate the heat from the fresh water system to the sea water. Sea water is passed through the intercooler, the oil cooler and then the heat exchanger in series flow. An engine driven fresh water circulating pump is usually incorporated in the system but the sea water pump may be either an independent unit or engine driven in tandem with the fresh water pump. The cooling systems may be arranged so that the sea water can be circulated through the engine jackets under emergency conditions. The details of the engine internal cooling arrangements will be apparent from the sectional arrangement in Figure 16–5.

The standard method of starting is by means of compressed air admitted to the engine cylinders at a pressure of up to 450 lb/in^2. The engine controls include a manually operated main starting air admission valve, which when opened under starting conditions admits a supply of air to the cylinders through individual cam operated timing valves and automatic non-return valves.

At entry to the engine fuel system the fuel oil passes through a twin compartment filter fitted with a changeover cock to enable one cartridge to be removed for cleaning without stopping the engine. In addition edge type filters are incorporated at the inlet to each fuel injector.

A forced feed system of lubricating is employed, the oil being contained in the engine bedplate. The oil pump is of the gear type, directly driven from the engine crankshaft. Oil is drawn by the pump from the bedplate and is passed through an oil cooler and then through a large full flow filter before being delivered to the lubricating oil busmain. A changeover cock permits the isolation of any one compartment from the operating circuit so that changing of the filter element can be carried out while the engine is running. Connexions from the pressure main are taken to the main, large end and camshaft bearings, governor bearings, camshaft driving chain and enclosed valve gear. Oil is also led up the connecting rods to the small end bearings and the pistons are splash lubricated. A hand priming pump is incorporated in the system to ensure that all parts can be supplied with oil before starting the engine and a tubular type cooler constructed with aluminium brass tubes and a cast iron body is normally provided.

For propulsion duties the engine employs a dry sump forced feed system of lubrication, the oil being contained in a separate tank. The oil pumps are of the gear type driven from the engine crankshaft. The pressure pump draws oil from the tank and passes it through an oil cooler and then the filter before delivery to the lubricating oil distribution main. After

FIG. 16–8—*Sheathed high pressure fuel pipe.*

circulating through the engine the oil drains into the engine bedplate where it is picked up by the scavenge pump and returned to the oil tank. Arrangements are made for changing over to a wet sump system for emergency operation.

The Allen engines are all adequately equipped for a completely "unmanned machinery space" requirement. The lubricating oil and fresh water systems are thermostatically controlled and sensing points are provided for comprehensive systems monitoring or data logging. To give protection against fire hazard Allen engines may be supplied with sheathed fuel injection pipes. In the event of failure of one or more pipes, the spill age of fuel is retained by a sheathing assembly over the pipe and transferred away from the engine to a duty tank. Here an alarm may be fitted to give warning of the failure. The cylinder on the left in Figure 16–8 shows the arrangement.

16.2. *Doxford "Seahorse"*

The Seahorse engine has been developed by the Doxford Hawthorn Research Services Limited, Doxford Engine Works, Pallion, Sunderland, who have kindly provided the material for this section.

This engine is of two stroke cycle, opposed piston configuration and is unique amongst medium speed engines in that the usual objectives of low height, compact power and light weight of the trunk piston design have been disregarded in favour of the ability to burn the heaviest fuels with complete reliability and to have the minimum number of cylinders in order to keep maintenance times and costs low. The resulting design follows closely the lines of the Doxford large bore engines and, as can be seen from the longitudinal and cross sections of the cylinder shown in Figure 16–9, it has a cross head and a piston rod sealing gland separating the cylinder skirt area from the crankcase running gear, so that there can be no contamination of the crankcase lubricating oil by combustion sludge. Leading technical particulars are given in the table.

Cylinders	4 to 7
Cylinder bore	580 mm
Piston stroke :	
Combined	1300 mm
Lower	880 mm
Upper	420 mm
Cylinder output	2500 bhp
Speed	300 rev/min
Mean lower piston speed	8·8 m/sec
B.M.E.P.	10·9 kg/cm^2
M.I.P.	12·0 kg/cm^2
Firing pressure	106 kg/cm^2
Piston area	2642 cm^2
Swept volume/cyl.	343·51 litres
BHP/cm^2 p.a.	0·473
BHP/litre	7·30

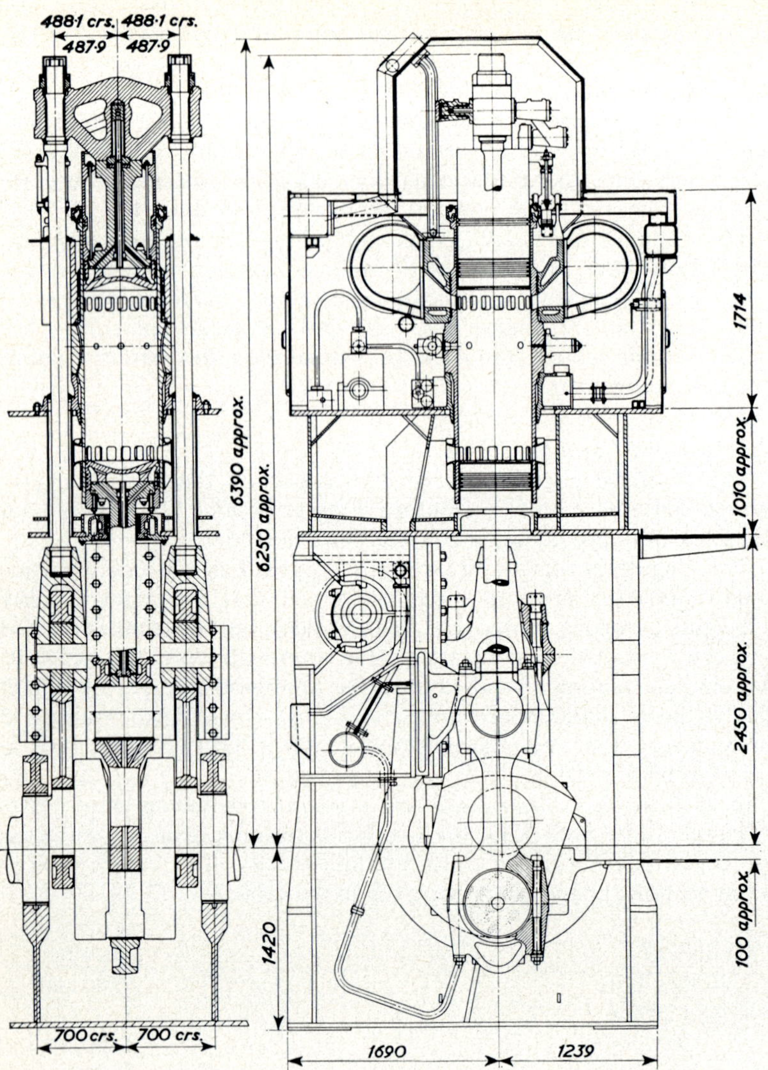

FIG. 16–9—*Section through Doxford "Seahorse" Engine*

Cylinder centres	1400 mm
Crankshaft main journal dia.	840 mm
Crankpin dia. (centre and side)	520 mm
Centre conn. rod length between centres	1520 mm
Side conn. rod length between centres	1300 mm
Exhaust crank lead	8 degrees.

The gas forces on the pistons and the internal inertia forces of the running gear are entirely contained by the centre crankshaft and side rod pin bearings, leaving the main bearings to carry only the weight of the running gear and a small amount of inertia force.

The frame is fabricated and incorporates cast steel diaphragms for supporting the main bearings.

The crankshaft is semi-built, from centre crank and pin forgings into which are shrunk bobbin pieces forming the side rod and main bearing journals. It will be seen from Figure 16–9 that the crank throw is offset from the centre of the webs to counteract the slight imbalance caused by the exhaust crank lead.

FIG. 16–10—*Side rods of "Seahorse".*

The running gear for each cylinder involves three sets of cross-heads and slippers. The upper ends of side rods are connected to the exhaust piston by a transverse beam consisting of a cast steel 'A' frame, whilst their lower ends terminate in inverted yokes which are attached to the cross-head slippers. The side connecting rods are two-armed steel castings with caps top and bottom tied by twin through bolts which also serve as the cap securing studs (Figure 16–10). The piston rod from the lower piston has an attached shoe which is secured to the crosshead pin. The centre connecting rod is of fixed centre design and the crosshead pin bears over its full length on the loaded side.

As will be seen from Figure 16–11, a train of four gears connects the crankshaft with a balance shaft extending along the back of the engine and running at crankshaft speed but rotating in the opposite direction. This is provided for engines of all cylinder numbers and results in virtually perfect balance.

The running gear is robust and relatively heavy, but because of the inherent good balance of the opposed piston arrangement and careful attention in the design, a running speed of 300 rev/min is achieved without

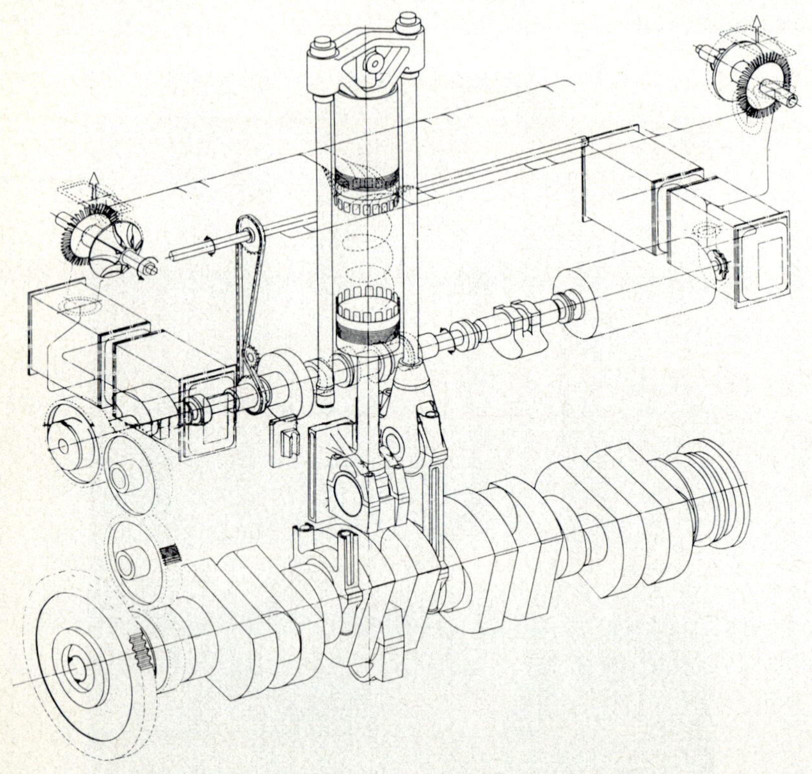

FIG. 16–11—*Arrangement of balance shaft and drive.*

difficulty. This results in the high output, for a medium speed engine, of 2500 bhp, per cylinder. As can be seen in Figure 16–12, the design is much smaller than a direct coupled engine of the same power; there is a considerable saving in space and the material saved in building the engine is sufficient to compensate for the additional cost of the necessary gearing.

The piston heads are single piece cast steel forgings. The upper piston is cooled by water which is supplied through swinging links connected by flexible joints. The lower pistons are cooled with oil which is supplied through channels formed in the crosshead guides and shoes and thence through passages in the crosshead and piston rod.

The cast iron liners are of single piece form with integral combustion chambers. Both the combustion chamber space and the exhaust port bars

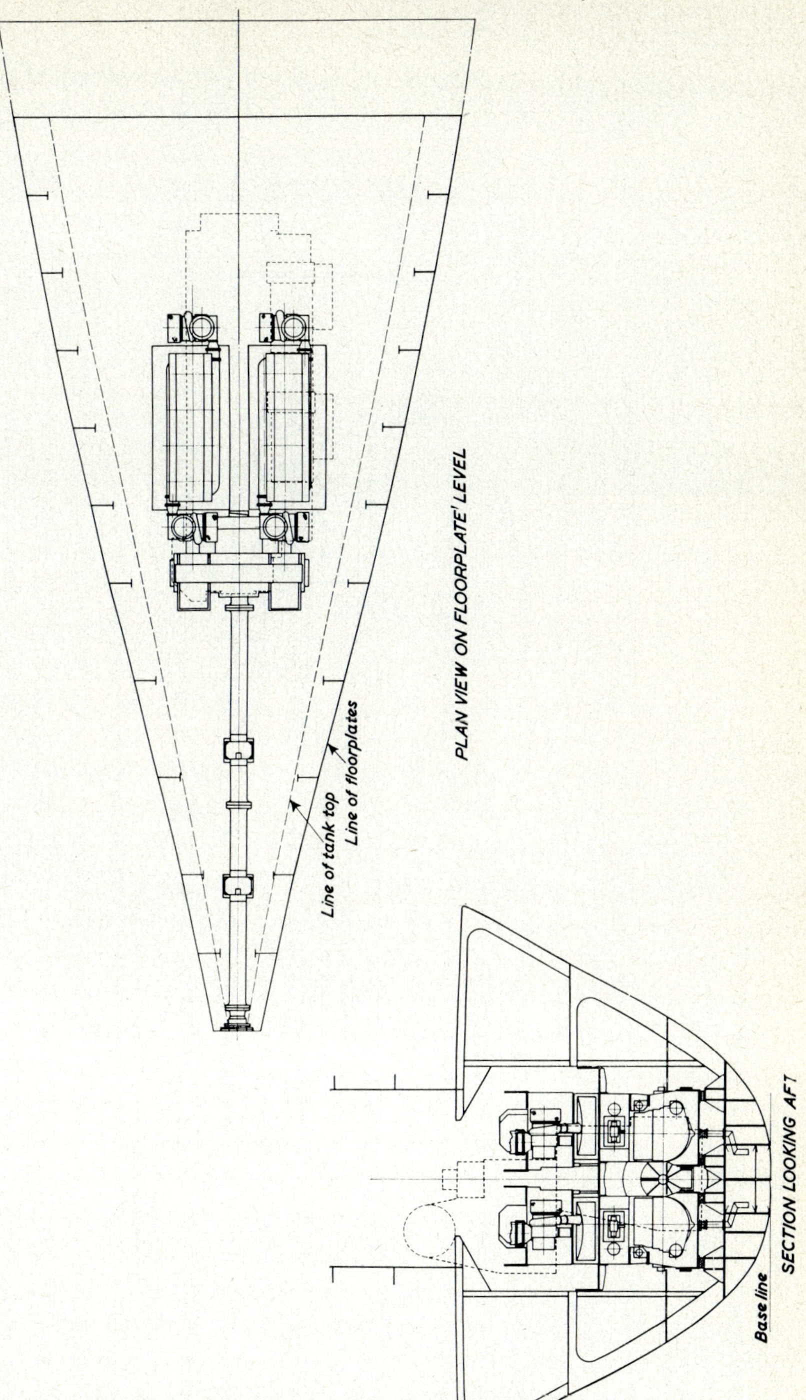

FIG. 16–12(a)—*Comparison between "Seahorse" and a low speed direct coupled engine—plan and section.*

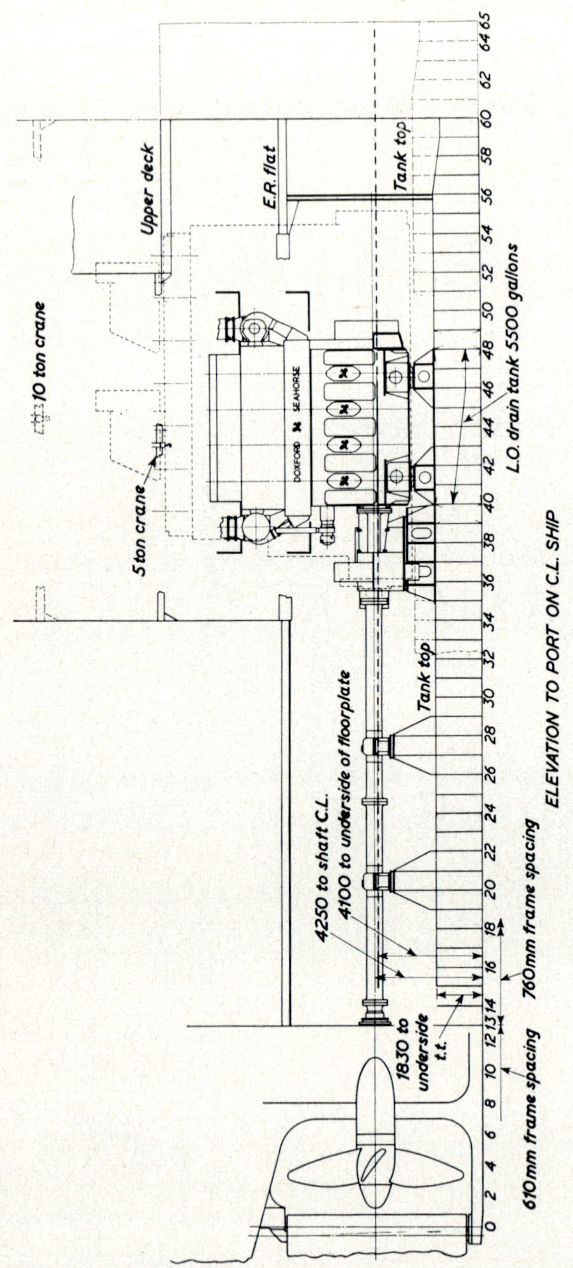

FIG. 16–12(b)—*Comparison between "Seahorse" and a low speed direct coupled engine.*

are intensively cooled by flow of water through drillings. There are eight lubricating oil points in the top and eight in the lower part of the liner, the lubricator shaft being driven by chain from the balance shaft.

The cylinder is charged by scavenge air entering through a ring of ports at the lower end and the exhaust is released by the ring of ports at the upper end.

The upper piston stroke is less than half that of the lower piston and in order to give favourable timing for exhaust blow-down followed by scavenging and charging of the cylinder, this upper piston is given a lead corresponding to a few degrees of crankshaft.

The combustion chamber has four injectors and the exhaust belts have two outlets leading to manifolds extending along both front and back of the engine, to a turbocharger arranged at each end.

Turbocharging is on the constant pressure system at a boost of about 2·4 atmospheres at full power. An auxiliary blower driven by gear train from the crankshaft and running at 22 000 rev/min delivers some 10 per cent of the total air required. This blower is driven through two Voith hydraulic couplings, one for each direction of rotation, mounted on the layshaft.

The structure has been designed to have very great bending and torsional stiffness to avoid any distortion in service due to ship working. This will permit mounting all sizes of the engine on four points only, using

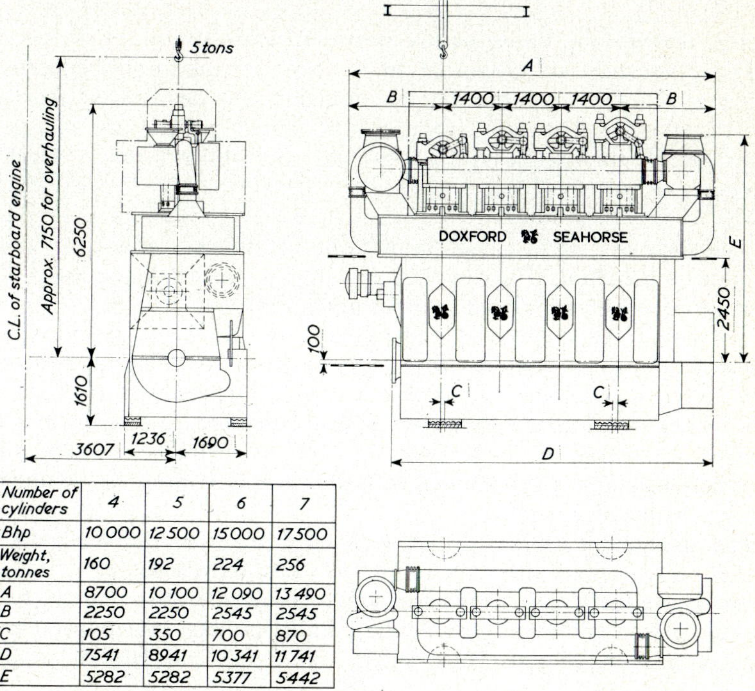

Number of cylinders	4	5	6	7
Bhp	10 000	12 500	15 000	17 500
Weight, tonnes	160	192	224	256
A	8700	10 100	12 090	13 490
B	2250	2250	2545	2545
C	105	350	700	870
D	7541	8941	10 341	11 741
E	5282	5282	5377	5442

FIG. 16–13—*Weight and overall dimensions of "Seahorse".*

a special mounting arrangement which allows for thermal expansion without misalignment. The overall dimensions are given in Figure 16–13 which also shows the position of these mounting points.

16.3. *M.A.N. RV/VV 52/55 type engine*

The RV/VV 52/55 type engine is manufactured by M.A.N. of Augsburg, West Germany who have kindly provided material for this section.

The M.A.N. RV/VV 52/55 is a high output medium speed engine which follows the design principles of the earlier RV/VV 40/54 type. This new and larger engine embodies the proven features of its predecessor which had accumulated a total of about 400 000 engine-hours of service at the time that the new engine was introduced. Development work has made it possible to make the new engine relatively more compact in respect of both weight and space. The following table lists specific engine data :

Bore	520 mm	20·47 in
Stroke	550 mm	21·65 in
Speed	430 rev/min	
Cylinder output	1000 hp	
Mean effective pressure	18 kg/cm²	256 lb/in²
Mean piston speed	7·9 m/sec	1552 ft/min
Power/weight ratio	8·5 kg/hp	18·7 lb/hp

A section through the Vee engine is shown in Figure 16–14.

The crankshaft is supported in a deep 'U' cast iron bedplate, the opening at the top, required for the insertion and bedding of the crankshaft, being closed by cast steel girders. These girders are secured by vertical and horizontal waisted bolts and the whole forms a rigid box section unit (Figure 16–15).

The crankshaft meets Classification Society requirements and has been classified for higher than the present output per cylinder to provide development potential. The bedplate also forms the crankcase and is fitted with large covers giving access to the running gear.

On top of the assembled bedplate are mounted the cylinder blocks which in the Vee engine form two separate banks extending over the entire engine length and which are secured to the framework by tie rods. This design results in an engine structure that is both flexurally and torsionally very rigid. Cast iron, the material chosen for the bedplate and the cylinder blocks, amongst other advantages, affords improved noise damping when compared with a steel structure.

The connecting rods are of articulated design and the designers state that their calculations show that this form results in a saving of 12 to 13 per cent in engine length compared with side-by-side rods. As can be seen from Figure 16–14, the design adopted permits pulling and re-installing the pistons without having to remove the connecting rod bearing from the crankshaft.

The cylinder liner has a large collar cast integral with its upper part.

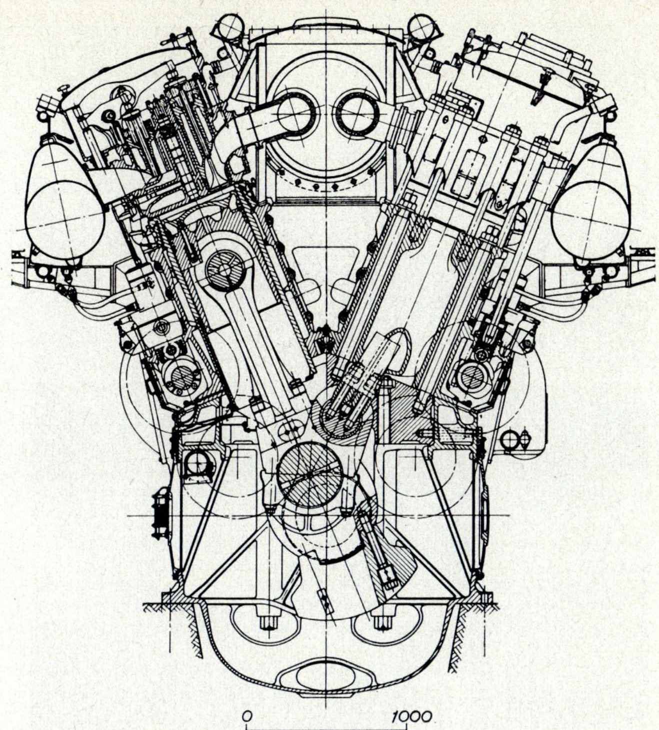

FIG. 16–14—*Cross section through M.A.N. VV.52/55 engine.*

This collar is designed to accommodate efficient cooling passages with the result that the temperature of the bore at the top piston ring reversal point is of the order of 160°C. The seating face of the cylinder liner resting on the cylinder block is at the lower end of the collar, an arrangement which completely avoids ovality caused by thermal deformation of the upper part of the liner.

The cylinder head design separates the mechanical from the thermal stresses, as far as is possible, by utilizing a comparatively thin combustion chamber top plate keeping the thermal stresses low, whilst the mechanical stresses are absorbed by a double cross formed by ribbing in the cool upper zone of the cylinder head. The cylinder head height to piston diameter ratio is almost 1·0 giving a stiff design (Figure 16–16).

All inlet and exhaust valves are in cages and exhaust valve cage seats are cooled (Figure 16–17), to maintain seat temperatures in a range which precludes the undesirable formation of vanadium and sodium deposits from the use of heavy fuel. The exhaust valves are fitted with rotators.

Cooling water and fuel are kept completely separate from the rocker arm lubricating oil. The water inlet and outlet of the exhaust valve cages are outside the oil compartment and fuel is supplied to the injector not from the top but from the side through the cylinder head.

Separate cylinder lubrication is featured, the lubricating oil being con-

FIG. 16–15—*Engine frame V6V 52/55.*

ducted through drillings in the liner wall from underneath so that it does not have to pass the hot portions of the cylinder liner where it could be in danger of carbonization or precipitation of additives. The use of this feature permits the supply of small metered quantities of fresh lubricating oil with characteristics to suit the fuel used. For instance, it is often sufficient to use a high grade mild alkaline oil only for cylinder lubrication. The cylinder lubrication is also of advantage in the event of failure of the lubricating oil supply.

The piston is of two-piece design incorporating a steel crown and a forged light metal trunk. The main advantages of this design are that only a small change in the piston clearance at the crown occurs between part load and full load due to the low heat expansion coefficient of steel, the excellent cooling of the ring belt and the possibility of using hardened piston ring grooves.

The crankshaft and connecting rod bearings are steel shells with centrifugally cast lead bronze and a thin electro-deposited overlay.

The temperatures in critical places are very low. The highest temperature on the crown of the piston is about 300°C, and the temperature in the top ring groove is only slightly above 100°C. Figure 16–18 shows the temperature pattern of an exhaust valve fitted with a valve rotator and it

FIG. 16–16—*Cylinder cover section VV.52/55.*

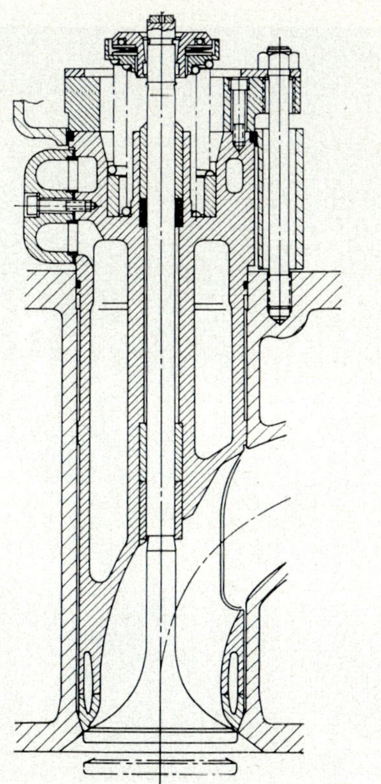

FIG. 16–17—*Exhaust valve VV.52/55.*

can be seen that it rotates at 4 rev/min. There are temperature fluctuations noticeable which depend on the position of the valve. Superimposed are fluctuations resulting from the valve operation cycle, which have a considerably higher frequency.

In the design of this engine great store has been set by easy accessibility to all components subject to wear, in order that the amount of maintenance work would be as low as possible.

Electrically heated necked-down bolts are used for mounting the cylinder heads. Warming up is controlled by a heating unit connected to

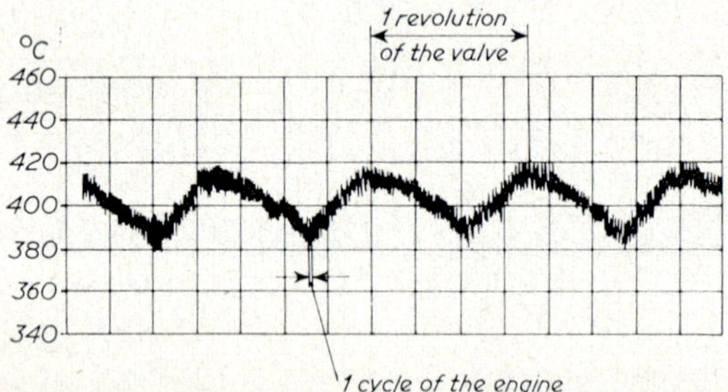

FIG. 16–18—*Temperature curve in the seat of the exhaust valve cone. VV.40/54.*

FIG. 16–19—*Fastening of cylinder cover bolts with heating pins VV.52/55.*

eight rods at the same time. The heating rods are inserted into the rifle drilled bores in the bolts and a pre-selected warming up time is set, after which the rods are automatically switched off (see Figure 16–19). The temperatures reached are far below the point of spontaneous ignition of any adjacent oils and well below temperatures at which changes could occur in the material structure of the bolt shanks.

The valve gear on the cylinder head can be taken off as a complete unit after the removal of only two nuts (Figure 16–20). The threaded studs which are then exposed are used for attaching a device which permits suspension of the cylinder head either horizontally or at an angle of $22\frac{1}{2}°$.

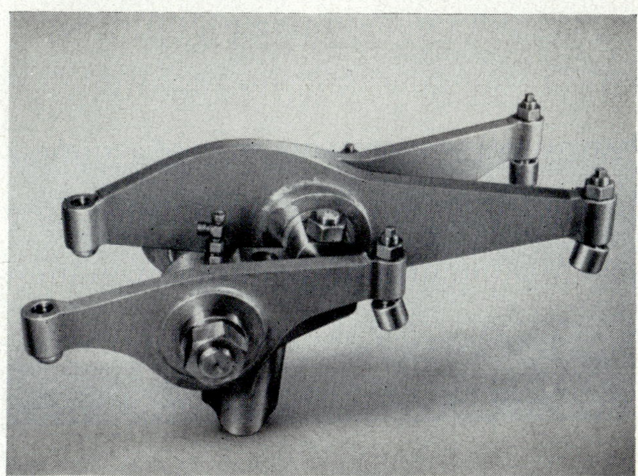

FIG. 16–20—*Valve rocker arm VV.52/55.*

As the inlet and exhaust valves are arranged in cages, the need to remove the cylinder head for inspection and repair of the valves is avoided. The flow of cooling water to and from the valve cages can be shut off so that it is unnecessary to drain engine cooling water when removing them. Should valve springs or valve rotators require attention, these parts can be exchanged without removing the valve, as a securing arrangement is provided to prevent it falling into the combustion chamber.

After removal of the connecting rod bolts with the aid of electric heating rods, the piston can be pulled (Figure 16–21). Before lifting the piston, a protective device is clamped to the connecting rod palm end. Similar to a cam, this device forces the connecting rod automatically into the correct position for removal relative to the cylinder liner, so that the contact surface is protected against damage (Figure 16–22). Furthermore, once the piston is out of the cylinder liner, this device prevents any deformation of the aluminium piston skirt caused by swinging of the connecting rod. The suspension device, which is screwed into the piston head, permits hanging of the piston either vertically or at the angle of inclination of the cylinder. It can remain attached to the piston while the

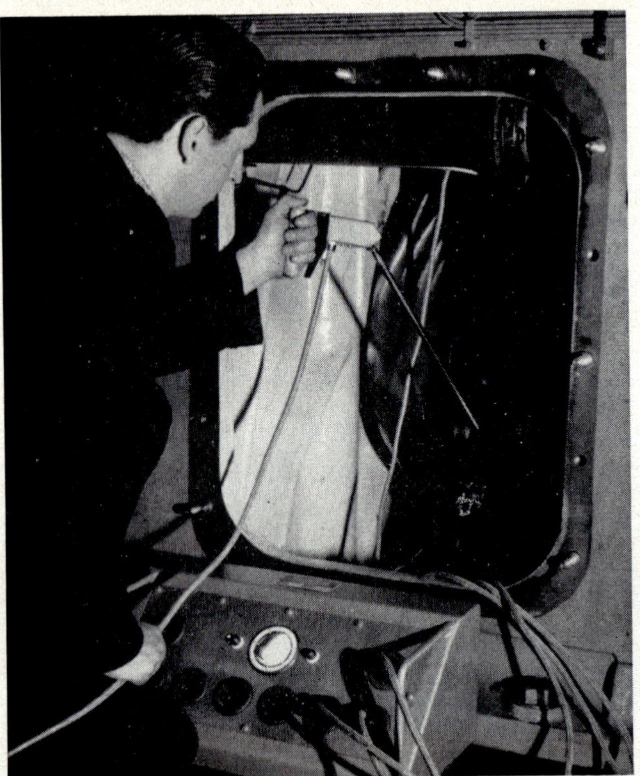

FIG. 16–21—*Fastening of connecting rod bolts.*

engine is being turned over. A guide rod, which can be screwed in, aids accurate and easy removal and installation (Figure 16–23).

Neither cylinder head nor connecting rod head need be removed to exchange the big end bearing shells as a special appliance is arranged to hold the pair of connecting rods out of the working area. Similarly, removal of piston and connecting rod is possible without disturbing the crankshaft bearing. This is of advantage as a well run-in bearing need not be removed unnecessarily. On account of connecting rod shaft and bearing being separated, the height required for removal of pistons and connecting rods is not inconsiderably lower, which is a further advantage. Auxiliary equipment has been designed which considerably facilitates the work to be carried out on the engine bearings, particular attention being paid to the transport of heavy components, e.g. bearing caps, from the crankcase. All these measures and maintenance facilities result in the times required for installing and removing essential engine components being short. Figure 16–24 shows a table of man hours required for a number of operations.

The turbocharging system has received special attention to minimize noise. The intakes can be equipped with efficient air silencers. In addition, the volute is encapsulated with sound proof lagging and at critical points of the charge air system absorption type silencers can be installed.

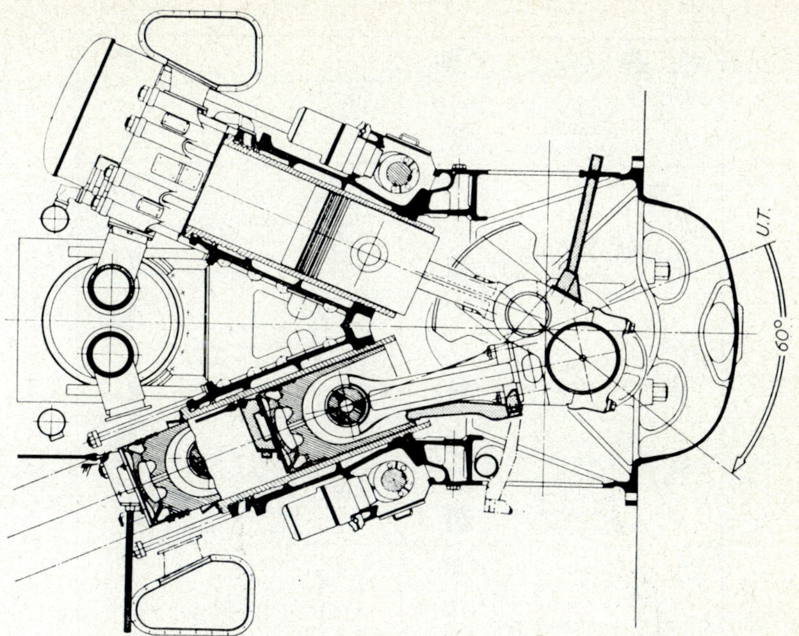

FIG. 16–22—*Removal of main piston.*

FIG. 16–23—*Inserting a piston VV.52/55.*

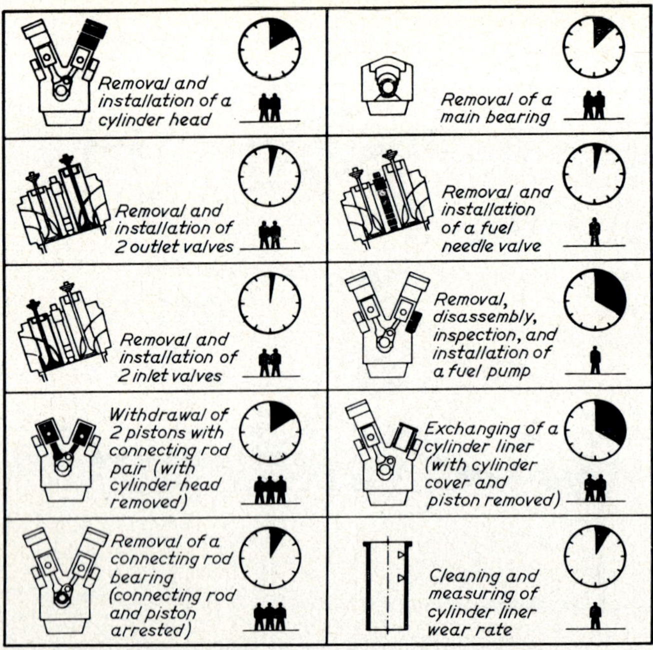

FIG. 16–24—*Man hours for maintenance work.*

16.4. *Mirrlees K Major*

The K Major range of engines is manufactured by Mirrlees Blackstone Limited, Hazel Grove, Stockport, Cheshire, who have kindly supplied the material for this section.

The Mirrlees K Major is manufactured in cylinders numbering 3, 5, 6, 7, 8 and 9 in line, 12, 14, 16 and 18 in 45° Vee form. It is designed to run on heavy fuel and to give a high power output from a well proven basic structure, the result of careful development of a conventional design.

The general appearance of the engine is shown in Figure 16–25 and its construction can be appreciated from Figures 16–26 and 16–27.

Technical details are given in the following table. It will be noticed that at the time of going to press the rating is being increased as a result of development. The manufacturers state that the scantlings of the crankshaft and all major parts are well in excess of the requirements of Classification Societies and potential exists for still further development in future.

The construction of the frame follows traditional lines, the crankshaft being supported by bearings formed in a bedplate. The bedplate is made of high grade cast iron and, as can be seen from Figure 16–27, is of deep section to resist internal bending moments and has integral longitudinal mounting girders embracing the seating bolt bosses. It carries a lubricating oil duct along its entire length through which the oil for lubrication and piston cooling is fed to each crankshaft bearing. The main bearings are of the thin wall type and consist of steel shells lined with copper lead with a thin lead-tin overlay. The main bearing caps are each secured by four bolts.

Bore	15 in	381 mm		
Stroke	18 in	457·2 mm		
		Stage 1 1970	Stage 2 1971	Stage 3 1972
Continuous rating				
BHP per cylinder		470	530	600
Speed rev/min		530	600	600
B.M.E.P. lb/in^2		220	220	250
kg/cm^2		15·5	15·5	17·6
Piston speed ft/min		1590	1800	1800
m/sec		8·1	9·15	9·15

FIG. 16–25—*Mirrlees KV. Major engines.*

The crankshaft is machined from a steel forging which is hardened and tempered to provide a wear resistant yet tough surface on all crankpins and journals. Drilled holes feed the lubricating oil from the main bearings to the large end bearings.

The upper half of the crankcase is formed by the column illustrated in Figure 16–28. This is a cast iron box-like structure which also incorporates the camshafts and timing gear. Inspection doors are provided on both sides for access to the running gear and on one side these doors carry

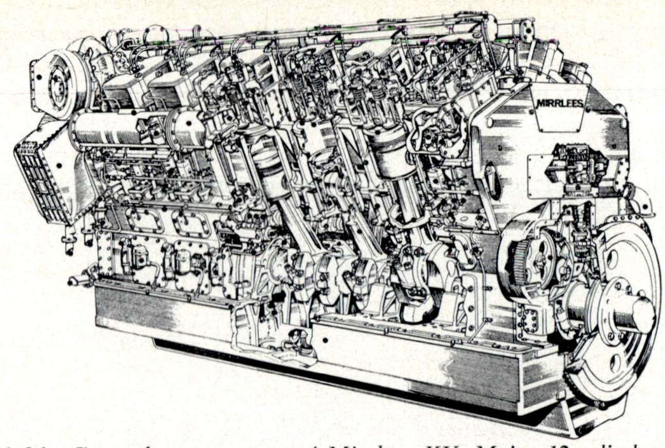

FIG. 16–26—*General arrangement of Mirrlees KV. Major 12-cylinder engine.*

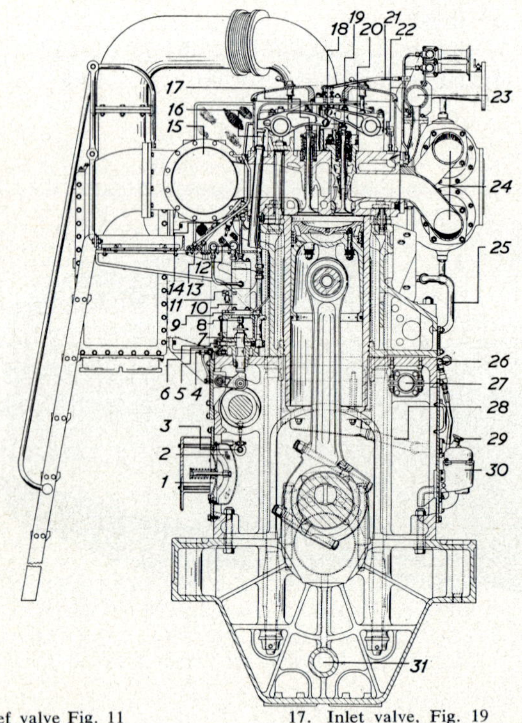

1. Crankcase relief valve Fig. 11
2. Gauze flame trap
3. Camshaft lub. oil manifold
4. Cam follower gear lub. oil manifold
5. Fuel drain pipe
6. Starting air manifold
7. Starting air controller, Fig. 18
8. Fuel pump tappet, Fig. 18
9. Fuel drip tray
10. Fuel injection pump, Fig. 38
11. Fuel pump control shaft
12. Fuel oil manifold
13. Valve gear lub. oil manifold
14. Pressure indicator cock
15. Air inlet manifold, Fig. 43
16. Push rod, Fig. 17
17. Inlet valve, Fig. 19
18. Link for exhaust valve levers, Fig. 19
19. Fuel injector, Fig. 39
20. Exhaust valve "Rotocap", Fig. 22
21. Injector nozzle cooling pipes
22. Exhaust outlet cage cooling pipes, Fig. 23
23. Water outlet manifold
24. Exhaust pipes, Fig. 44
25. Water pipes to turbocharger
26. Lub. oil manifold to filters
27. Lubricating oil pipe
28. Piston cooling oil drain pipe
29. Piston cooling oil thermometer, Fig. 8
30. Lubricating oil by-pass filter, Fig. 12
31. Main lubricating oil gallery, Fig. 8

FIG. 16–27 (a) and (b)—*Sections through Mirrlees K Major engine.*

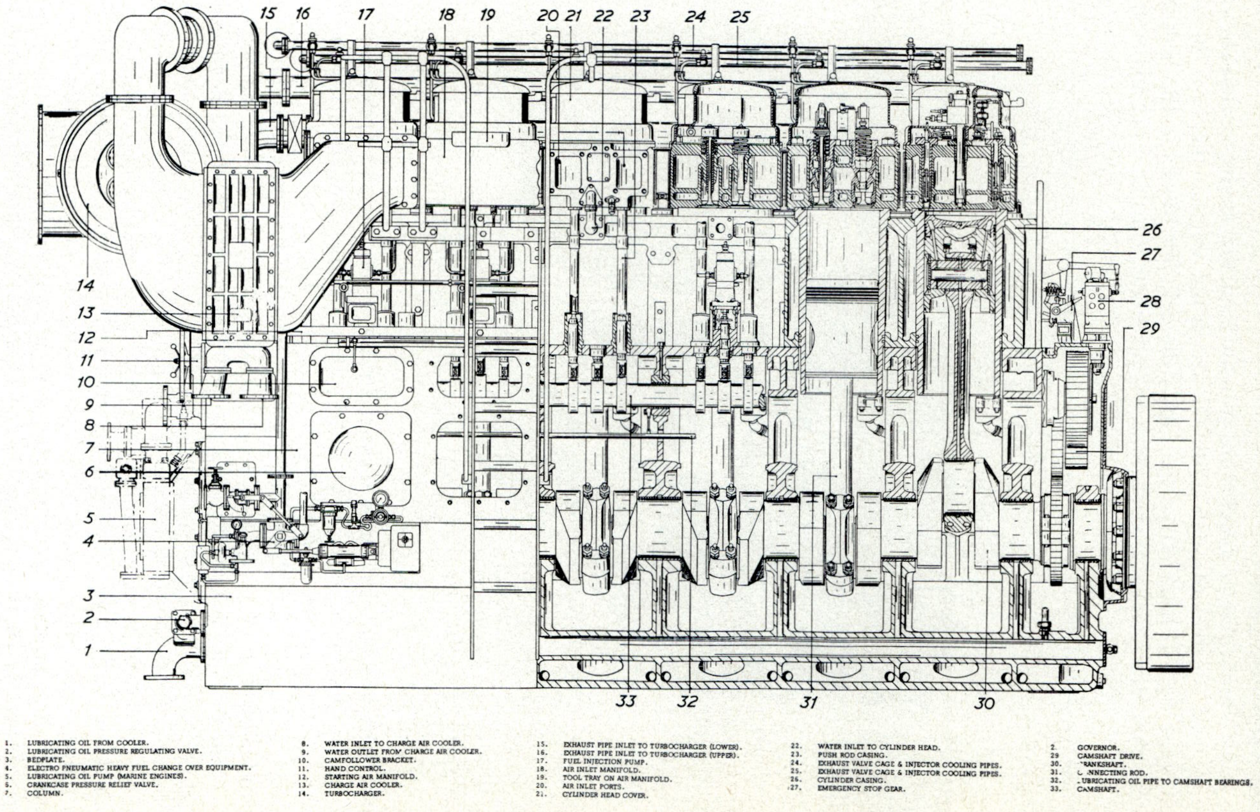
1. LUBRICATING OIL FROM COOLER.
2. LUBRICATING OIL PRESSURE REGULATING VALVE.
3. BEDPLATE.
4. ELECTRO PNEUMATIC HEAVY FUEL CHANGE OVER EQUIPMENT.
5. LUBRICATING OIL PUMP (MARINE ENGINES).
6. CRANKCASE PRESSURE RELIEF VALVE.
7. COLUMN.
8. WATER INLET TO CHARGE AIR COOLER.
9. WATER OUTLET FROM CHARGE AIR COOLER.
10. CAMFOLLOWER BRACKET.
11. HAND CONTROL.
12. STARTING AIR MANIFOLD.
13. CHARGE AIR COOLER.
14. TURBOCHARGER.
15. EXHAUST PIPE INLET TO TURBOCHARGER (LOWER).
16. EXHAUST PIPE INLET TO TURBOCHARGER (UPPER).
17. FUEL INJECTION PUMP.
18. AIR INLET MANIFOLD.
19. TOOL TRAY ON AIR MANIFOLD.
20. AIR INLET PORTS.
21. CYLINDER HEAD COVER.
22. WATER INLET TO CYLINDER HEAD.
23. PUSH ROD CASING.
24. EXHAUST VALVE CAGE & INJECTOR COOLING PIPES.
25. EXHAUST VALVE CAGE & INJECTOR COOLING PIPES.
26. CYLINDER CASING.
27. EMERGENCY STOP GEAR.
28. GOVERNOR.
29. CAMSHAFT DRIVE.
30. CRANKSHAFT.
31. CONNECTING ROD.
32. LUBRICATING OIL PIPE TO CAMSHAFT BEARINGS.
33. CAMSHAFT.

crankcase explosion valves. The design allows for the inspection and maintenance of main bearings and large end bearings without disturbing other main component parts.

Cast iron cylinder blocks are mounted on the engine crankcase and are secured by steel through bolts. On the in-line engines the steel through bolts pass right through from the top of the cylinder blocks to the bedplate where they are located at each side of each main bearing. Each cylinder is

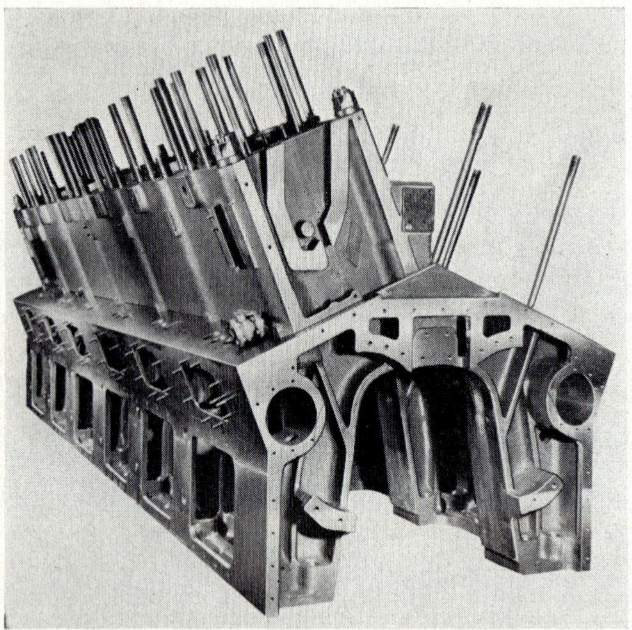

FIG. 16–28—*Cylinder casing and column, KV Major.*

fitted with a cast iron liner located by the flange at the upper end and sealed at the lower end with rubber rings which permit expansion. The liners and block are cooled by fresh water which is introduced at the lowest point of the block, the outlet being at the highest point of the cylinder head.

Side-by-side connecting rods on each crankpin are of the fixed centre type machined from steel stampings of 'H' section. The bottom end bearing housing is split obliquely to enable the piston and connecting rod assembly to be removed through the cylinder bore. The bearing cap is secured by four alloy steel bolts, the bearings being thin shell type, similar in construction to the main bearing. Oil for cooling the pistons is conducted up a bore in the centre of the connecting rod, a small quantity of it being used for lubrication of the top end bush. This bush is made from aluminium alloy which, along with other good properties as a bearing material, has the virtue of resisting corrosive attack from the additives in modern lubricating oils.

Individual cylinder heads are provided for each cylinder. They are made of close-grained cast iron and are deep in relation to the cylinder bore to provide rigidity. A deck plate is cast internally a short distance above the flame plate constraining the cooling water to flow rapidly over the latter, so intensifying the cooling effect. The joint between the cylinder head and cylinder liner is by a copper ring seal and each head is secured by eight alloy steel studs. There are two inlet valves and two exhaust valves in each head.

The two exhaust valves are mounted in cages which can be removed for maintenance without disturbing the cylinder head. These cages have water cooled seats and spindle guides and also the valve spindles are fitted with valve rotators to enable heavy fuel to be burned without incurring unduly short maintenance intervals.

The flow lubrication of the valve guide takes advantage of this rotation by using the valve itself as a timing device. Two flats are provided on the valve spindle which periodically line up with oil inlet and outlet drillings in the guide as the valve rotates. The linear positioning of the flats only allows the oil to pass when the valve is open and thus the oil space around the valve is only pressurized when the exhaust pulse pressure is present in the valve cage gas passage, the oil acting as a seal against gas penetration up the stem and the exhaust pressure preventing leakage of oil from the guide. This intermittent pressure lubrication of the valve stem makes it possible to use a very small clearance between the stem and the guide bore without any risk of valve sticking and this helps the heat transfer from the valve to the water cooled guide. In addition, the danger of stem or guide bore corrosion at low load running conditions is avoided.

The fuel injector is located in the centre of the cylinder head and has a water-cooled nozzle. The valve rocker gear is enclosed by an individual cover on each cylinder head designed so that the injector and high pressure fuel pipes are external to the cover to avoid any possible fuel dilution of the lubricating oil (Figure 16–29).

Cooling water is supplied to the exhaust valve cages and fuel injectors from a separate fresh water circuit.

Electron beam welding is used in the construction of the coolant passages of both the water cooled exhaust valve cage seat and the water cooled injector nozzle. The use of this technique has permitted the design of neat passages close to the surfaces to be cooled without any possibility of water leaks.

The pistons are of two piece construction designed on the lines described in Chapter 5 article 5.10. and illustrated in Figure 5–8. A well cooled heat resisting alloy steel crown, shaped to minimize thermal and mechanical stresses, carries all the compression rings in a suspended ring belt and is mounted on a cast iron skirt which carries the oil control rings and the gudgeon pin.

Individual fuel pumps are provided for each cylinder. Each pump is mounted on its own tappet block and actuated by a cam roller attached to a fulcrum lever. Fuel is provided under pressure to the inlet side of the pump and flows round an annular cavity in the pump body and out to a return manifold. Heated heavy fuel is thus continuously circulated and an even temperature of the pump is ensured under all conditions.

A hydraulic servo type governor is driven from the camshaft drive gears and can be provided with either manual or pneumatic speed control. An entirely separate overspeed trip governor is also fitted to the engine.

One exhaust gas driven turbocharger is used on an in-line engine and one on each bank for a Vee engine. These can be mounted at either the

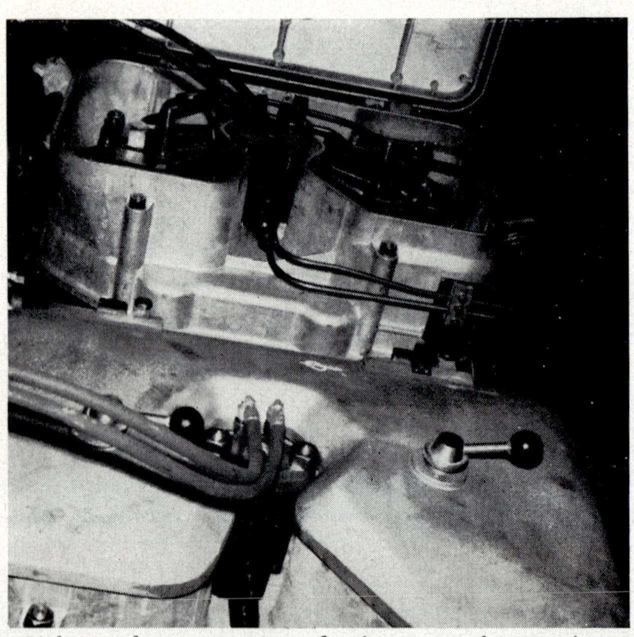

FIG. 16–29—*Valve rocker gear covers showing external connexions of fuel and cooling water pipes to injectors, K. Major.*

forward or the aft end of the engine and can be provided with a combined air intake filter and silencer or, alternatively, connected to air inlet ducting. Air from the turbocharger compressor passes to the intercooler and then to the engine intake manifold.

On eight cylinder in-line engines and on each bank of sixteen-cylinder Vee engines the exhausts are grouped in four pairs of two cylinders firing at 360° intervals. However, the four exhaust systems do not enter the turbine by the usual four entry casing but are combined into two streams by "pulse converters". Each pulse converter consists of a 'Y' piece connector in which the internal passages are carefully proportioned so that the pressure energy in each pulse is almost wholly converted into kinetic energy as it passes through. As the amplitude of the pressure wave is greatly reduced beyond the junction, it is unable to enter the complementary exhaust system to interfere with the scavenging of other cylinders. The kinetic energy from the conversion of the pulses is added to the total energy of the exhaust gas leaving each pulse converter. The two streams enter the turbine through a two entry casing.

The engine is started by admitting compressed air to the cylinders. For uni-directional engines the air supply is timed by the fuel pump tappet. In the case of direct reversing engines, starting air timing is controlled by interaction between the fuel pump and exhaust valve tappets.

Reversal is accomplished by withdrawing the cam followers by means of a linkage which can be seen in Figure 16–27, sliding the camshaft axially to bring into operation the appropriate air and exhaust cams and then replacing the followers. The fuel pump cams are symmetrical and the same cams are used for ahead or astern running. Interlock arrangements are made to prevent fuel being injected until the engine is turning by starting air in the desired direction of rotation.

The engine can be operated from a control position at the forward end or from a remote control panel. The manoeuvring gear is pneumatically operated but mechanical means are provided to enable the operation to be effected in the event of an air failure.

The engine operates on a dry sump lubrication system with forced feed to all-important running gear. Separate motor driven pumps are employed. One pump acts as a scavenge pump drawing the oil from the bedplate and delivering it to a drain tank, the other pump feeds the engine pressure system, drawing oil from the tank and passing it through lubricating oil filters and coolers to the engine.

16.5. *Nohab Polar F Engines*

The Nohab Polar F range of engines is manufactured by Nohab, Trollhättan Sweden who have kindly provided the material for this section. The engines are also manufactured in Great Britain under licence by British Polar Engines Limited, Glasgow.

The manufacturers state that this engine has been designed with a wide range of purposes in mind. It is used for power generation, ashore and afloat, for the propulsion of small vessels and, when geared in multiple engine installations, for the propulsion of large vessels. It is a 4-stroke cycle, trunk piston engine of 250 mm bore and 300 mm stroke. At 750 rev/min the mean piston speed is 7·5 m/s. It is at present marketed to carry a bmep of 14·2 kg/cm^2 (200 lb/in^2). The engine is produced as an in-line version from 3 to 6 cylinders and in a Vee version of 8, 12 and 16-cylinder units.

A Vee 12 engine is pictured in Figure 16–30 and cross sections of the in-line and Vee versions are shown in Figures 16–31 and 16–32. Figure 16–33 shows performance curves for the engine.

The cylinder block and crankcase form a one-piece casting constituting the frame of the engine. As a result of development work to meet the high bmep's now offered, this casting is made in S.G. iron. There are no tie bolts, the walls of the casting being designed to carry the gas forces.

The crankshaft is supported by underslung main bearings and is fitted with counter weights and a torsional vibration damper if these are necessary. The bearings are steel backed and lined with lead bronze with a thin layer of lead-tin-copper electro-deposited.

The connecting rods are drop forged toughened steel, the large end

FIG. 16–30—*Nohab Polar-F 12-cylinder Vee engine.*

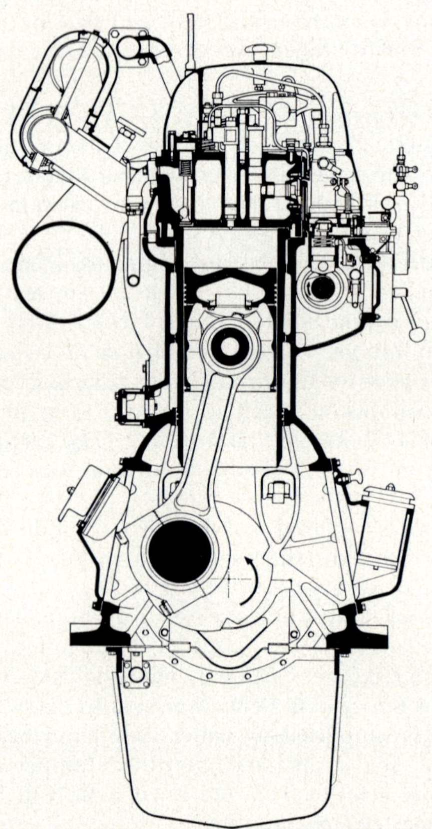

FIG. 16–31—*Section through F type in-line engine.*

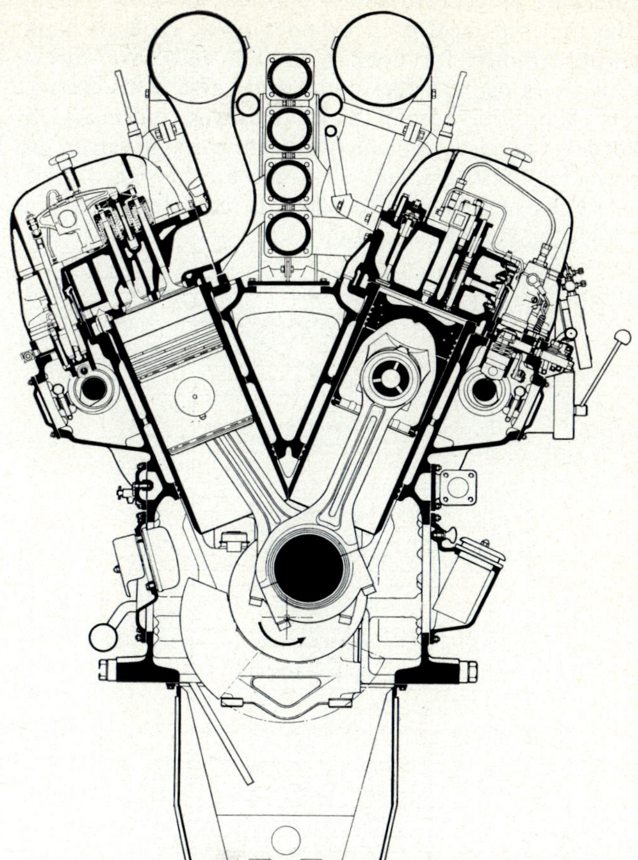

Fig. 16–32—*Section through F type Vee engine.*

bearings being split obliquely to facilitate dismantling of the shells and withdrawal of the pistons.

The pistons are made of light metal alloy with cast in coils for cooling. Four compression rings and two oil scraper rings are used. The top two rings are chromium plated and the topmost groove has a cast iron insert. The bearings in the piston are made of tin bronze and pressure lubricated.

For each cylinder there is a separate cylinder head made of cast iron. Two inlet and two exhaust valves are fitted for each head and all four valves are identical and made of high alloy steel with valve seat and top stellited. The valve seats in the cylinder heads are replaceable.

The cast iron cylinder liners are clamped at the flange at the top between the cylinder head and the cylinder block. The lower end is sealed by rubber rings allowing expansion of the liner. In addition to being supported at the top and at the bottom, each liner has a ring of circumferential supports about its mid-length so that it makes contact with the cylinder block at three points, giving it increased rigidity.

The camshaft is gear driven supported in white metal lined steel bushes at the upper part of each cylinder block on each bank. The gear-wheels are rubber lined for noise reduction and have surface hardened teeth. The followers of both fuel pump and valve cams are in the form of roller tappets. The motion from the valve cam is transmitted by push rods to the rocker levers and yokes which operate pairs of valves. All the mechanism is pressure lubricated and all joints are hardened. The light alloy rocker box covers are shaped to separate completely the fuel injection equipment from the valve mechanism.

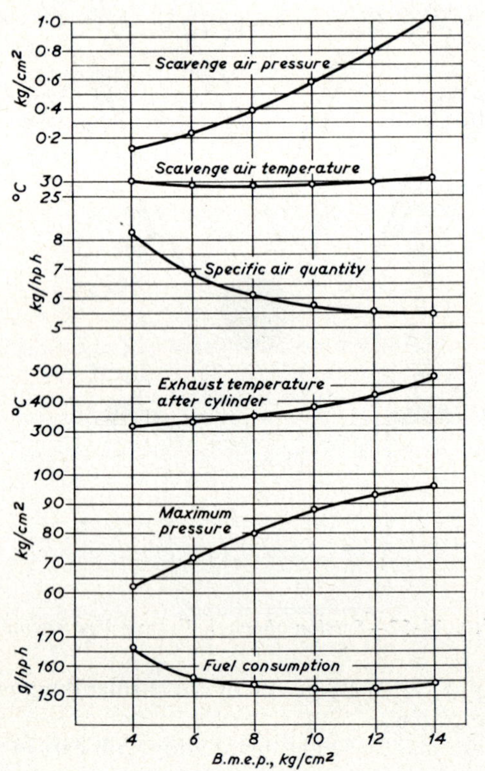

FIG. 16–33—*Performance curves, 6-cylinder F-engine at constant speed 750 rev/min.*

The fuel injector needles are set to open at 4000 lb/in² pressure and, for use at high b.m.e.p., a constant pressure unloading delivery valve system has been developed to avoid secondary injections and risk of cavitation. The injector nozzles are uncooled for distillate fuels and cooled for residuals. The F engines are normally running on distillate fuels but development testing has been carried out with residual fuels up to 3500 Red. 1 and the engine is offered to run on heavy fuels up to about 1000 Red. 1. The camshaft follower gear operating fuel pumps is robustly designed to withstand the loads on pumping the viscous fuel at about 75°C as the normal fuel temperature for 200 Red. 1. The problem of vanadium deposits on exhaust valve seats is met by cooling the seats directly and at the same time arranging the turbocharger match to give a high throughput of air. Furthermore, valve rotocaps are fitted. These measures maintain the valve seat temperatures at about 430 to 450°C on full load, which experience has shown to give a sufficient margin to avoid deposits and subsequent cracking and burning.

In order to maintain crankcase conditions satisfactorily alkaline, a lubricating oil having a T.B.N. of about 30 should be used for heavy fuel service. Replenishment of the oil consumed will then result in a level of about 15 T.B.N. being attained in continuous operation. The engine is fitted with bypass centrifugal type lubricating oil filters to keep the content of insolubles below the desirable 2 or 3 per cent. The main lubricating oil filter, the lubricating oil cooler and thermostat are separately mounted, as also is the fresh water cooler and thermostat.

A Brown-Boveri turbocharger and air filter are fitted, together with a GEA charge air intercooler. The lubricating oil pump, fresh water pump and sea water pump are arranged to be direct driven from the engine.

The engine is started by compressed air which is admitted to the cylinders via a distributor. The engines are uni-directional, not arranged for direct reversing, but can be built to rotate in either direction.

An hydraulic governor with pneumatic or electric remote control is used for speed regulation.

16.6. *Ruston Paxman RK Engine*

Ruston Paxman Diesels Limited, Lincoln, England, a management company of English Electric Diesels Limited, are responsible for the design and manufacture of a number of medium speed engines of various types covering a power range from 750 bhp to 8000 bhp. They have kindly provided the material for this section which describes the RK engine series of the range.

The engine designated RK has a bore and stroke of 10 in and 12 in respectively and is manufactured in Vee form with 8, 12 and 16-cylinder configuration and as a 6-cylinder in-line engine. The engine operates on a pressure charged four-stroke cycle and is available in both charge cooled and non-charged cooled versions. Technical details are given in the following table :

Bore		10 in	254 mm	
Stroke		12 in	304 mm	
Rev/min		900		
Piston speed		1800 ft/min	9·15 m/s	
B.M.E.P.		205 lb/in²	14·4 kg/cm²	
Firing pressure		1400 lb/in²	98 kg/cm²	
No. of cylinders	6 in-line	8 Vee	12 Vee	16 Vee
Maximum continuous rating bhp	1320	1760	2640	3520
Specific fuel consumption lb/bhp/h	0·356	0·351	0·344	0·346

Overall length	13 ft 0 in	9 ft $10\frac{3}{4}$ in	12 ft $8\frac{1}{4}$ in	18 ft $2\frac{3}{4}$ in
	3962 mm	3016 mm	3867 mm	5556 mm
Overall width	4 ft 6 in	5 ft 6in	5 ft 6 in	5 ft 6 in
	1372 mm	1676 mm	1676 mm	1676 mm
Overall height	8 ft $10\frac{1}{2}$ in	7 ft $8\frac{1}{2}$ in	7 ft $8\frac{1}{2}$ in	7 ft $8\frac{1}{2}$ in
	2705 mm	2350 mm	2350 mm	2350 mm

A longitudinal section of the engine is shown in Figure 16–34 and a

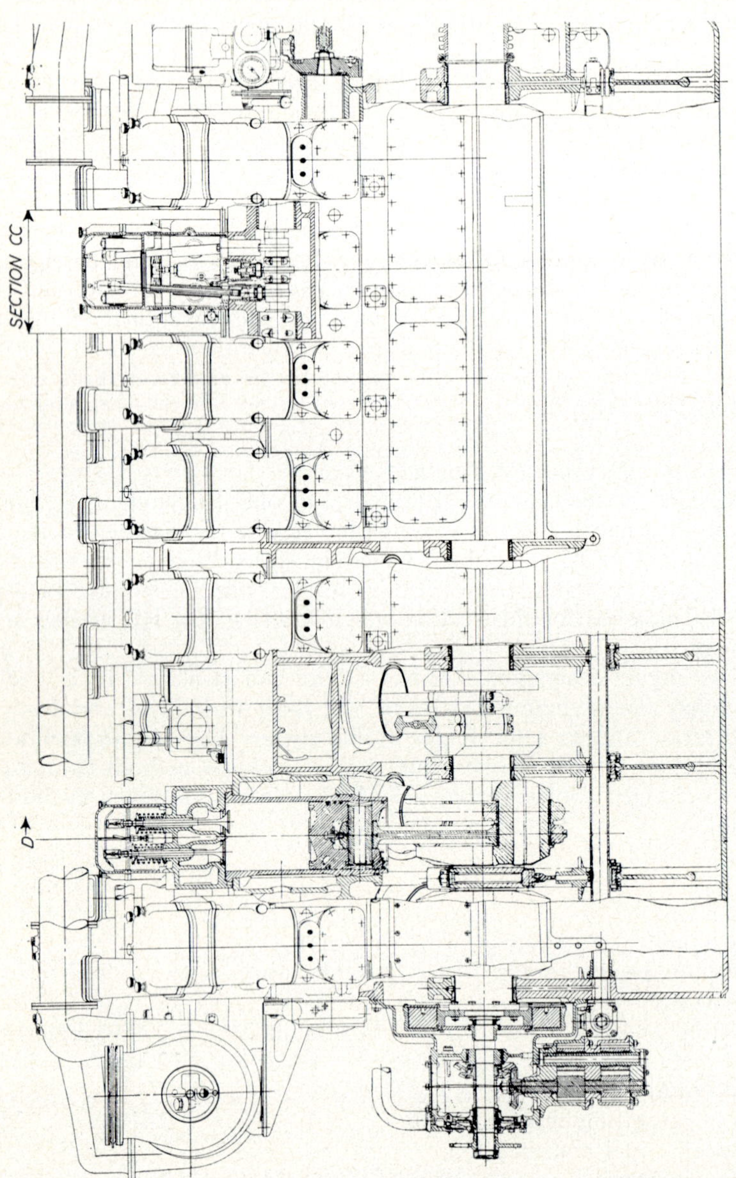

FIG. 16–34—*Longitudinal section through Ruston Paxman RK engine.*

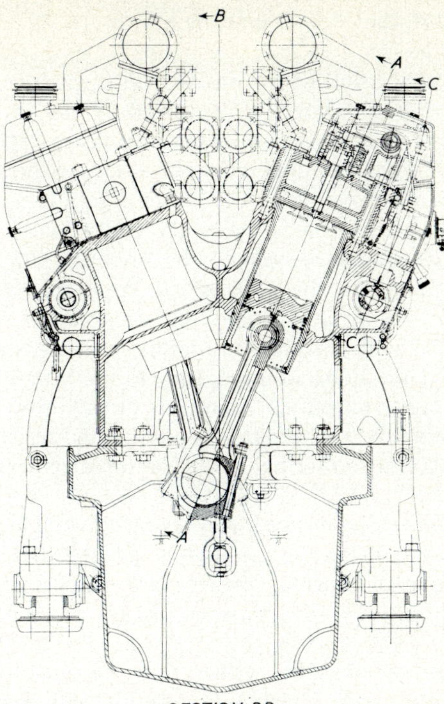

FIG. 16–35—*Transverse section through Ruston Paxman RK engine.*

FIG. 16–36—*Ruston Paxman RK engine.*

transverse section of the Vee engine in Figure 16–35. A complete engine is shown in Figure 16–36.

The main engine structure is formed from a cast iron crankcase bolted to a cast iron bedplate jointed at the level of the crankshaft axis. The design of these two castings results from extensive rig testing and direct stress measurements on running engines, allied to improved methods of stress calculation made possible by the use of computers.

The structure is designed to carry the firing loads without the use of through bolts. During development to the present ratings special attention was given to the design of the arch in the crankcase and to the bedplate bearing diaphragms to ensure acceptable stress levels and adequate stiffness. The crankcase and bedplate assembly is designed to be sufficiently rigid for all known methods of mounting to suit various applications. In particular these include mounting by means of side girders on to a concrete block or a rigid structure. Alternatively the engine together with its driven machine may be flexibly mounted at either three or four points according to the engine type.

Both in-line and Vee engines incorporate a single piece alloy crankshaft forged by the continuous grain flow process before full machining. The main and big end bearing surfaces are induction hardened and the shaft is located longitudinally between thrust pads at the centre main bearing.

The crankpins are bored to reduce rotating out-of-balance forces and oil passages from the main bearings to the big end bearings are provided by oblique, off-set drillings. Bolted on balance weights are provided to reduce the internal loads due to rotating couples and on the eight-cylinder Vee form engine a system of balance weights rotating at twice engine speed is employed to provide secondary balance. There are no external out-of-balance forces on the 6, 12 and 16-cylinder engines and any out-of-balance force associated with the 8-cylinder engine is reduced to a minimum.

Viscous torsional vibration dampers are fitted to the 12- and 16-cylinder engines as required by operating conditions, chiefly for variable speed engines.

Stamped steel connecting rods are employed, the rod design being common to in-line and Vee versions of the engine, the Vee engine utilizing a side by side rod arrangement in conjunction with off-set cylinder banks. The large end bearing employs a split arranged normal to the connecting rod centre line and a patented arrangement of inclined large end bolts. This arrangement prevents fretting of the abutment faces and allows the connecting rod to be withdrawn through the cylinder bore.

Overlay plated lead bronze on thick steel shells are utilized for main bearings and overlay plated lead bronze on thin steel shells for crank pin bearings. The small end bearing is a full sleeve type thin shell steel bearing with a lead bronze lining.

The pistons are manufactured from low expansion aluminium alloy castings and are attached to the connecting rods by fully floating gudgeon pins. Details of the piston rings are shown in Figure 16–37. Each piston is fitted with three compression rings, and two slotted oil control rings are positioned one above the gudgeon pin and the other near the bottom of the piston skirt. The top compression ring is chromium plated on the surface which bears against the cylinder wall, and an 'Alfin' insert of austenitic iron is provided for this ring. Cooling oil is fed into the centre of the piston, via a slipper, from the top of the connecting rod, and thence via two transverse cored passages to a cored annular space behind the top compression ring. Drain drillings from this space to the bottom of the piston are arranged to maintain the space half full of oil.

Cast iron cylinder liners are carried in the main crankcase structure of the engine. Separate cast iron cylinder heads seat on the liner flange. The heads are attached to the crankcase by six equally spaced bolts which provide even loading to the joint ring between head and liner and ensure that the correct transfer of gas loading to the crankcase structure is achieved.

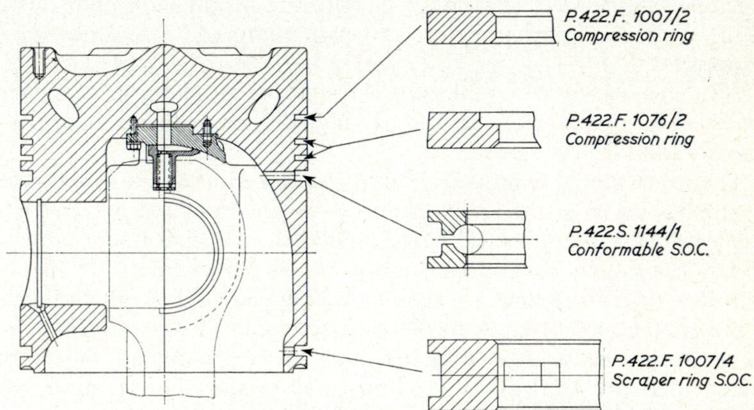

FIG. 16–37—*Piston and piston ring pack, RK engine.*

The cylinder head is designed to withstand high mechanical and thermal loading. To achieve this the outer walls of the head have been positioned as nearly as possible above the cylinder liner flange and joint rings, and an intermediate deck has been used immediately above the flameplate. The provision of this deck, together with thickening of the top deck, give mechanical strength and stiffness to the head and permit the flameplate to be reduced to a practical minimum to reduce thermal loading. The intermediate deck also controls the distribution and velocity of coolant over the top of the flameplate and round the central fuel injector.

Coolant is transferred from the crankcase water jacket to each cylinder head by a single flexible connexion at a point near the exhaust port.

All valves are provided with renewable wear-resisting seat inserts.

Each cylinder head houses two inlet and two exhaust valves operated in pairs by a bridge piece, rocking lever and push rod system from a low level camshaft which is housed directly in the crankcase structure. Charge air and exhaust systems on the engine are located on the same side of the engine. On Vee engines this conveniently allows the induction and exhaust systems to be located between banks, allowing efficient induction and exhaust energy utilization.

On the in-line version of the engine the camshaft is driven by a system of chains and gears from the flywheel end of the crankshaft. On Vee engines the camshaft is driven by a train of hardened and ground gears from the flywheel end of the crankshaft. Balancer gear, driven from the free end of the engine is incorporated in the 8-cylinder version.

Individual fuel injection pumps located on the crankcase are actuated through tappets directly from the camshaft. The fuel pumps are calibrated and pre-set to maintain even distribution of load over all cylinders. Fuel is delivered direct to the combustion chambers through low inertia nozzles.

Auxiliaries are driven from the free end of the engine through a spring drive and spur gears. Various arrangements of pumps are possible to suit different engine applications, but the standard arrangement includes one water pump for engine coolant circulation, one water pump for the oil cooler and charge-air cooler and a lubricating oil pressure pump. It is also possible to drive a lubricating oil scavenge pump and a fuel transfer pump if required.

An extension of the crankshaft through the auxiliary drive casing may be used as a power take-off for driving larger auxiliaries such as compressors, radiator fans etc.

The lubricating oil pressure pump mentioned above is of the gear type and supplies oil from the engine sump to engine-mounted oil coolers under the control of a thermostatic valve, and thence to fine filters with renewable felt cartridges, to a main supply rail in the bedplate. High pressure (70 lb/in^2) oil is supplied to the main, big-end and small-end bearings, and the pistons. It also goes to the camshaft and all gear drives. The valve gear is lubricated at low pressure and the cams by an oil bath. The cylinder liners are lubricated by oil mist and splash from the main bearings and pistons. Main oil pressure is regulated by a relief valve situated on the delivery side of the pump which discharges into the sump, and the system may be primed before starting by means of a semi-rotary hand pump.

Two entirely separate coolant circulation systems are used. Treated fresh water is used for the engine water jackets in conjunction with a heat exchanger or radiator. Raw water (sea water in the case of marine application) is used for circulation through the charge coolers and oil coolers.

A variety of mounting positions for exhaust driven turbochargers and charge air coolers is available to enable the range of engines to be readily adapted to any application.

The engine is capable of operation on distillate fuels and special engine builds are available for operation on residual fuels up to a maximum viscosity of 1000 Red. 1 at 100°F.

16.7. *S.E.M.T.-Pielstick PC3–480 Engine*

S.E.M.T.-Pielstick engines are designed and developed by Société D'Etudes de Machines Thermiques, 2 Quai de Seine, 93 Saint-Denis, France. They are manufactured under licence in Great Britain by Crossley Premier Engines Limited. S.E.M.T. have kindly provided the material for this section.

S.E.M.T.-Pielstick engines were the first medium speed engines to be installed in large numbers in ocean-going vessels and running on residual fuels. The PC range started with the PC 1 version having an output of 250 hp per cylinder which, with continued development, grew into the PC 2 engine producing 500 hp/cylinder. A larger bore design, the PC 3–480 engine, based on the experience with the PC 2, is described here and many of its features have been embodied in turn in the latest development of the PC 2 known as the PC 2–5.

The PC 3-480 engine has a bore of 480 mm and a stroke of 520 mm. It runs at 460 rev/min and is currently rated at 850 metric horse power per cylinder, but it is stated by the designers that this will be increased to 950 metric horse power per cylinder. The engine has a 45° Vee form for the 12, 14, 16 and 18 cylinders and an in-line form for 6, 8 and 9 cylinders. The main features are given in the following table and Fig. 16–38 shows a cross section through the engine.

Bore	480 mm	18·89 in
Stroke	520 mm	20·47 in
Swept volume	94·1 dm³	5740 in³
Maximum continuous power per cylinder	625 kW	850 metric hp
	or 838 British hp at first stage	
	700 kW	950 metric hp
	or 937 British hp at second stage	
B.M.E.P. at 460 rev/min and 850 mhp/cyl.	17·7 kg/cm² (252 lb/in²) at first stage	
	19·8 kg/cm² (281 lb/in²) at second stage	
Mean piston velocity at 460 rev/min	7·97 m/s	1570 ft/min

The following table gives the rated powers and respective weights of each engine type:

Engines	*Powers metric hp*	*Weights metric tons*	*Weight/power kg/m hp (850 hp)*
12 PC3 V 480	10 200	110	10·8
14 PC3 V 480	11 900	125	10·5
16 PC3 V 480	13 600	140	10·3
18 PC3 V 480	15 300	160	10·1

The basis of the engine structure is a frame comprised of rolled, cast and drop forged steel components welded together to form a rigid box shaped structure in which the crankshaft is supported in an underslung manner and to which the individual cylinders are fitted. The side walls of the lower part of the frame form the crankcase and are fitted with circular inspection doors having explosion safety valves. The frame is extended at the sides to carry the camshafts and at one end to form the timing gear case. The crankcase is closed at the bottom by a light plate oil sump.

The main bearings are of completely underslung design. The rigid main

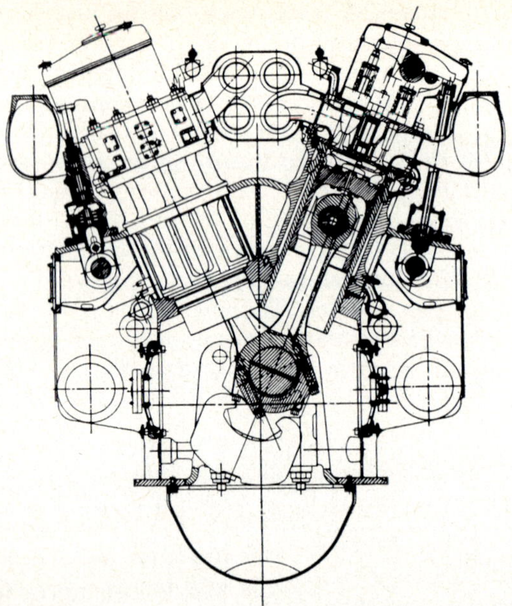

FIG. 16–38—*Cross section through S.E.M.T.-Pielstick PC3-480 engine.*

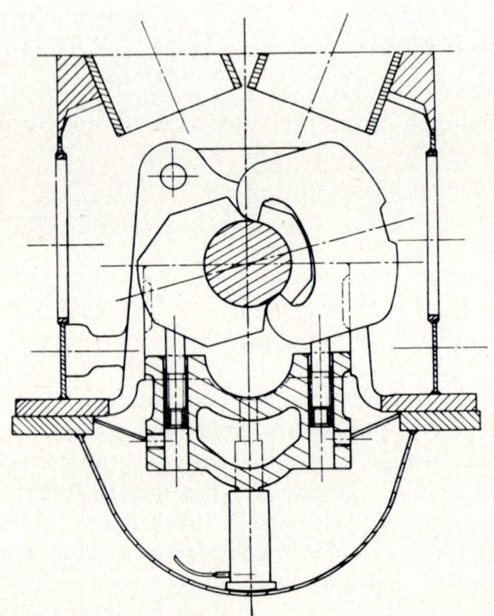

FIG. 16–39—*Dismantling of main bearings PC3-480.*

bearing cap is held up into the frame by two long bolts and the frame is clamped sideways on to it by further bolts; all these are hydraulically tightened. To inspect the main bearing, the cap is lowered into the sump by a hydraulic jack, as shown in Fig. 16–39. Any pair of main bearing shells can be inspected without disturbing balance weights or connecting rods and running gear.

The main bearing shells are of the thin wall type consisting of mild steel shells lined with copper lead and coated with a thin lead tin overlay.

The single piece crankshaft is forged from chromium molybdenum alloy steel and has a coupling flange at both ends. The full output can be taken from either end of the engine and the free end flange is used for mounting a barring gear disk and vibration damper. Holes bored through the crankshaft journals, webs and pins convey lubricating oil from the main bearings to the connecting rods and pistons. Balance weights are retained on the crank webs by means of dovetailed slots in the webs against which the balance weights are secured by a single centrally located screw in compression.

The camshafts, one for each bank of the Vee engines, are driven by a double reduction gear train, the first train also being used to drive the governor and such auxiliaries as lubricating oil and cooling water pumps if required. The individual inlet, exhaust and fuel pump cams for each cylinder are shrunk on to the shaft and located by pins.

The camshaft is underslung in the side compartment of the frame. It is supported at each cylinder by two plain bearings directly fastened to each injection pump support which is mounted on top of the frame. The reactions from the pumping loads are thus conveyed directly between the camshaft and the pump. Dismantling of the camshafts is carried out by sideways removal from the engine.

When the engine is required to be directly reversible, this is accomplished by moving the camshafts axially by means of hydraulic jacks and levers. All the cams have double profiles, one corresponding to ahead and the other to astern operation.

Cylinder assemblies are each comprised of a cylinder head and a liner within a water jacket. The design ensures that no cooling water can come into contact with the main frame. Each assembly is held in position in the main frame by eight tie bolts running from the top of the cylinder head and clamping the water jacket and liner between the cylinder head and the lower part of the frame.

The cylinder liner is of centrifugal cast iron and is located by a flange at its upper end which is retained between the water jacket and the cylinder head. The cooling water is conducted close to the bore by hyperbolic drillings in the upper part of the liner. The temperature of the bore surface at the highest point reached by the top piston ring is 160°C. The seal between the water jacket and the liner at the lower end is made by 'O' rings and similar 'O' rings seal between the lowest part of the liner and the crankcase as shown in Fig. 16–40. There is an annular space between the rings sealing the water spaces and the rings sealing to the crankcase so that in the event of leaks developing the water is drained away outside the main frame and cannot pollute the crankcase. Similarly a leak on the crankcase side would not cause oil to enter the cooling water.

The pistons are of two piece construction with a thin steel crown fastened to the top of a light alloy skirt by eight studs. The dished crown incorporates four peripheral pockets to clear the valve heads during the scavenging phase of the cycle. It is cooled by lubricating oil conveyed up the connecting rod and through the piston pin, circulating round the upper part of the ring belt and the crown before being allowed to fall back into the crankcase. The temperature of the top ring groove is 150°C. There are four compression rings and two oil scraper rings located above the gudgeon

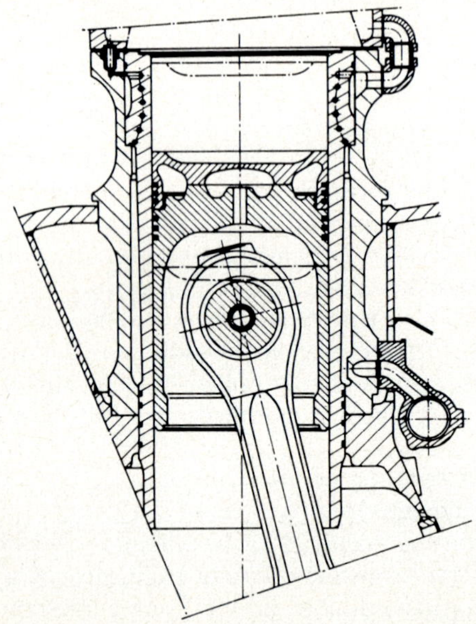

FIG. 16-40—*Bore cooling of cylinder liner.*

pin. Only the top two compression rings are carried by the crown, the grooves for these being induction hardened to resist wear when running on heavy fuel.

The side-by-side connecting rods are of 'I' section in drop forged steel. The cap is fastened by four bolts, two of which are fitted and the joint face is arranged obliquely to permit withdrawal through the cylinder bore. Thin wall shell bearings of a similar design to the main bearings are used for the large end. The small end is fitted with a phosphor bronze bush.

The individual cylinder heads are of cast iron and are of symmetrical octagonal shape. An intermediate deck is used to give two-stage water circulation. Each head has two inlet valves which are identical and which seat directly in the cylinder head. The valve seat surface is coated with hard heat resisting material.

The exhaust valves, of which there are two symmetrical sets to each cylinder, are contained in removable cages which can be dismantled without removing the cylinder head. The cages are secured by long tie bolts having long sleeve nuts to provide resilience by virtue of their elastic length. This arrangement absorbs the elongation resulting from expansion of the valve cage arms as they reach operating temperature (Fig. 16–41). The exhaust valves are fitted with rotators which are placed at the lower end of the valve springs. For engines operating on heavy fuel with a high vanadium content, the valves can be of the cooled seat type, without rotators, cooling water being conveyed down the centre of the stem of the valve to cool the head and then returned by an annular passage in the stem (Fig. 16–42). In all cases both valve and cage seats are made in heat resisting metal and the valve guide is fitted with two cast iron bushes. The valve body is water cooled by a bypass circuit from the cylinder head.

The injector is water cooled by a separate cooling system which can be

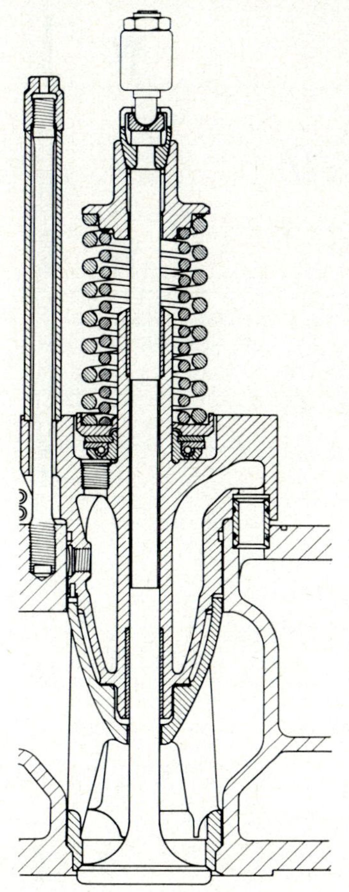

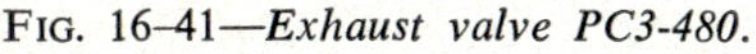

FIG. 16–41—*Exhaust valve PC3-480.*

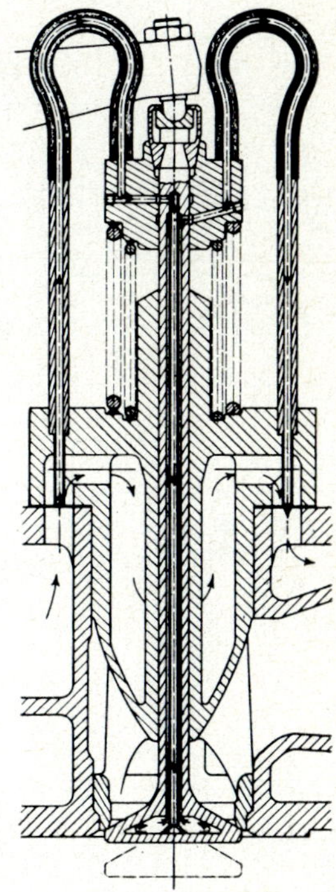

FIG. 16–42—*Water cooled exhaust valve.*

controlled thermostatically to a different level from the jacket cooling water. A sight flow return indicator facilitates checking for the presence of fuel in the water system. The delivery pipe, from pump to injector, is of double wall, co-axial tubing to prevent contamination of the lubricating oil by the fuel, should the delivery pipe fail.

Individual flange mounted jerk type injector pumps are provided, one for each cylinder. Their control racks are actuated by a linkage operated by a speed regulation governor driven from the camshaft drive gears. A separate overspeed trip is also provided.

There is one turbocharger for each bank of cylinders. These two turbochargers can be mounted at either end of the engine and the air

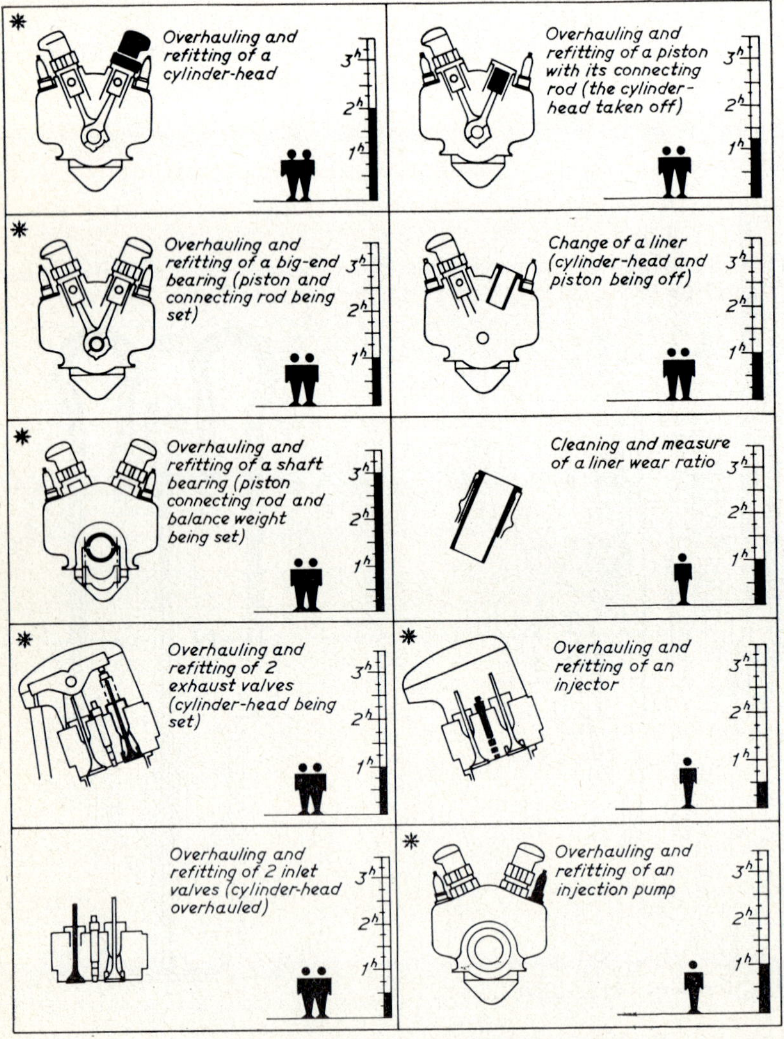

FIG. 16–43—*Man hours for overhauling operations.*

coolers for them are mounted at the other end with ducting in between. The arrangement is compact and gives effective cooling of the charged air; furthermore, absorption silencing can be applied in ducting.

Lubricating oil is arranged to drain by gravity from the engine crankcase into the double bottom tank. From the tank the oil is taken by a pump, which may be separately driven or engine driven, and passed to the oil cooler which is thermostatically controlled, thence via a pressure controlled valve to the filter. The pressure controlling valve by-passes any excess flow back to the service tank. After passing the filter the oil is conducted to a manifold along the engine from which there are tappings to the main bearings through which the oil passes to the internal system where it is distributed to force feed lubricating oil to all the moving parts in the crankcase and also to the piston cooling system. The valve rocker gear is lubricated by a separate circuit which incorporates a small pump, again either engine driven or independent, drawing oil from a separate tank, the steady level of which is maintained by make-up lubricating oil from the main system. This separate valve rocker system safeguards against pollution of the main lubricating oil system if water or fuel leaks should occur under the cylinder head covers.

In the design of this engine attention has been paid to accessibility for ease of maintenance and Figure 16–43 shows the man hours required for a number of maintenance operations.

16.8. *Stork-Werkspoor TM410*

The TM410 engine is manufactured by Stork-Werkspoor Diesel, Czaar Peterstraat 213, Amsterdam, who have kindly provided the material for this section.

The TM410 is a four-stroke engine, built in in-line and Vee-form, and can be built directly reversible. The main data is given in the following table :

Cylinder dimensions :

Bore	410 mm	16·2 in
Stroke	470 mm	18·6 in
Cylinder centre distance	700 mm	27·6 in

Continuous service ratings :

	initial		*present*	
Output bhp/cyl	500		600	
Speed rev/min	500		550	
B.M.E.P. kg/cm² (lb/in²)	14·5	(206)	16	(255)
Mean piston speed m/sec (ft/min)	7·8	(1540)	8·6	(1700)
Maximum cylinder pressure kg/cm² (lb/in²)	92	(1310)	112	(1590)

Dimensions and weights:

Length (from front to coupling flange)		Weight (metric tons)
6 cylinder in line	6020 mm (19 ft 9 in)	52
8 cylinder in line	7920 mm (25 ft 10 in)	68
9 cylinder in line	8620 mm (28 ft 2 in)	76
12 cylinder Vee	6005 mm (19 ft 8 in)	95
16 cylinder Vee	7925 mm (26 ft)	125
18 cylinder Vee	8495 mm (26 ft 10 in)	140
20 cylinder Vee	9375 mm (30 ft 9 in)	155

In the philosophy on which this design was based the most important requirement was quality in terms of reliability and maintenance cost, and the second was that the engine could be produced at moderate cost. To be as free as possible to satisfy these two requirement, a third one, low weight, was considered of less importance. As a result the TM410 is, at its initial rating, heavier than most of its competitors, but the manufacturers state that uprating by further development will bring the specific weight down to a normal four-stroke level.

The construction of Vee form and in-line engines is much the same. Figure 16–44 shows a cross section of the Vee form engine. The main

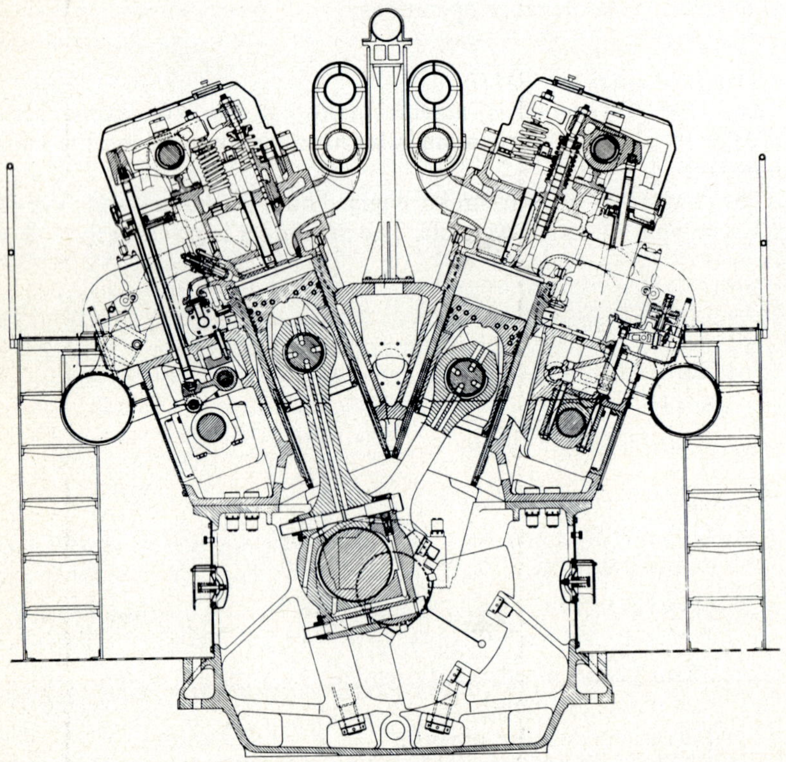

FIG. 16–44—*Cross section of Stork-Werkspoor TM.410 Vee engine.*

structure consists of a U-shape bedplate and a cylinder block, both pearlitic iron castings. These two main components are connected by alloy steel tie rods, which are accommodated in columns cast integrally in the bedplate and which prevent high tensile stresses occurring in the cast iron. Only two rows of tie rods are applied, even in the Vee engine, their main function being to keep the tensile stresses in the bedplate low. In the Vee engine a number of short bolts are fitted to prevent relative movement of the cylinder block and the bedplate. The main bearing caps are fitted on the bedplate with serrated joints and contribute considerably to the rigidity obtained.

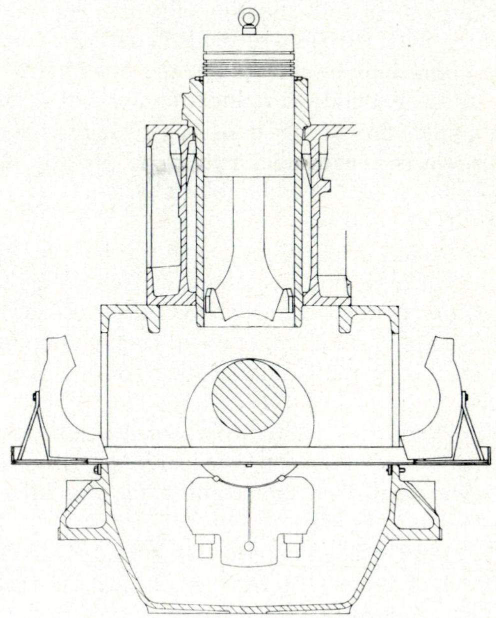

FIG. 16–45—*Dismantling piston and bottom end bearing TM.410.*

Large inspection openings in the bedplate give access to the crankcase. The inspection covers are of light alloy and are fitted with relief valves.

The carbon steel crankshaft is a fully machined one-piece forging. Counterweights are fitted on all crankwebs by means of a construction (see cross section, Figure 16–44) employing two keys, two hydraulically stressed studs and a split to allow for small machining errors. Main and big-end bearings are steel shells with copper-lead lining and lead-tin plating.

The cast iron cylinder block is provided with supports for the underslung camshaft bearings and for the cam follower casings. The camshaft is ground to a single diameter over its entire length and the cams are hydraulically shrunk on by means of tapered bushes. The whole camshaft can be removed sideways. Directly reversible engines are provided with double cams and with hydraulic reversing gear which moves the camshaft in an axial direction.

The connecting rods of the Vee engine are side by side on the crankpin and have large, very rigid big end bearing housings. To enable the rods to be withdrawn through the cylinder bore the big end is split in two planes by three serrated joints (Figure 16–45). For maintenance purposes the two bearing caps are removed along a horizontal slide, (Figure 16–45) and the two bolts which secure each big end assembly are fitted at right angles to the centre line of the connecting rod and are tightened hydraulically. The small end is stepped to provide enlarged bearing areas in the direction of the firing load in both rod and piston.

The light alloy piston has a cast-in top ring carrier and a cast-in cooling oil tube. The piston pin is fully floating. Piston pin, connecting rod and big-end are provided with two sets of drilled passages, one for the cooling oil flow from the crankshaft to the piston, the other back from the piston to the underside of the big-end, to reduce the amount of splash oil.

In the thick upper rim of the cast iron cylinder liner, cooling water passages are drilled in a hyperboloid pattern to provide intensive cooling. This cooled rim rests on the cylinder block without any restraints which may interfere with its circularity.

Two holes are drilled in each liner from the bottom, each feeding one cylinder lubricating oil quill halfway along the liner length. The provision of cylinder lubricators permits the cylinder oil to be adapted to the fuel used and, in conjunction with the piston ring configuration, creates a cylinder oil film that is virtually independent of the speed at which the engine is operated.

The cast iron cylinder head has two exhaust valves which are in cages and two inlet valves which seat on hardened cast iron inserts mounted directly in the cylinder head. It also accommodates the injector and starting air valve.

Cooling of the cylinder head combustion surface is obtained by guiding the cooling water, which enters the head by drilled passages in the bottom periphery, over the combustion plate by means of a horizontal division plate. Cooling of the exhaust valves, which have stellited seats, is achieved by water cooled valve cages (see Figure 16–46) in which the total water flow of the engine is forced through the cooling space, which is very close to the seat. The valve stems of both inlet and exhaust valves are force lubricated.

The exhaust valve rocker is elliptical and goes completely around the inlet valve rocker which is a simple forked lever. The arrangement allows access to the injector by the space left between the end of the inlet valve rocker and the inside of the exhaust valve rocker. Both rockers have a common fulcrum shaft and only two nuts have to be loosened to remove the whole mechanism.

A special feature is valve gear lubrication giving very small quantities of oil on the right places. The valve rockers are fitted with grease-filled needle bearings, requiring no oil supply at all. The other requirements are satisfied by impulse oiling equipment. Once in 20 minutes the whole system is subjected to a pressure of 40 kg/cm^2 (570 lb/in^2) and metering blocks carry a controlled quantity to each of the valve gear contact faces and to the valve stems. The oil used for valve gear lubrication purposes is drained separately and contamination of the crankcase oil by leaking fuel, fuel injector cooling water, or exhaust gas passing the exhaust valve stem, is ruled out. As there is no risk of crankcase contamination, there is no need to keep the injector outside the valve gear cover which is of simple design.

The fuel injection pumps are designed for oil pressures up to 1,100 kg/cm^2 (15,600 lb/in^2). The injectors are water cooled.

The pilot starting air valves are actuated by the fuel cams and not from

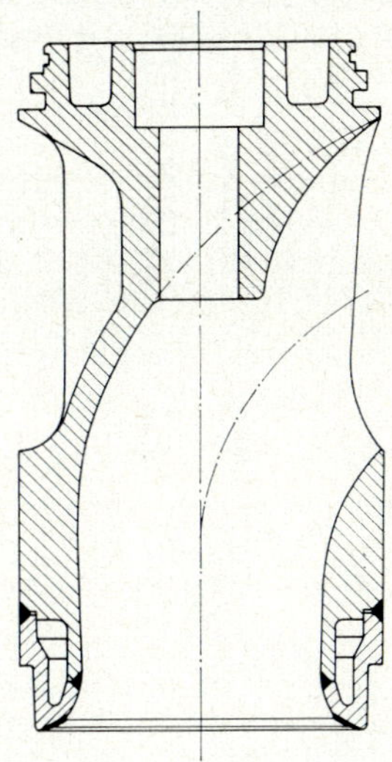

FIG. 16–46—*Valve cage seat cooling TM.410.*

a central air distributor. This entails short pipe-lengths between pilot valve and starting air valve with consequently small delay, a more economic use of starting air and also a more powerful braking effect when air is applied to reverse the rotation of an engine that is driven by the propeller when the ship is still moving ahead.

The mechanism for the starting air pilot valves is also used to stop the engine at overspeed. In this case the mechanical overpeed trip actuates a pneumatic valve causing air pressure in auxiliary cylinders to keep every fuel injection pump lifted after one last injection stroke. Figure 16–47 shows the fuel injection pump drive gear with the air starting pilot valve and the pump lifting pistons mentioned. Safety devices for low lubrication oil pressure and for low cooling water flow are integrated in the engine systems and are thus independent of the external electrical alarm system.

Turbocharging is applied using a pulse system exhaust. The 12-cylinder engine has one turbocharger for each bank of cylinders and they may be mounted at either end of the engine.

The 16-cylinder has one turbocharger at each end of the engine. The 18-cylinder is equipped with three and the 20-cylinder with four turbo-blowers.

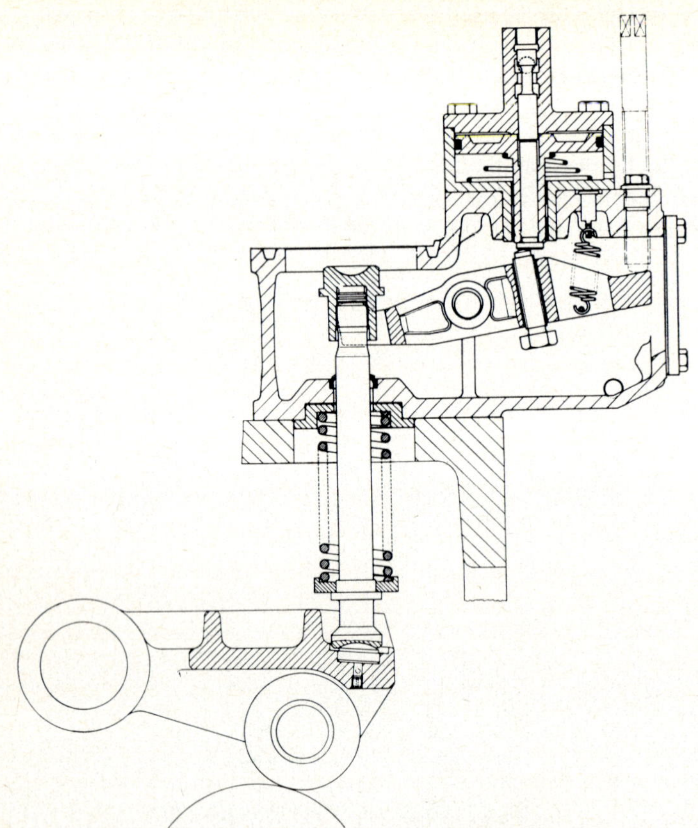

FIG. 16–47—*Fuel pump operating gear TM.410.*

16.9 *The Sulzer Z40 Type Medium Speed Two- and Four-Stroke Engines*

Sulzer Brothers, Winterthur, Switzerland manufacture a wide range of engines of both low speed and medium speed types. They have kindly supplied the material for this section describing their Z40 series, which is of unusual interest in that it has been designed to be built in both two-stroke and four-stroke forms. The change from one version to the other requires only simple modification as the main parts are identical for both. The important exceptions are:

> cylinder liners; camshaft drive gears; cams; turbocharging and inter-cooling equipment; crank sequence of crankshaft-dimensions, however, are identical.

The engines are primarily intended as ship propulsion units but are equally suitable for power generation. The two-stroke system is well suited to applications where its ability to cope with considerable speed drops at constant bmep is of advantage. Moreover, the two-stroke version has a higher output reserve than the four-stroke version. The four-stroke engine has a somewhat higher piston speed and is suitable in applications where a better acceleration capacity and the highest possible engine speed is of advantage.

The table below gives the principal technical particulars of Z40 engines:

Design	trunk piston direct injection turbocharged intercooled
Cylinder configurations	in-line or V-form
Bore, mm	400
Stroke, mm	480
Output per cylinder (MCR), bhp	600

	2-stroke	*4-stroke*
Nominal speed, rev/min	445	500
Mean piston speed, m/sec	7·12	8·0
Mean effective pressure, kg/cm²	10·1	18·0
Firing pressure, kg/cm²	97	115
Type denomination and number of cylinders	*Z40/48*	*ZB40/48*
(a) in-line engines	6, 9, 12	6, 8
	ZV40/48	*ZVB40/48*

FIG. 16–48—*Sulzer 6Z40/48 prototype engine.*

(b) V-engines	12, 16	10, 12, 16
Weight per bhp approx. kg/bhp		
(a) in-line engines	13·1–12·5	12·8–12·5
(b) V-engines	11·1–10·9	11·3–10·5

Figure 16–48 shows the prototype two-stroke in-line engine. Its cross-section is shown in Figure 16–49 and cross-sections of the four-stroke in-line and Vee engines are shown in Figure 16–50 and Figure 16–51 respectively. The design will permit substantial increases in specific output in the course of further development; attention having been paid particularly to the thermally loaded components: piston, cylinder liner and cylinder head, as will be apparent from the following descriptions.

A rotating piston is provided as standard equipment for the Z40 engine (Figure 16–52). Besides the usual reciprocating motion, the piston is also made to perform a slow rotating motion around its longitudinal axis. This innovation requires a spherical design of the top end bearing and a driving mechanism to generate the rotating motion. The only disadvantage of the design is a slightly higher manufacturing cost, which is by far outweighed by the following advantages:

1. At each stroke there is always a newly oil-wetted portion of the piston sleeve surface in contact with the pressure side of the cylinder liner, thus practically eliminating any danger of piston seizure.
2. Due to the rotation of the piston rings, local heating of the working surface of the cylinder liner, caused by blow-by at the ring gap, can be avoided. This results in a greatly reduced risk of seizure, due to uniform thermal loading. Furthermore, rotation of the piston rings considerably improves their lubrication and reduces their wear rate, which now becomes uniform.
3. The oil scraper rings rotate in a similar manner to the piston rings; the running-in is thereby improved and better sealing properties ensue.
4. Since the upper connecting rod bearing is of spherical shape, the piston is stressed symmetrically and edge loading, as occurs with a gudgeon pin, is avoided even at higher load. The piston can align itself in the liner along its longitudinal axis at all times.
Furthermore, the piston body is not prone to deformation and can be made, therefore, of circular shape, which simplifies manufacture.
5. Due to the fact that no unsymmetrical deformation occurs, the piston clearances can be kept small, resulting in improved guiding of the piston and reduced wear of piston rings, ring grooves and cylinder and low oil consumption.

To rotate the piston, the connecting rod is fitted with two pawls in a position slightly out of the centre of the sphere of the small end bearing (Figure 16–53). When the connecting rod swings relative to the piston, the pawls impart an intermittent rotating motion to a toothed rim; this is transmitted to the piston by means of an annular spring (Figure 16–53). This layout permits the reduction to a very low level of the forces required for the rotation of the piston. Due to the energy stored in the spring, it operates

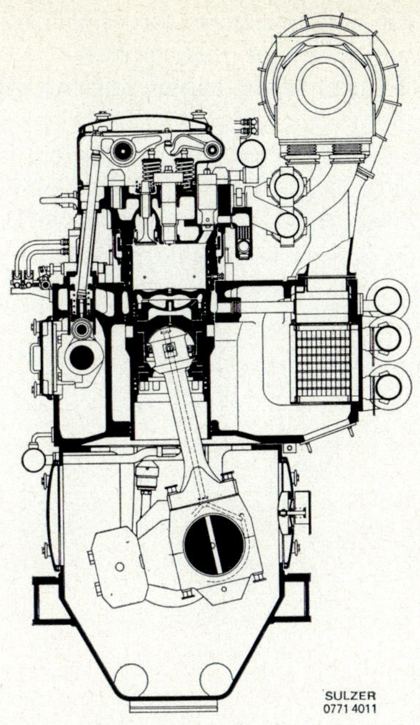

FIG. 16–49—*Cross section of Sulzer 6Z40/48 prototype engine.*

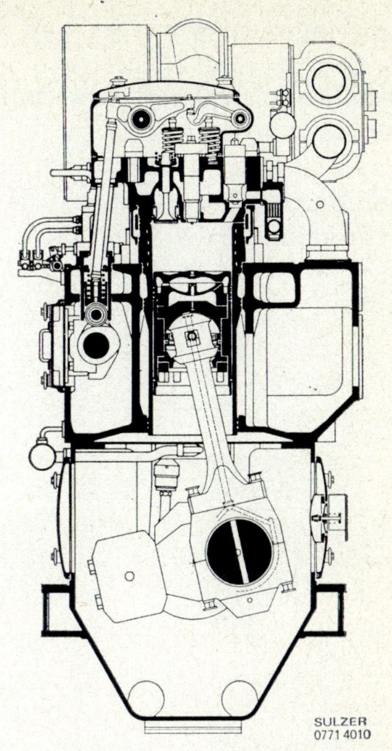

FIG. 16–50—*Cross section of four-stroke in-line engine Z40 series.*

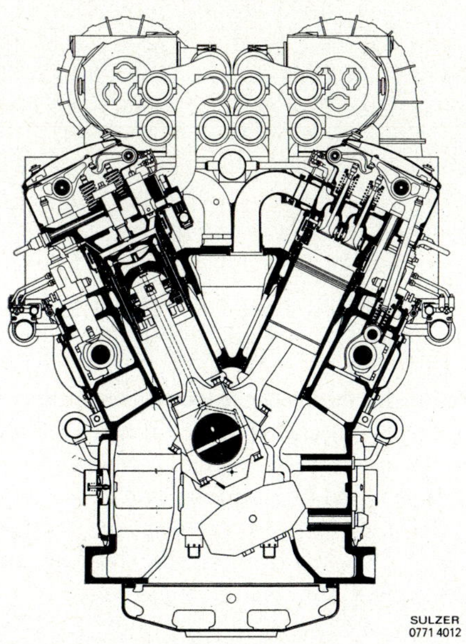

FIG. 16–51—*Cross section of four-stroke Vee engine Z40 series.*

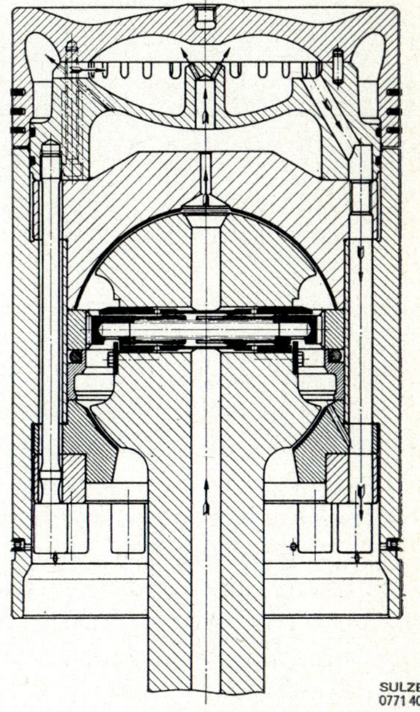

FIG. 16–52—*Rotating piston.*

at that moment when the load from the gas and the mass forces is low. It has been proved by practical service experience that this indexing mechanism works without appreciable wear. Figure 16–54 shows a pawl after 9,000 working hours.

In addition to the design details already described, the piston features a steel crown which carries the compression rings. The upper ring groove is chromium-plated. Particular attention was paid to the cooling of the piston, whereby the vicinity of the top ring groove is internally cooled by a spray of oil from numerous radial holes and in addition, the small wall

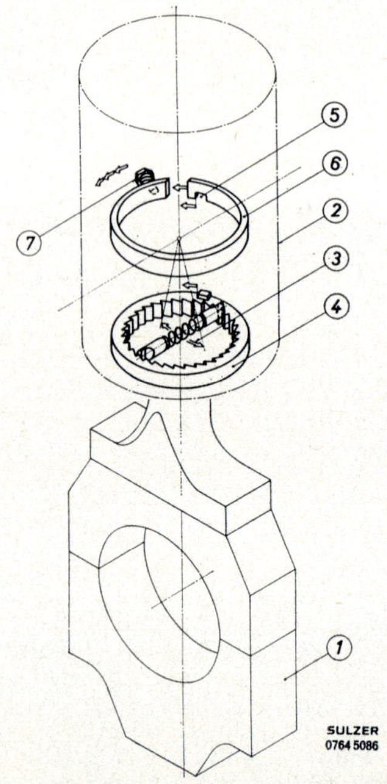

FIG. 16–53—*Driving mechanism of rotating piston.*

thickness of the piston crown above the top ring groove ensures a reduced heat flow in this area. The top of the piston is cooled by oil splashing in the central chamber.

The top collar of the liner, where the highest temperatures and gas pressures prevail, is thick-walled but has a special 'bore cooling' arrangement (Figure 16–55) consisting of a system of holes drilled tangentially at an oblique angle into the cylinder wall, so that the cooling water is led as closely as possible to the heated inner liner surface. In this way, the temperature of the liner running surface is low, thus maintaining the oil film and creating favourable operating conditions for the piston rings, and at the same time keeping thermal stresses at a low level. The cold part of the cylinder outside the bores is relatively thick-walled and embraces the inner portion like a shrunk-on ring, thereby minimizing the pulsation stress set up by the gas pressure. Due to its thickness, this upper collar of the liner is extremely rigid and provides support for the cylinder head.

For the two-stroke version, cylinder liners with inlet ports arranged all around the bore are fitted to obtain satisfactory uniflow scavenging. On both the two- and four-stroke versions, the same forced liner lubrication system is provided.

The cylinder head is designed with a double bottom plate (Figure 16–55). The lower flame plate portion is extremely thin walled to keep thermal stresses low. It is supported by the thicker upper plate through the four valve channels, as well as by the outside wall, the latter being so formed

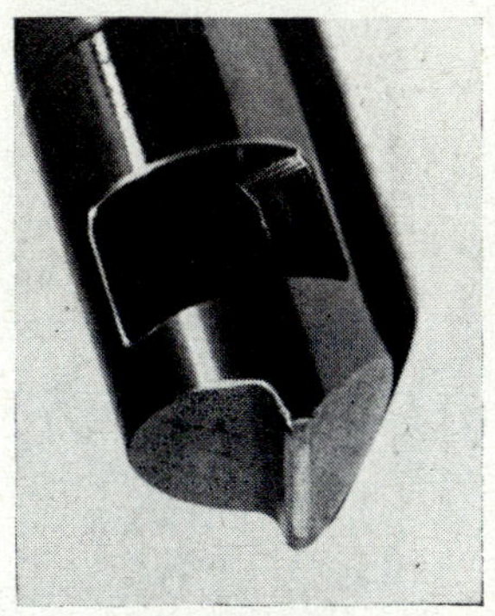

FIG. 16–54—*Pawl of rotating piston after 9000 operating hours.*

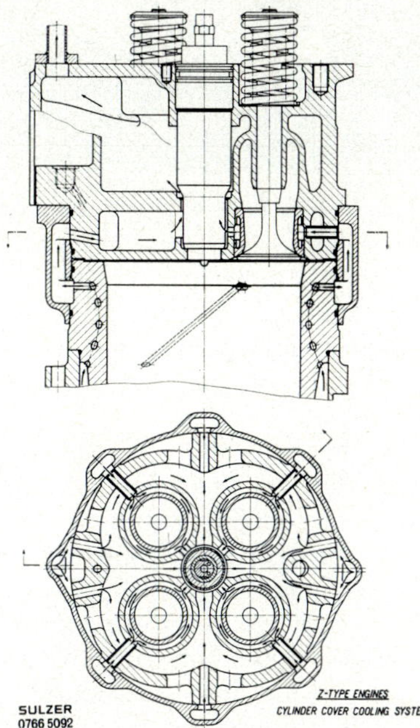

FIG. 16–55—*"Bore cooling" of cylinder liner and section through cylinder head.*

that the unsupported areas of the flame plate are as small as possible in order to keep down the pressure stresses in it.

The cylinder head is made of cast iron and carries four identical poppet valves. In the two-stroke version all four valves are used as exhaust valves. and in the four-stroke version two valves are used as inlet and two as exhaust valves. The injection valve, located at the centre, is fitted with a water-cooled nozzle. The starting valve is pneumatically controlled. The pressed in seats of the poppet valves are forced-water-cooled, thus ensuring low seat temperature, an imperative requirement when operating on heavy fuel. The cooling water leaving the cylinder liner is led through a jacket into the cylinder head. Part of it is fed radially between the two bottom

FIG. 16–56—*Hydraulic tightening of main bearing bolts.*

plates toward the centre and the other part cools the four valve seats. From there on, the water is led through a central passage around the injection valve and cools the upper chamber of the cover. The four poppet valves are made from alloy steel and their seats are specially treated to withstand attack by heavy fuel residues.

For the in-line engine the crankcase is made of cast steel cross members welded to steel plate walls. The crankcase and cast iron block are assembled by tie rods, thus placing the engine structure under pre-stress.

For the V-version a one-piece cast iron engine frame with an under-slung crankshaft has been adopted.

The crankshaft is a single piece forging and has large crankpin and journal diameters. Counter-weights are bolted on to the crankwebs to achieve favourable bearing loads and vibration-free running. The crank-shaft scantlings allow for future higher mean effective pressures. It is supported in trimetal bearings with white metal linings. In the in-line engine, the main bearing cap is held down by pressure bolts designed as hydraulic jacks. The oil pressure needed for pre-stressing is generated by a pump placed outside the engine (Figure 16–56) facilitating the maintenance of correct static pre-load of the bearing caps and bearing clearances. Moreover, the two pressure bolts are used to feed the running gear with lubricating oil and the piston with cooling oil, thus obviating the necessity of a separate line to the main bearing cap.

In the V-version with the underslung crankshaft, hydraulic jacks are also used for pre-tensioning of the bearing cap studs to obtain correct and even tightening.

The big end of the connecting rod is designed as a separate part bolted

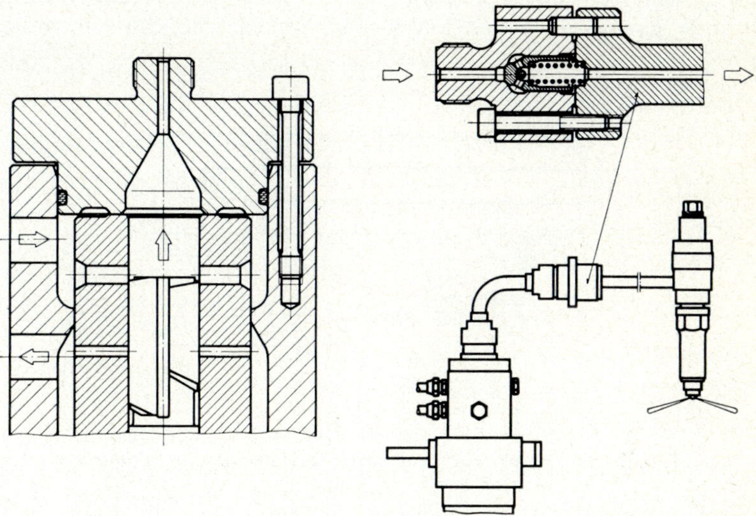

FIG. 16–57—*Section through top part of fuel injection pump and non-return valve.*

to the bottom end of the shank. This arrangement has the following advantages:

(1) The withdrawal height of the piston is considerably lower than with a big end split diagonally.

(2) The piston can be withdrawn without having to dismantle the big end bearing.

(3) The fitting of a shim between the head and the shaft of the connecting rod permits easy adaptation of the compression ratio to special conditions, if necessary.

(4) This design does not imply any limitations in the dimensioning of the crankshaft journal diameter.

The fuel injection pumps are of the helix-controlled type (Figure 16–57); an individual pump is provided for each cylinder.

The pump casing has two separate chambers, one connected to the inlet and the other to the spill pipe. Priming is carried out from the upper chamber and the excess fuel not required for injection returns to the lower chamber. In this way the fuel leaving the injection pump under high pressure is kept separate from the suction system so that harmful high pressure pulsations on the suction side of the pump are avoided. With the engine at rest, the two chambers are inter-connected by the control edges of the plunger so that fuel may circulate through the pump cylinder and when operating on heavy fuel, preheat the injection pump.

As may be seen from Figure 16–57, the fuel pump has no built-in relief valve. In optimizing the injection system in respect of smokeless combustion, low fuel consumption and cavitation-free operation, a non-return valve has been provided between the injection pump and the injector.

The fuel feed to the injection valve takes place laterally through a solid tube having its connexion located outside the cylinder head, thus

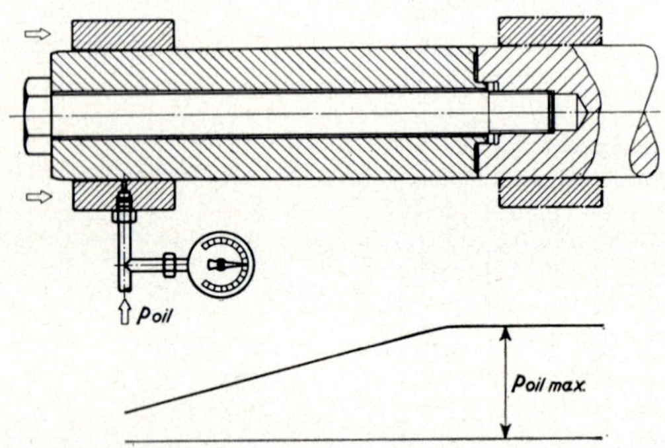

FIG. 16–58—*Hydraulic cam fitting and dismantling procedure.*

completely eliminating any danger of the lubricating oil being contaminated by fuel oil.

To avoid any cracks, the high pressure fuel lines are made from drilled solid bars, which is possible because of their short length.

The velocity curve of the fuel cam has a declining characteristic during the pumping stroke to keep the maximum injection pressure as low as possible.

The control of the four valves in the cylinder head is effected by push rods, main and auxiliary rocking levers. On the engine control side there are three cams per cylinder—two for operating the poppet valves and one for the injection pump.

The cams are mounted on the camshaft by the hydraulic oil injection SKF method. This design results in a plain, strong and keyless camshaft and permits the timing of the cams to be adjusted easily.

For assembly and dismantling, as shown in Figure 16–58, a short length of slightly tapered shafting is screwed on to the end of the camshaft.

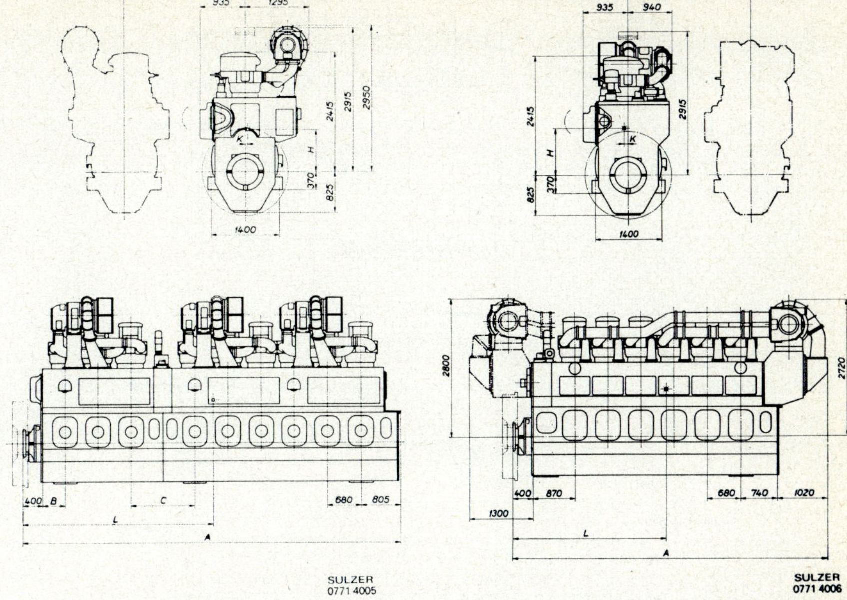

FIG. 16–59—*Main dimensions of two-stroke in-line engines Z40.*

FIG. 16–60—*Main dimensions of four-stroke in-line engines ZB 40.*

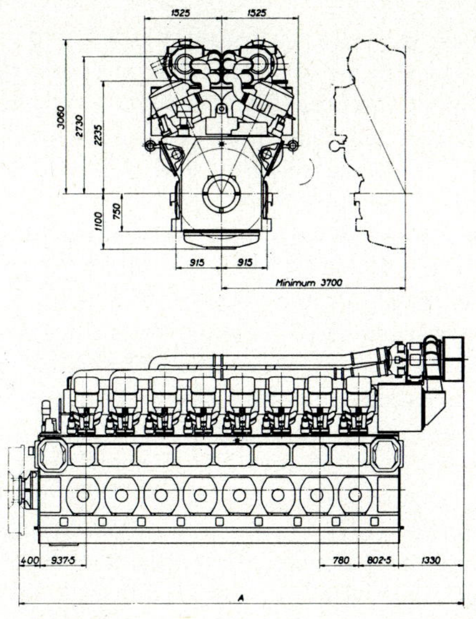

FIG. 16–61—*Main dimensions of four-stroke Vee engines ZVB 40.*

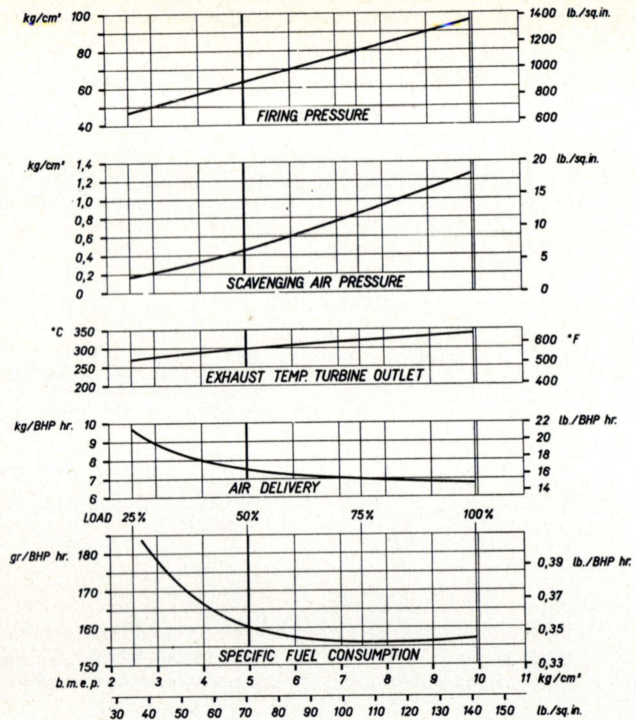

FIG. 16–62—*Test results of 6Z40/48 engine.*

The Z40 engines are pulse pressure-charged and intercooled. On the two-stroke in-line engines one turbocharger and one intercooler are provided for each group of three cylinders, whereas on the four-stroke in-line engines one turbocharger and one intercooler only are required, which can be either at the free or at the flywheel end. For the four-cycle V-engines two units are placed at the same end.

Pressure charger arrangements and engine main dimensions are shown on Figure 16–59 for the Z40/48 types, on Figure 16–60 for the ZB40/48 types and on Figure 16–61 for the ZVB40/48 engines.

The test bed performance of a two-stroke 6 cylinder in-line prototype engine is shown in Fig 16–62. Outstanding characteristics are the low exhaust temperature amounting to only 340°C and the relatively low firing pressure at full load of only 97 kg/cm².

Some of the results of the test bed measurements on the 6ZB40/48 engine are shown in Figure 16–63 for Diesel oil operation.

The Z-engines are currently offered with a maximum rating of 600 bhp per cylinder at speeds of 445 and 500 rev/min for the two-stroke and four-stroke versions respectively. Development continues and includes endurance tests at still higher outputs.

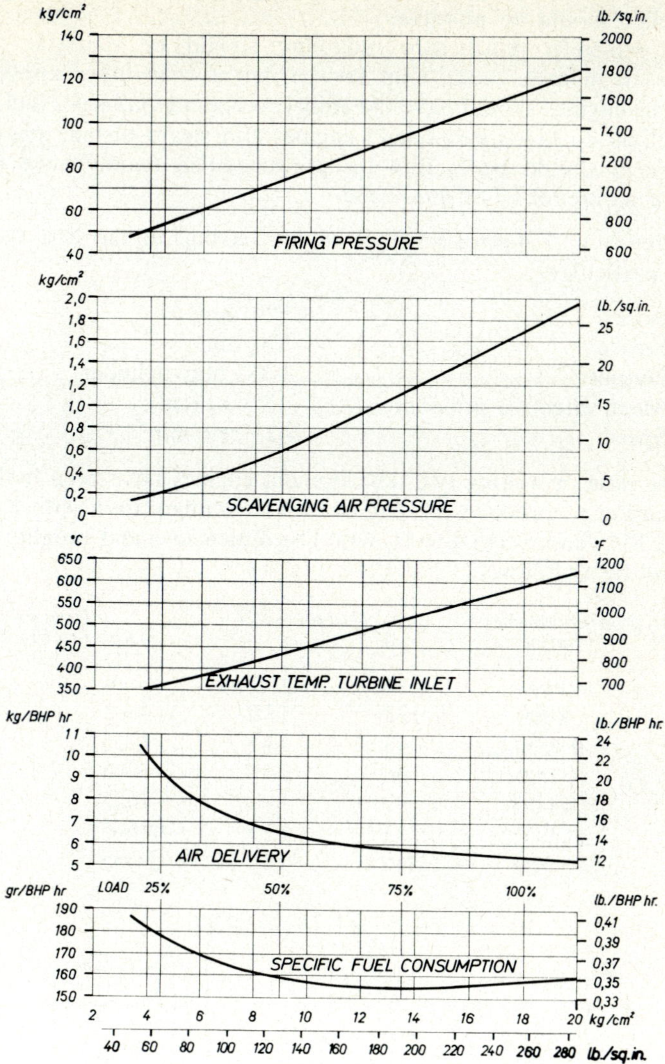

FIG. 16–63—*Test results of 6ZB 40/48 engine.*

16.10 *UDAB 520/570 Engine*

The two major Swedish Diesel engine manufacturers, Messrs. Götaverken and Nohab, co-operated in the formation of a mutually owned development company registered as United Diesel A.B., Gothenburg, Sweden. The outcome of their collaboration in the design of a medium speed engine is the UDAB 520/570 engine. The description of this engine which follows is based on material kindly supplied by UDAB.

The design characteristics have been based on the following points:

1. Output up to 40,000 bhp from two engines geared to one common

propeller should be possible;
2. Completely satisfactory reliability should be available for use above all in trans-ocean ship service, but also in other applications;
3. The engines should have the ability to operate on heavy fuel equally as well as do low speed Diesel engines and steam turbine plants;
4. There should be facility for pre-scheduled maintenance routines with a minimum of off-hire time.

The result is a 4-stroke Diesel engine having, in the first stage, the following particulars:

Bore	520 mm
Stroke	570 mm
Output	1000 bhp/cylinder
Mean effective pressure	17·5 kg/cm^2
Speed	425 rev/min

At the time of writing two development engines have been built and it is the intention to produce the engine in Vee configuration with 8, 10, 12, 14, 16 and 18 cylinders (Figure 16–64). The dimensions and weights of these engines will be as follows:

No. of cylinders	*Continuous output*	*Weight*		*Dimensions (Refer to Fig. 16–64)*	
		total		L_1	L_2
	bhp	*tons*	*kg/bhp*	*mm*	*mm*
8	8000	122	15·2	4805	5570
10	10 000	148	14·8	5755	6250
12	12 000	168	14·0	6705	7470
14	14 000	197	14·0	7655	8420
16	16 000	213	13·3	8605	9370
18	18 000	240	13·3	9555	10 320

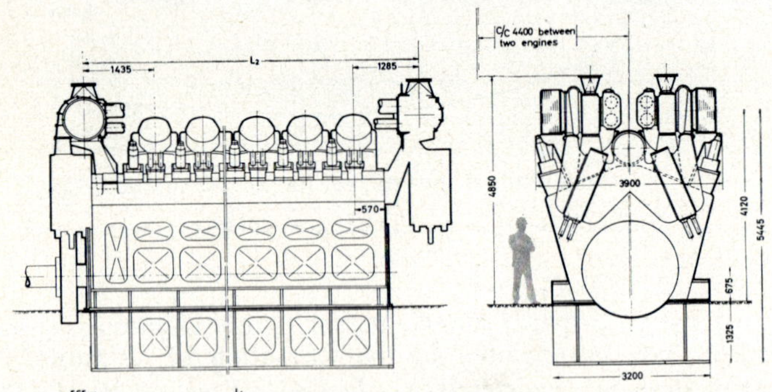

FIG. 16–64—*Main dimensions of UDAB 520/570 Vee engine.*

The UDAB engine is a monobloc engine with an underslung-type crankshaft, the main bearing caps being keyed by serrations. The crankcase frame, fabricated of steel plate with cast steel brackets for cylinder head studs and cast steel main bearing housings, is a single unit even for 18-

cylinder engines. All joints are butt welds or TK-welds machined for complete penetration.

The monobloc design means that the number of machined surfaces has been reduced to a minimum. There is also a minimum number of bolts. The crankcase frame is placed on top of the engine base (Figure 16–65) and the two components are held together by heavy bolts, thus forming a rigid unit that can be mounted on four brackets, two on each side of the base. These brackets can, for instance, be placed on extensions of ordinary hull frames, thus giving an extremely simple installation. The engine is in this way practically independent of any deformation of the hull.

Large covers are provided, as can be seen in Figure 16–65, both on

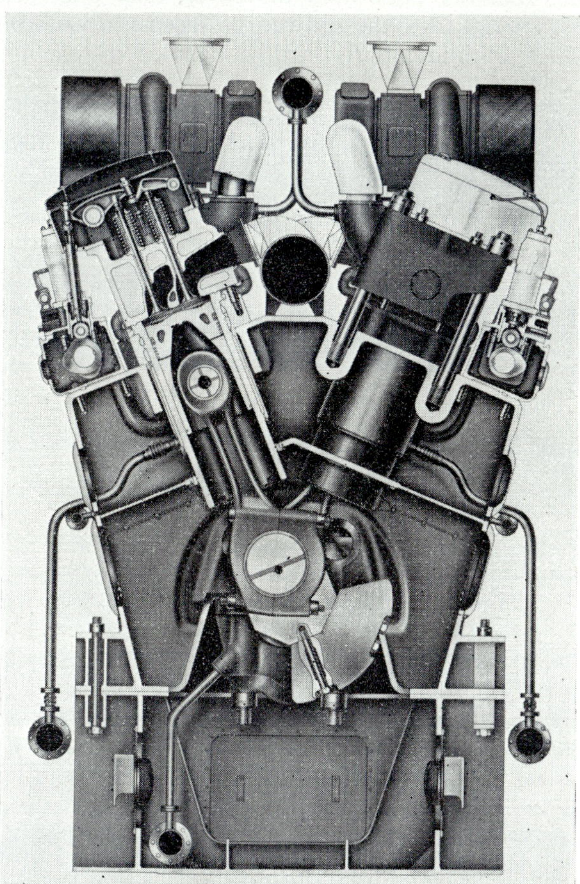

FIG. 16–65—*Transverse section of UDAB Vee engine.*

the crankcase and the base, giving access for inspection of crankshaft bearings. The covers on the base are fitted with explosion relief valves.

The side-by-side principle was chosen for the connecting rods (Figure

16–66) in spite of the consequent slight increase in engine length and the somewhat heavier crankshaft scantlings required because of the increased bending moments. These drawbacks were considered to be less disadvantageous than those of the articulated master design.

The disadvantages of master and slave are that the height of the engine must be increased, because the journal on the master for the slave requires some space. The design will also be much more complicated and expensive, and there is a much greater risk of stress concentrations. Furthermore, assembly and dismantling are more complicated. Finally, it is very difficult to arrange for the passage through the journal and the pin of a sufficient quantity of cooling oil to the piston on the slave rod.

The cylinder assembly includes the cover, liner, cooling jacket, piston and connecting rod. Six bolts hold the cover, liner and jacket together. By means of a special tool, the rod can be fastened to the bottom of the liner after which all these parts can be handled as one unit having a weight of about 5000 kg.

Cylinder assembly units can be serviced by exchanging for reconditioned units so reducing the time for work on the engine itself. Overhead hoists have been designed to make it possible to do this in a minimum of time either in port or at sea with only one engine stopped.

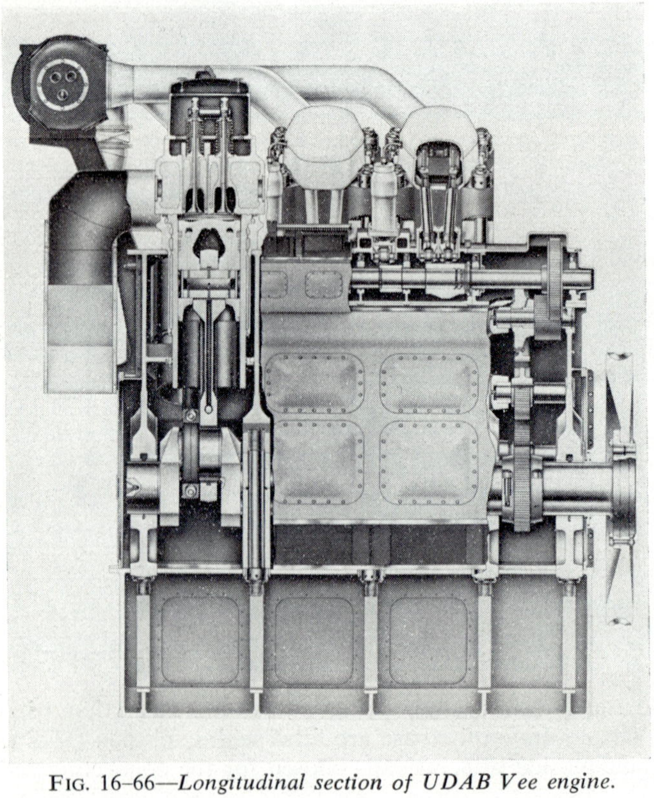

FIG. 16–66—*Longitudinal section of UDAB Vee engine.*

The cover is of chromium-alloyed cast steel. The inlet and outlet ducts have been designed for least possible flow resistance. The gas pressure forces are transferred to the cylinder block through four studs. Six other bolts ensure close sealing between liner, jacket, and cover and facilitate the handling of these parts as one unit.

The exhaust valves are mounted in detachable cages which are water cooled including the seating. Valve rotators are fitted.

The cylinder liner is of vanadium-titanium alloyed cast iron. The detachable cooling jacket is of cast steel and is separate for each cylinder. The jacket, strongbacking the top of the liner, is also designed to provide high cooling water velocity in the channels around the combustion chamber. When exchanging a cylinder assembly, it is possible to drain only the assembly concerned, thus reducing work and the loss of treated cooling water to a minimum. To ensure good cooling in highly turbocharged engines, it is necessary to check or clean the water chambers of the cylinder at the same intervals as for piston overhaul. This is facilitated by the detachable jackets.

Initial trials have been conducted using light metal pistons as shown in Figure 16–65, but for running on heavy fuel two-piece pistons have been designed. These have a steel crown carrying the upper two rings and a piston trunk of light metal. Temperatures attained by these pistons are very low, the temperature at the bottom of the top ring groove being well below 150°C at an m.e.p. even of 21 kg/cm^2.

The connecting rod big end is split obliquely in order that it can be removed upwards through the liner when piston overhauls are due.

The camshafts are mounted on top of the crankcase frame in separate boxes. Thus, very short pushrods have been obtained, reducing the inertia forces and diminishing the risk of vibrations.

In addition, by placing the camshafts in separate boxes instead of in the upper part of the crankcase frame, discontinuities, which could disturb the stress flow in the frame, have been avoided. The assembly and machining of components are also simplified.

The arrangement further facilitates the use of a separate lubrication system for the cylinder tops and the camshafts, thus avoiding contamination of the crankcase lubricant which may be caused by leakage from the fuel or water-cooling systems.

The fuel pumps, one for each cylinder, are located on the top of the camshaft boxes. The length of the high-pressure line is reduced to a minimum, diminishing the risk of breakage and the danger in the event of such a breakage.

The top platform is level with the top of the cylinder block. The inlet and exhaust manifolds are nested in the Vee, leaving unrestricted access to the fuel pumps, the racks, the fuel lines and their connexions. It is also easy to reach the valve mechanism and the fuel injectors.

The flexible connexions of the water pipes to the jackets can easily be reached through the upper openings on the cylinder block. The connexions for liner lubrication and the bolts on the big end are accessible through the lower openings (Figure 16–65).

CHAPTER SEVENTEEN

Some High Speed Engines

17.1. *Foden FD6 Marine Engine*

The Foden FD range is manufactured by Fodens Ltd. of Elworth Works, Sandbach, Cheshire, who have kindly provided the material for this section.

Foden diesel engines were designed originally for the propulsion of heavy commercial road vehicles. The Mark I version went into full production in 1950 following a design and development period of about eight years. Even before it was announced commercially it had attracted the attention of the Royal Navy and an FD6 engine was delivered to the Admiralty for type test. The initial type test was entirely satisfactory but a vehicle engine required considerable modification to make it completely suitable for marine use. Once this had been done the Royal Navy adopted it as a standard range engine. From then on the development to the present Mark VI and Mark VII engines has been strongly influenced by marine requirements.

A power range from 81 b.h.p. to 375 b.h.p. is covered at 1,800 rev/min as a continuous rating and extends up to 440 b.h.p. at 2,000 rev/min as an intermittent rating. The engines are suitable for propulsion or power generation. (See Figures 17–2 and 17–3).

The basic design can be seen in Figure 17–1. It is a 2-stroke, exhaust valved, uniflow engine using the Kadenacy principle and having a 2-lobe roots blower to provide the scavenging and charging air. The frame of the engine is of aluminium construction which, being non-magnetic, was attractive to the Navy for minesweeper applications. The use of this material, together with the 2-stroke principle, has resulted in an engine of high specific output, small size and light weight.

The range of engines comprises 4, 6 and 12 cylinder models with exhaust turbocharged models of the 6 and 12 cylinder engines. The mechanically blown engines are designated Mark VI, whilst the exhaust turbocharged versions are designated Mark VII. The 12 cylinder engine in each Mark is in effect two 6 cylinder engines with contra rotating crankshafts geared together, mounted side by side in a common crankcase. Between the different models there is considerable interchangeability of main working components, such as connecting rods, pistons, cylinder liners, cylinder heads, etc.

The Mark VI engines are uni-flow blower scavenge engines, the roots type blowers feeding air into a common air chest from where it is delivered to the cylinders through the air ports in the cylinder liners which are designed to impart a high degree of swirl. The two exhaust valves for each

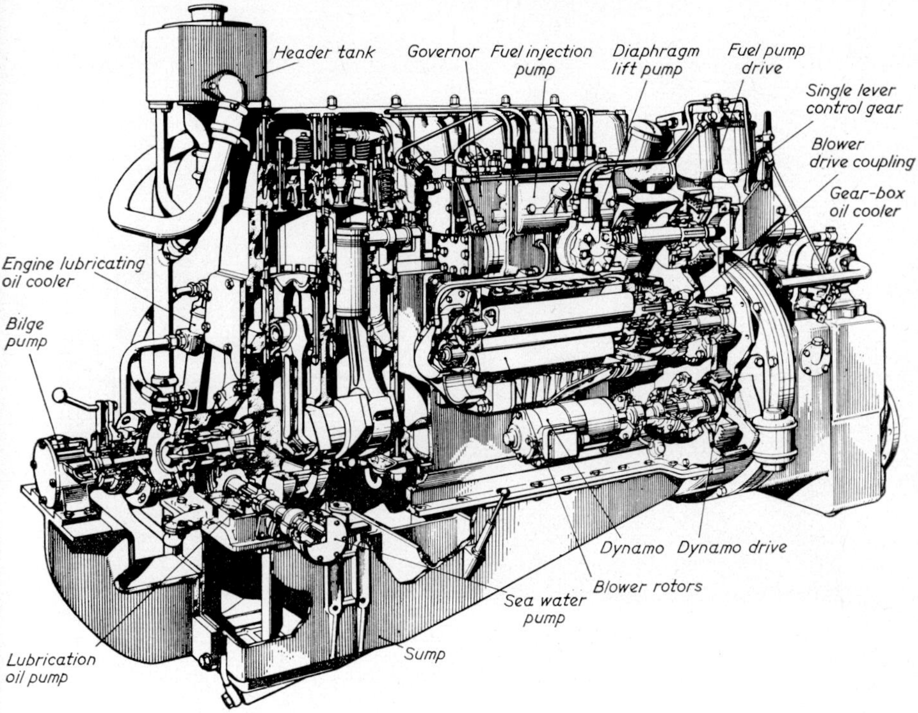

Header tank
Governor
Fuel injection pump
Diaphragm lift pump
Fuel pump drive
Single lever control gear
Blower drive coupling
Gear-box oil cooler
Engine lubricating oil cooler
Bilge pump
Dynamo
Dynamo drive
Blower rotors
Sea water pump
Sump
Lubrication oil pump

FIG. 17–1—*Foden Mk. VI engine.*

FIG. 17–2—*FD 6 Mk. VII propulsion.*

FIG. 17–3—*250 kw marine generating set FD 12 Mk. VII.*

cylinder are located in the cylinder head and fuel is injected through a single hole nozzle.

The Mark VII model is basically the Mark VI engine with the addition of an exhaust turbocharger and intercooler. The air passes from the turbocharger through the intercooler to a reduced size roots type mechanical blower which is retained on this engine only for starting and to ensure stability with change of load.

The crankcase, air chest and cylinder block form a single piece aluminium alloy casting having three separate horizontal compartments and which in consequence is very rigid. It extends from the cylinder head to well below the crankshaft centre line. The crankshafts, which are nickel steel forgings are carried in aluminium-tin thin shell bearings underslung in the crankcase. The lower part of the crankcase is closed by a detachable oil sump designed so that the engines can operate under wide angles of rake and list.

Each cylinder liner is attached to its individual cylinder head by six studs and the heads are connected to the frame by long studs carried through to the main bearing. These studs take the gas loads and the aluminium casting is not subjected to tensile loads from this source.

The twin exhaust valves are made of nimonic 80 material and are operated through short push rods and overhead rockers from a hardened steel camshaft carried in six bearings located high in the cylinder block. The tappets and push rods are carried in the heads and are removable with them as a unit. The injectors can be removed without taking off the rocker boxes. There is no possibility of lubricating oil being contaminated by fuel through leakage.

Each connecting rod is of H section and is a high tensile steel forging. The big end is split obliquely so that it can pass through the cylinder bore for withdrawal. Lubricating oil is fed through drilled passages in the crankshaft to lubricate the large end bearing and then by a drilling up the connecting rod to feed the small end and also to spray from the top of the rod on to the inside of the piston crown for cooling purposes.

The pistons are made of cast iron. They are long, as is typically the case with 2-stroke trunk pistons and which helps in the performance of the cross head function. Each piston carries a fire ring at the top of the crown and three compression rings lower down. At the bottom of the piston skirt there are two oil control rings. The gudgeon pin is of nitrided steel.

The timing gears are located at the aft end of the engine and are of straight spur profile ground on the surface. They drive the camshaft, fuel pump, blower and dynamo. The location near to the crankshaft node minimizes the transmission to them of torsional vibration.

The fuel pump is a monoblock unit complete with built-in filter, hydraulic governor and diaphragm type fuel lift pump. It is mounted on top of the blower and the camshaft is fed under pressure from the engine lubricating oil system.

A lubricating oil pump of the gear type is driven from the cross shaft at the forward end of the engine, the oil being delivered through a heat exchanger which is cooled by water from the fresh water circuit. This arrangement is adopted so that the thermostat also controls the lubricating oil temperature, resulting in rapid warm up and improved fuel consumption under low load conditions. A fuel flow filter is fitted in the lubricating oil circuit.

Water circulation is by means of a centrifugal pump, also driven at the forward end of the engine. The heat exchanger is mounted on the engine. A positive displacement pump is employed to circulate sea water through the tube stacks of the heat exchangers and intercoolers.

The turbocharger is a single stage radial flow turbine driving a single stage centrifugal compressor. It is of the Holset type described in Chapter 19. The shaft runs in plain bearings which are pressure lubricated from the engine oil gallery and the oil returns under gravity to the engine sump.

An interesting feature of the engine is the rather unusual character of the noise it emits. Airborne noise from intake and exhaust is naturally of higher frequency than that of four-stroke engines of similar speed and number of cylinders, but this is not a difficult matter to silence. Noise emanating from the engine structure, however, is very different from most four-stroke engines, and essentially not much noise comes from the lower half of the engine; also the intensity is relatively low owing to the low rate of change of pressure rise in the cylinders during combustion.

This is accounted for partly by the high swirl of the air entering the cylinders causing minimum ignition delay and partly by the rigid construction.

17.2. *Gardner 6 LW*

The Gardner 6 LW engine is one of a range of engines marketed by

Gardner Engines (Sales) Ltd., Patricroft, Eccles, Manchester, who have kindly provided the material for this section.

The sturdy construction and low fuel consumption for which Gardner engines are renowned are the result of highly developing a classical design of naturally aspirated, four stroke cycle, direct injection type.

The LW series marine engines have a $4\frac{1}{4}$ in (107·95mm) bore and a 6 in (152·4 mm) stroke. The 6 LW model, taken here as an example, is the six cylinder engine of the range. It is rated, under normal atmospheric temperature and pressure, at 84 b.h.p. at 1,300 rev/min for heavy duty commercial craft, 94 b.h.p. at 1,500 rev/min for yachts, cruisers and similar craft and at 107 b.h.p. at 1,700 rev/min for high speed craft. The general appearance is shown in Figure 17–4 and details of its performance appear in Figure 17–5.

FIG. 17–4—*Gardner 6 LW marine engine.*

Figure 17–6 is a drawing showing the constructional details of the engine. The basic framework consists of a single piece crankcase, with a bolted on sump, carrying two cylinder blocks each comprising three cylinders and surmounted by two matching cylinder heads.

These cylinder heads are made from iron castings and each one carries the overhead valves, valve levers and fuel sprayers for three cylinders. Hard alloy iron inserts, highly resistant to wear or deformation, are pressed into the heads to form valve seats. Each cylinder has one inlet valve and one exhaust valve. The inlet valve is larger than the exhaust valve and carries a masked portion which, in conjunction with the carefully controlled shape of the inlet passage, produces the correct degree and direction of air movement within the cylinder. Means are provided for decompressing by slightly lifting all inlet valves; this being essential for

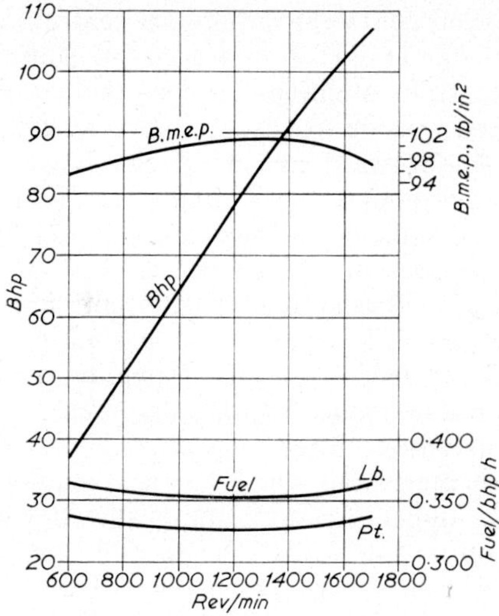

FIG. 17–5—*Gardner 6 LW marine engine performance curves.*

hand starting and a most desirable feature to facilitate turning the engine during servicing adjustments.

The cylinder blocks are cast in the same material as the cylinder heads and are fitted with renewable dry type liners. Coolant flow passages are designed to give the best possible distribution in both block and head. The coolant is transferred from block to head via cadmium plated spigot tubes and synthetic rubber rings arranged to prevent it making contact with the joint faces so that it cannot cause them to corrode.

The cylinder blocks are bolted to the crankcase by long high tensile bolts which extend through the crankcase and at their lower ends form the main bearing cap studs.

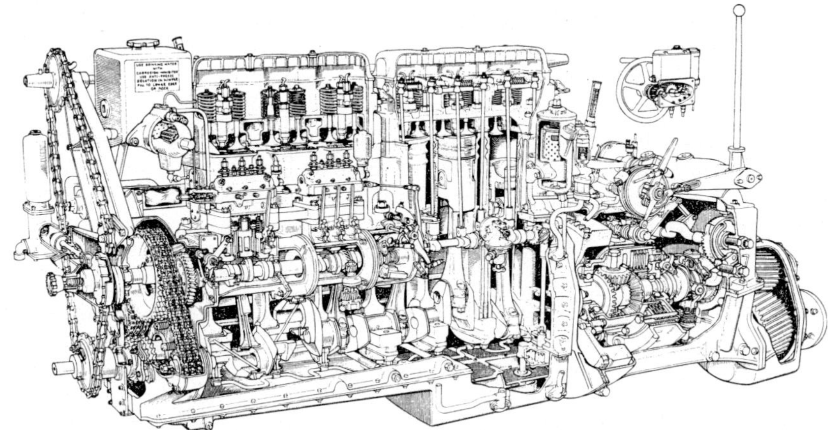

FIG. 17–6—*Gardner 6 LW marine Engine with reversing and reducing gear.*

The crankcase casting can be supplied in either aluminium alloy or cast iron. It extends well below the crankshaft centre line and its lower face is closed by a cast aluminium or cast iron oil sump which adds to its rigidity. When made from aluminium alloy, the crankcase is generously ribbed both inside and outside to ensure complete support for each crankshaft main bearing. The forward end of the crankcase contains the drive to the valve camshaft, fuel pumps, water circulating pump and alternator.

The lubricating oil sump is attached to the lower face of the crankcase and contains the oil reservoir, surge baffles and a gauze screen which prevents large particles of foreign matter from entering the oil pump.

The pistons are cast in a medium silicon aluminium alloy. The design embodies an exceptional length of skirt, providing a large area of contact with the cylinder walls to reduce loading and wear, increase heat transfer and secure quiet operation. The hemispherical combustion chamber is formed in the top of the piston which carries two pressure rings and one oil control ring. The upper pressure ring is plated with chromium on its periphery and also on its side faces. The second pressure ring is plated on its periphery only. The hollow gudgeon pin is free to rotate in the small end of the connecting rod and also in the piston. It is located by aluminium pads fitted in the ends of the pin. The axis of the pin is slightly offset from the piston centre line.

The connecting rods are produced from die stampings of chromium molybdenum steel and are of "H" section. They are machined and polished all over. The big-end is of two-bolt design and the rods are rifle drilled from end to end to provide lubrication for the special copper-alloy small-end bearing. The big-end is fitted with pre-finished thin wall bearings lined with copper lead overlay plated.

The crankshaft is machined all over from a die stamping of chromium molybdenum steel. It has hollow bored journals and crankpins joined by stiff webs. It is not locally or surface hardened and is carried in seven pre-finished copper-lead overlay plated steel shell bearings with an additional roller bearing at the forward end. The forward end of the shaft carries the chain sprocket for the timing and auxiliary drives and also the special ratchet sprocket for the hand starting gear. Also at the forward end is a friction-type torsional vibration damper. Oil is fed from the main bearings to the crankpins by means of steel tubes pressed into holes drilled across the shaft from journal to pin.

The valve camshaft is chain driven from the forward end of the crankshaft. The inlet and exhaust cams for each cylinder are a combined unit secured to the camshaft by 60° pointed set-screws. The cams operate tappets carried in cast-iron guides fitted to the crankcase and the valve levers are operated by the tappets and ball-ended push rods of light tubular steel.

A gear-type pump draws oil from the sump and delivers to an external spring loaded relief valve which maintains a constant pressure in the whole system. Approximately half the total oil delivered from the pump is by-passed by this valve and returns to the sump after having flood lubricated the fuel pump cambox, governor and timing case. The oil which is not diverted by the pressure relief valve passes through a paper element filter to a passage cast in the crankcase connecting with an oil distribution pipe which feeds all main bearings. The big end bearings are lubricated by means of the steel pipes leading from the journal bearings, some of this oil being forced through the central holes in the connecting rods to the small ends and gudgeon pins. The overhead valves, operating levers and push rods are fed with oil taken from a tapping on the pipe leading from the filter to main bearings. From this tapping the oil is led through a small pipe to the hollow valve lever shafts which are drilled to permit oil to reach the lever fulcrums. The levers are drilled vertically and horizontally; the vertical hole leading oil to the top edge of the lever, which is "flatted" to a predetermined width to lead the correct amount of oil to valve tip and stem. The horizontal holes carry oil to lubricate the ball ends of the push rods. All oil delivered to the overhead gear returns to the sump via the push rod passages in the cylinder and through holes in the tappet guides.

The engine is designed for fresh water cooling, using either a sea water cooled heat exchanger or an outboard mounted keel cooler. With either arrangement an engine mounted fresh water header tank provides the necessary head of coolant which is circulated through the system by the integrally mounted centrifugal type water pump.

Temperatures are automatically maintained within desirable limits under all loads and conditions by a dual wax-type thermostat, incorporating a primary and secondary element each set to operate a control valve at a pre-determined temperature. The primary unit maintains the approximate full load running temperature of 140°–150°F. In the unlikely event of failure of the primary element the secondary valve is caused to open by the resulting rise in coolant temperature and prevents the engine attaining dangerously high temperatures.

Flange mounted CAV type BPF fuel pumps are fitted; operated by cams and spring loaded tappets. Individual fuel rams are provided with a hand-operated lever for priming the system and testing the action of the sprayers.

The fuel pump camshaft is driven through a helical gear meshing with a similar gear mounted on the valve camshaft. The gear is free to slide on a helical spline on the fuel pump camshaft and, by means of a yoke coupled to the speed control lever, is moved axially whenever the lever is operated. The timing of injection is thus varied automatically with engine speed.

17.3. *Kelvin TSC8 Engine*

The Kelvin TSC8 engine is manufactured by the Kelvin Marine Division of English Electric Diesels Limited, Kelvin Works, 254 Dobbies Loan, Glasgow, who have kindly provided the material for this section.

The Kelvin T range of engines, of which the 8 cylinder model, designated Model T8, is the largest, was introduced in 1959. It is the culmination of continuous development over 64 years of a series of engines which, almost from the beginning, have been designed and built primarily as marine propulsion units.

The T8 engine was initially designed as a naturally aspirated unit, but since it was fully anticipated that much higher performance versions of it would later be required, more than adequate safety factors were built into it. Thus the turbocharged TS8 engine came into being in 1962 and, more recently, the TSC8 model, which not only has its charge boosted by a turbocharger but also cooled to improve breathing efficiency. The power outputs, available at the propeller, of these three sub-types are:

T8 Engine	240 h.p. at 1,000 engine rev/min
TS8 Engine	320 h.p. at 1,000 engine rev/min
TSC8 Engine	400 h.p. at 1,150 engine rev/min

The following statistics and description apply to the TSC8 engines:

Bore	$6\frac{1}{2}$ in
Stroke	$7\frac{1}{4}$ in
Swept volume	1,920 cu.in
B.m.e.p. at max. rev/min ...	152 p.s.i.
Fuel consumption rate at full speed	18·5 Imp. gall/hr
Specific fuel consumption ...	0·355 lb/b.h.p./hr
Firing cycle	4-stroke
Firing order	14768523 (standard rotation) 13258674 (reverse rotation)
Compression ratio	14·0 : 1
Combustion system	Direct injection
Overall length	122·5 in (including gearbox)
Overall width	42·25 in
Overall height	63·25 in (including gearbox)
Weight–dry	11,200 lb (including gearbox)

The general appearance of the engine can be seen in Figure 17–7.

There is a separate cylinder head for each cylinder and each one can be removed without disturbing the others. Each head carries two valves, one inlet and one exhaust, and each valve is controlled by duplex valve

FIG. 17–7—*Kelvin TSC8 engine.*

springs. Both valves seat on renewable inserts in the cylinder head, made from special alloy cast iron. The exhaust valves are Stellite faced. A de-compressor is fitted on the valve cover to operate on the exhaust valve.

To cope with the increased b.m.e.p. of TSC8 engines, a cylinder head joint made from solid metal has been introduced. This withstands firing pressures better than the normal copper-asbestos type of gasket used on earlier Model T engines.

The cylinder blocks are cast in pairs, so as to be adaptable for 4-cylinder, 6-cylinder and 8-cylinder engines, but the pairs are rigidly bolted together endwise to form, in effect, a continuous cylinder block. Each bore is fitted with a centrifugally-cast slip-in wet type cylinder liner to give even cooling. The liners are sealed at the bottom by two rings made from a synthetic silicone compound which can withstand temperature and chemical effects.

The crankcase and sump are both of robust one-piece construction and are well bolted together to provide rigid support for the crankshaft. Additional rigidity, in the plane of each main bearing, is obtained by the use of horizontal screws which bind the main bearing caps securely inside their recesses in the crankcase. These screws are additional to the normal set-bolts which clamp the caps vertically.

The aluminium pistons each carry five rings, three compression and two oil-control. The top-most ring is chromium-plated on the periphery to give a low rate of wear and works in a special insert, cast integral with the piston, to reduce groove wear. Piston temperatures are controlled by jets in the crankcase which spray oil to the undersides of the crowns, in which are formed the central-displacement type combustion spaces.

The connecting rods are of conventional design, but are particularly rigid. The big end cap on each is secured by four bolts. The big end bearings, like the main bearings, are of the thin-wall steel-backed type with a copper-lead bearing surface. When new, the bearings also have a flash of lead-tin alloy to ease the running-in period. Piston-connecting rod assemblies can be withdrawn upwards through the cylinder bores.

The crankshaft is of tempered alloy steel with ample bearing areas for long life: re-grinding is not normally required for several years. It is not counter-balanced, because it is well supported by the crankcase-sump structure and because rotational speeds are kept to a moderate level. A torsional-vibration damper, of the viscous type, is fitted at the forward end.

The timing gears are driven from a pinion at the rear of the crankshaft at a point adjacent to the flywheel. In this position they leave the front of the engine adaptable for the inclusion of auxiliary drives.

A casing, containing a train of gear wheels, has been arranged at the front of the engine to provide essential drives for the sea water and fresh water pumps and also optional auxiliary shaft drives to suit the various requirements of operators. The auxiliary shafts can be fitted in many different assembly combinations and include drives for winches, dynamos, additional pumps and a variety of pulleys.

The fuel injection pump has 11 mm dia. elements, as against the 10 mm elements used on the lower-rated engines, for the sake of improved injection characteristics. The governor is of the all-speed hydraulic type and the injectors have four-hole nozzles.

The turbocharger is mounted directly on the exhaust manifold. It comprises a centrifugal air compressor driven by an inward-flow exhaust gas turbine. The compressor and turbine rotors are mounted on the same shaft and this runs in plain bearings.

An intercooler is interposed between the turbocharger and the inlet manifold. It is cooled by sea water passing through internal tubes which carry fins for the dissipation of the heat in the air.

The engine is cooled directly by fresh water and indirectly by sea water which removes the heat in the fresh water through the medium of a large heat exchanger. A wax-type thermostat controls the actual temperature of the fresh water within prescribed limits. The fresh water also passes through the jackets of the exhaust manifold and the engine oil cooler. The sea water system has been re-arranged, as compared with that on the TS8 engines, because of the introduction of the intercooler. Zinc anodes are incorporated in this system to minimize corrosion.

Particular attention has been paid to filtration, both of the fuel oil and the lubricating oil. It has been found, from experience, that the presence of water in the fuel oil is quite common and that it can have a damaging effect on the injection system. A sedimenter unit, in which the water separates away from the fuel oil by gravity, is therefore fitted on the engine, so that water can be drained off periodically before reaching the feed pumps. Dirt in the fuel is dealt with by a paper-element type of filter, which gives a high degree of filtration, between the feed pumps and the injection pump. This latter is of the duplex type, so that one element can be renewed while the other remains in circuit to keep the engine running. The main lubricating oil filter is of the full-flow type and an additional smaller filter gives added protection to the turbocharger bearings.

Starting is normally by electric starter, but other types can be fitted.

17.4. *Paxman Valenta RP200 Engine*

The Paxman Valenta series is part of a range of engines manufactured by Ruston Paxman Diesels Limited, Paxman Works, Colchester, Essex, who have kindly provided the material for this section.

The Paxman Valenta range has been developed particularly for application to high speed craft, naval patrol vessels and marine generating sets. It is a compact prime mover of high specific output and is based on a family of similar engines designed for heavier duties. To achieve the compact design many items of equipment are packed inside the overall engine dimensions, as can be seen from the illustration in Figure 17–8. The table below gives some details of the engine which is built entirely in Vee form.

Number of cylinders	8, 12, 16
Bore and stroke	7·75 × 8·5 in (197 × 216 mm)
Cycle	4-stroke
Compression ratio	13:1
Full load speed range	750–1,500 rev/min

Mean piston speed—

at 1,500 rev/min 2,125 ft/min (10·8 m/sec)

RATINGS	ENGINE SPEED rev/min	LOAD BMEP lb/in²	kg/cm²	ENGINE TYPE 8-RP200 bhp	12-RP200 bhp	16-RP200 bhp
High speed craft						
Maximum	1600	255	17·9	1650	2475	3300
Intermittent	1550	240	16·8	1500	2250	3000
Continuous	1500	223	15·6	1350	2025	2700

FIG. 17–8—*Paxman Valenta RP200 engine.*

The construction will be understood from a study of Figure 17–9.

The housing is fabricated from high grade steel plates and steel castings. The underslung crankshaft is carried in main bearing caps located against side thrust by fitting faces. Lateral bolts screwed into the caps through the housing skirt increase the rigidity of this assembly. The housing carries wet type steel cylinder liners, surmounted by individual cylinder heads. The underside of the housing is enclosed by a sheet metal oil sump complying with Classification Societies' requirements for angles of pitch and roll. Crankcase relief doors are fitted.

The forged steel hardened crankshaft is carried in main bearings of the steel shell type, lined with aluminium tin. It is located by independent thrust rings at the drive end bearing. The crankshaft is flanged at the drive end for the attachment of the flywheel and driven machinery by set bolts, the drive being taken by hollow dowel pins. A viscous torsional vibration

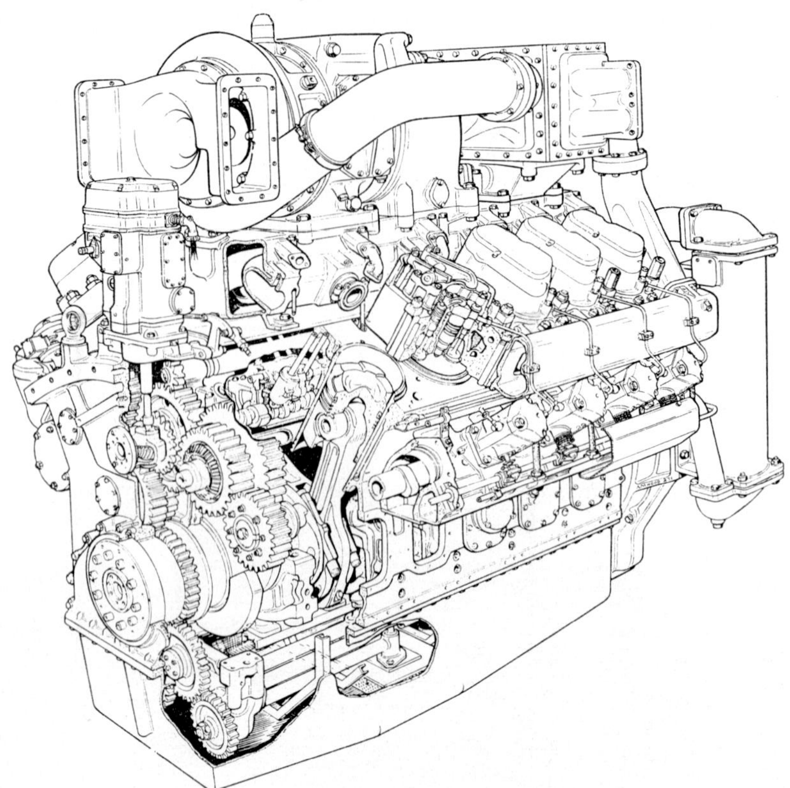

FIG. 17–9—*Paxman Valenta RP200 engine.*

damper (two on 16 cylinder engines) is bolted to the flange at the free end of the crankshaft. The main drive for the camshaft, governor, injection, oil and water pumps is from a train of hardened gears located at the drive end of the engine. Eight cylinder engines have a gear driven secondary balancing system and the usual primary balance weights on the crankshaft are fitted to all engines.

Fork and blade connecting rods have been employed in Paxman designs for over 35 years. A large end block attached to the forked rod houses a replaceable steel back aluminium tin lined, lead-tin flashed bearing shell, which runs on the crankpin. A similar bearing shell but lined with lead bronze in the blade rod runs on a hard chromium plated outer surface on the large end block. The joint faces of the bearing block and blade rod are serrated for location purposes. The rods and bearing blocks are drilled for lubrication of the small ends and for piston cooling.

Aluminium alloy oil cooled pistons are each fitted with three compression and one oil control ring above the fully floating gudgeon pin which is located by circlips. The top pressure ring is fitted in a cast iron ring groove insert.

The single unit cylinder heads each house two inlet and two exhaust valves seating on replaceable inserts, and a central injector. Each pair of valves is operated by one push rod and rocker by a bridge piece. Hard chromium plated wet type liners of seamless steel tube flanged at the head end, carry a copper plated steel ring above and below the flange for sealing the cylinder head against gas and water leakage; the lower end of the liner is sealed by three rubber rings. The water side surface of the liner is protected by chromium plating.

The camshaft is located in a trough formed by the two banks of cylinders. It is gear driven from the crankshaft and runs in pressure lubricated cast iron housings. A separate tappet housing for each line of cylinders is fitted above the camshaft and contains four roller type cam followers for operating the tubular push rods.

Single element injection pumps are mounted on a cambox attached to each side of the engine below the air manifolds. The fuel injection pump plungers are lubricated from the engine pressure system via a filter which also supplies the pump camshaft. Both fuel feed and injection pumps are driven from the gear train at the drive end of the engine. The combustion system employs direct injection of fuel by a central multi-hole injector into an open toroidal cavity in the top of the piston. Short pipes are fitted between the pumps and injectors which may be withdrawn from their housings by releasing a clamp. The injector connexion is designed so that any fuel leakage occurs outside the valve covers.

An all speed governor, driven from the engine gear train, is located above the cam trough at the drive end of the engine. The governor is of the centrifugal weight type, operating through a hydraulic servo-mechanism and controls the effective stroke of the injection pump plungers.

One water cooled exhaust gas driven turbo-blower is located above the vee of 8 and 12 cylinder engines. Two turbo-blowers are fitted to 16 cylinder engines. The turbo-blowers are mounted directly on to a water cooled exhaust manifold within the vee of the engine. Water cooling the exhaust manifold reduces radiant heat and dispenses with all expansion joints and hot surfaces which create a fire risk in unattended operation.

The intercooler for cooling the charge air from the turbo-blower is cooled by sea water.

Two engine driven gear type lubricating oil pumps are housed inside the sump at the drive end of the engine. These operate in parallel through engine mounted oil coolers to the engine main, large and small ends, camshaft, valve gear and gear train bearings via full-flow filters. Oil at reduced pressure is supplied to the water pump bearings. The filters and coolers are protected by relief and safety devices.

Water is circulated through the oil cooler, engine jackets, exhaust manifold and turbo-blower cooling system with controlled flow round the injectors and valve seats, by a single pump built into the free end of the engine and driven from the drive end gear train by a tubular shaft. The pump supplies fresh water to the engine mounted oil cooler (one per bank—12 and 16 cylinder engines) and to the inlet manifolds at the top of each water jacket. Outlet manifolds from the turbo-blowers are brought together to provide a common inlet to a thermostatic control valve mounted above and attached to the pump.

17.5. *Perkins T6.3543 Marine Engine*

The Perkins T6.3543 marine engine is manufactured by Perkins Engines Limited, Peterborough, England, who have kindly provided the material for this section.

Although it is the policy of the Perkins Engines Group to base its marine engines on the vehicle version of a particular model, the final aim is to produce a true marine engine and not a marinized unit.

By following this policy, not only is it possible for Perkins to achieve a cost advantage by being able to use the large scale quantity production facilities set up to produce vehicle, agricultural and industrial engines, but also to take advantage of the immense mileage and number of hours built up on vehicle engines during development and early production.

This compares most advantageously with the much slower build-up of hours possible in marine field testing and contributes materially to the reliability of the unit, which is an all-important feature of marine engines. To underline the fact that the end result is a marine engine and not a marinized unit, Perkins will always change the basic engine to match the marine requirements, should there be any conflict of specification. (Figures 17–10 and 17–11 show sectional arrangements of the engine).

In general, as far as the basic engine unit is concerned, the design

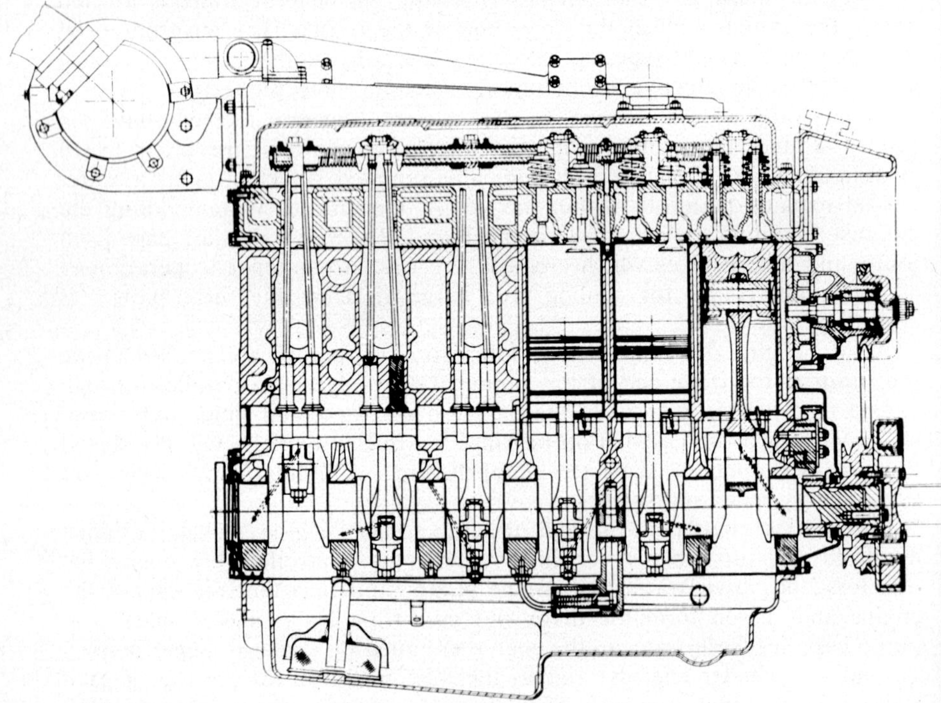

FIG. 17–10—*Perkins T6 3543 marine engine.*

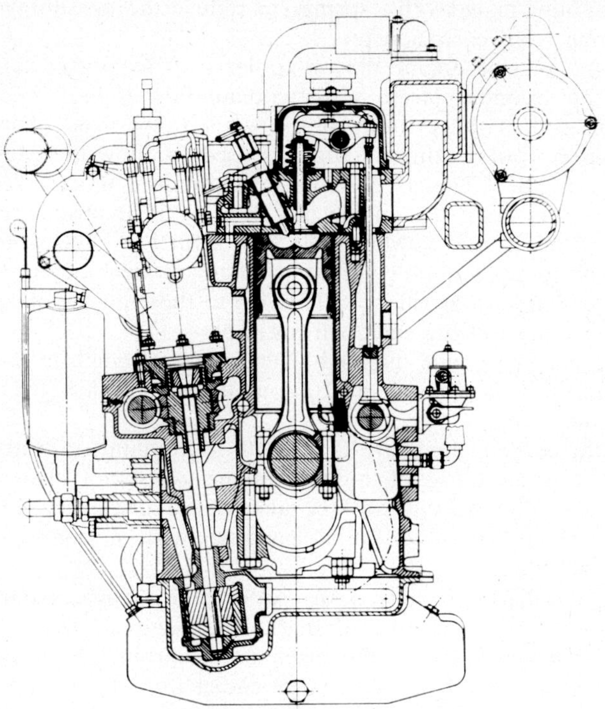

FIG. 17–11—*Perkins T6 3543 marine engine.*

requirements of a modern high speed marine diesel are similar to those required for an automotive engine and thus conflicts are few.

For example, the rating of the T6.3543 vehicle engine is 155 b.h.p. at 2,600 rev/min, whilst the marine engine is sold at these ratings:

145 s.h.p. at 2,000 rev/min continuous rating for commercial craft.

175 s.h.p. at 2,450 rev/min intermittent (1 hour) rating for pleasure craft or light duty high performance commercial craft.

245 s.h.p. at 2,600 rev/min special high performance rating for approved planing hulls only. The requirement for an approved hull is that it shall be capable of cruising on the plane with the engine throttled back to 145 s.h.p.

The 245 s.h.p. engine has a lower compression ratio of 13·25 : 1 which gives acceptable maximum cylinder pressure while still allowing the use of excess boost air to keep cycle temperature down.

Cold starting is affected by the use of a low compression ratio but it will still meet a requirement for marine engines of 15°F using only engine-mounted heat start equipment.

In the case of the new T6.3543 marine engine the changes to the basic engine are:

as in the vehicle engine.

1. Cadmium plated valve springs to reduce the possibility of fracture arising from corrosion pits.
2. In the marine engine the oil cooler uses sea water as the coolant so the secondary fresh water coolant flow is fed directly into the front end of the cylinder block water rail instead of into the rear
3. A special lubricating oil sump is used to allow the engine to sit as deep in the bilges as possible to maintain a more level shaft line. With this sump the engine can operate at an installation angle of 17° down at the flywheel end with a further increase to 20° when under way or at a continuous angle of heel of 30° when used for example in auxiliary sailers. Up to these angles of operation the connecting rod big ends will not dip.
4. The marine engine uses a higher fatigue strength 60 hour nitrided crankshaft in place of the 20 hour treated crankshaft in the vehicle engine.

   Main bearing shells are faced with aluminium tin bearing surfaces but use an alloy of higher fatigue resistance than the vehicle engine. The bearing surface also has a 0·0007 in thick tin lead overlay to improve initial bedding with a consequent greater fatigue life.

Engine auxiliaries include a sea water pump driven from the end of the auxiliary drive power take-off shaft. The pump is sized to maintain a sea water discharge below 130°F even under extreme tropical conditions. The coolant heat exchanger, air charge cooler and engine and gearbox oil coolers are sea water cooled. A cast iron sea water jacketed exhaust manifold is fitted. This manifold has twin exhaust passages leading to a twin volute turbocharger which permits the maintenance of useful boost down to lower throttle opening for the operation of winch or trawl gear. The engine can be built for either left or right hand rotation.

The engine is also built in a horizontal configuration with either left hand or right hand rotation.

The cylinder block and crankcase form a single piece casting with a deep skirt around the underslung crankshaft bearings and the cylinder block is stiff as tappet covers have been eliminated to reduce noise and give improved cylinder head gasket sealing. The cooling passages are designed to give full cooling up to the top position of the top piston ring travel. Cast iron, press-fitting dry liners are used, having a top flange which incorporates a raised fire ring to protect the gasket. Careful attention has been given to the cooling passages around the top of the cylinder and in the cylinder head to minimize thermal distortion. Oil is fed to the connecting rod bearing through drillings in the crankshaft which are arranged so that the centre main journal is not drilled. The big end bearings are plain and ungrooved.

The pistons have cored passages under the crown forming a cooling annulus to make effective use of the piston cooling oil which comes from fixed jets in the crankcase; this arrangement eliminates the need for oil drillings up the connecting rods and explains why plain big end bearings can be used. Wedge form small end bosses are employed to keep the bearing pressures low and the pistons are fitted with two compression rings and two oil control rings, one above the pin and one at the bottom of the piston skirt. The top compression ring is chromed and barrel-faced with a 6° wedge and negative ovality, and this is carried in an armoured insert groove of cast iron bonded to the aluminium alloy piston. The second ring, which is also chromed, is internally stepped with negative ovality, and the upper oil control ring is a conformable scraper of insert design with expanding spring. The bottom ring is a slotted oil control ring.

A large lubricating oil pump is used to provide the capacity needed for the six piston cooling jets in the crankcase. The oil cooler is mounted on the side of the engine and forms a unit with the oil filter. The cooler incorporates a bypass valve to safeguard against undue restrictions and a dump valve is fitted in the cylinder block to relieve the cooler of excess pressure under cold start conditions.

The fuel injection system consists of a distributor type pump and a mechanical governor. Fuelling at low speeds is regulated by a leaf spring controlled rising metering valve. The injectors, which are carried in copper sleeves in the cylinder head, are set to 210 atmospheres.

The specific fuel consumption is in the region of 0·37 lb/b.h.p./hour (168 gm/b.h.p./hour).

The bare engine without flywheel, flywheel housing, starter motor or after cooler weighs 912 lb (412 kg).

17.6. *Rolls-Royce D Range Engines*

The D range engines are manufactured by Rolls-Royce Limited, Oil Engine Division, Shrewsbury, who have kindly provided the material for this section.

The Rolls-Royce D Range diesel is representative of recent, high-speed, compact marine power units.

It is a turbocharged, 4 cycle, 90 degree Vee eight, liquid cooled, direct injection engine. A naturally aspirated version is available. The bore and stroke are 6·625 in (168·3 mm) and 7·25 in (184·15 mm) respectively giving a cubic capacity of 2,000 cubic inches (32·75 litres). The maximum rated power is 750 h.p. at 1,800 rev/min, the b.m.e.p. being 165 p.s.i. (11·6 kg/cm^2) and mean piston speed 2,174 ft/min (11·11 m/sec).

Construction details of this engine are indicated in the table below and in Figure 17–12. Performance data of the turbocharged and inter-cooled engine is given in Figure 17–13 and its weight and dimensions appear in Figure 17–14.

Type	8 cylinder 90° Vee 4 stroke liquid cooled
Combustion system	Direct injection with open chamber
Bore	6·625 in (168·275 mm)
Stroke	7·25 in (184·150 mm)
Capacity (swept volume) ...	2,000 cu. in (32·75 litres)

Maximum governed rev/min ...	1,800
Compression ratio	Naturally aspirated 15·5:1 Turbocharged 13·5:1
Rotation	Anti clockwise—viewed on flywheel
Engine firing order	A1, B1, A4, B4, B2, A3, B3, A2
Injection pump firing order ...	1, 2, 7, 8, 4, 5, 6, 3
Valve timing	Inlet opens 22° B.T.D.C. Inlet closes 50° A.B.D.C. Exhaust opens 70° B.B.D.C. Exhaust closes 20° A.T.D.C.
Injection timing	Naturally aspirated 36° B.T.D.C. Turbocharged 32° B.T.D.C.
Injection opening pressure ...	Atmospheres 200

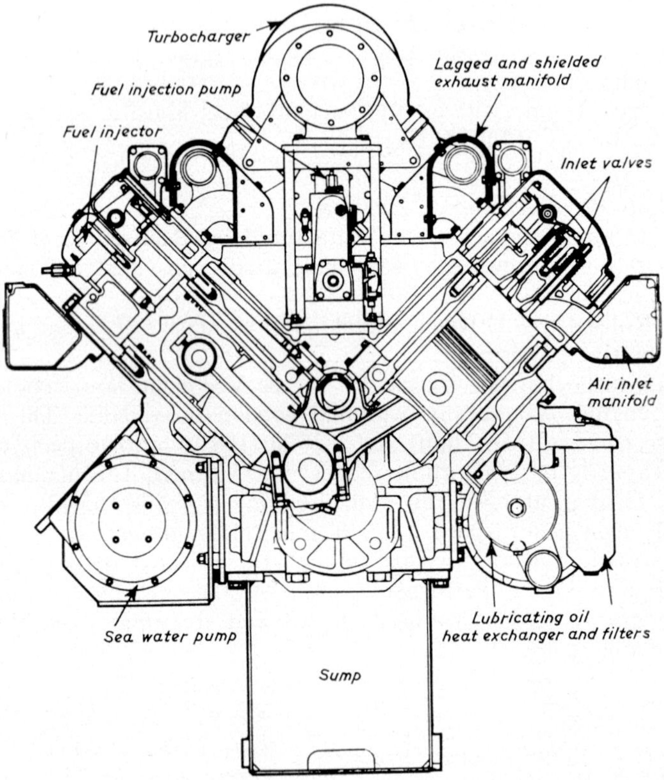

FIG. 17–12—*Rolls-Royce D range engine.*

Typical operating conditions for turbocharged and intercooled engine (DV8TCW) at both intermittent and continuous ratings are:

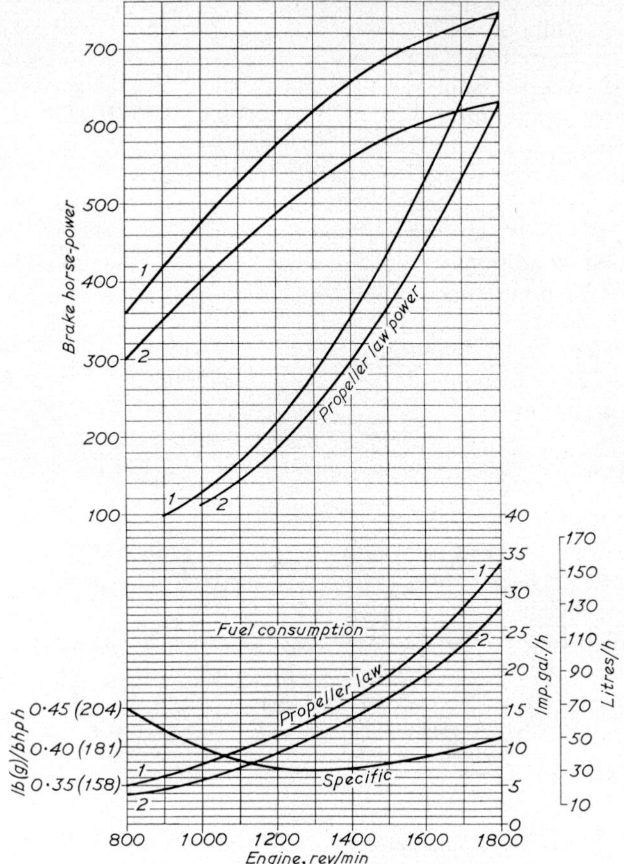

FIG. 17–13—*Rolls-Royce D range engine performance curves.*

	Intermittent Rating	*Continuous Rating*
Rev/min	1,800	1,800
Horsepower (gross)	750	635
B.m.e.p.		
p.s.i.	165	139·7
kg/cm^2	11·6	9·82
Fuel consumption		
lb/b.h.p./hr	0·38	0·38
gm/b.h.p./hr	172	172
Mean piston speed		
ft/min	2,174	2,174
m/sec	11·11	11·11
Combustion air requirements		
cfm	1,650	1,500
cu. m/sec	0·779	0·708

Engine ratings are quoted at conditions of:

Air inlet temperature	85°F (29·4°C)	
Air inlet restriction	10 in water gauge	
Exhaust back pressure	2 in Hg (static)	
Sea water temperature	75°F (23·9°C)	
Barometer	29·5 in Hg	
Cooling water at entry to charge air cooler	75°F (23·9°C)	
Derating for other climatic operating conditions	As BS.649	
Exhaust temperature at turbocharger outlet		
Water cooled manifolds ...	435°C	400°C
Air cooled manifolds ...	500°C	450°C
Exhaust gas flow		
cfm	4,500	3,900
m/sec	2·124	1·84

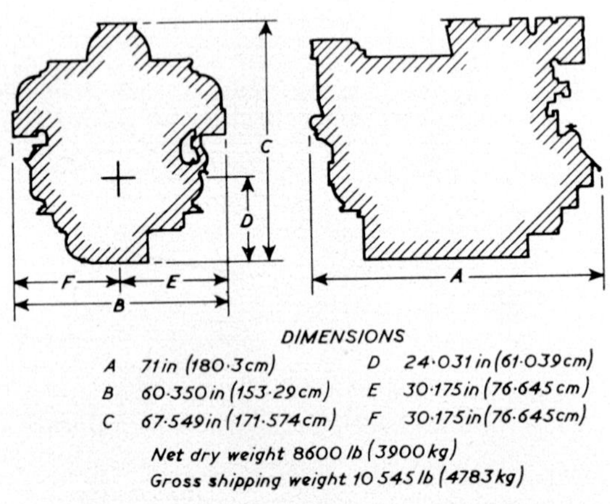

FIG. 17–14—*Rolls-Royce D range engine weight and dimensions.*

The basic engine frame is a monobloc cast-iron structure, combining cylinder blocks and crankcase. This same casting also incorporates the gearcase at the drive end. Main bearing caps are underslung and cross tied to the crankcase skirt by lateral bolts. Cover plates are fitted to both sides of the crankcase and permit internal inspection and removal of the connecting rod bearings. The underside of the crankcase is enclosed by a fabricated steel sump.

The flanged cylinder liners are a "slip-fit" in the crankcase and have three synthetic rubber seals at the lower end. The liners are differentially hardened by an induction process during manufacture in order to meet the design requirements of a relatively ductile upper flange, together with a hardened and tempered bore.

The chrome-molybdenum crankshaft, is forged from a single billet and undergoes a 60 hour nitride hardening process. After nitriding the crank-pins and journals are lapped to restore their prehardened size and to remove the superficial skin or "white-layer". It runs in five lead-bronze steel-backed shell bearings and is located by thrust bearings about the centre bearing. An additional bearing, adjacent to the flywheel provides additional support for a proportion of the weight of driven machinery such as marine gearing or the rotating parts of a single bearing generator used in connexion with a diesel electric propulsion or generating system.

Individual, alloy cast-iron cylinder heads each carry two inlet and two exhaust valves, which operate in pairs from single push rods which actuate T-shaped bridge pieces via the rocker arms. Stellite valve faces are utilized and the exhaust valve stems are sodium filled. Positive valve rotation is a feature. A spigot on the cylinder head joint face engages with a recess in the cylinder liner flange and gas sealing at this joint is accomplished with a hard copper sealing ring. Fuel injectors are centrally mounted, outside the valve rocker covers, for ease of service access and avoidance of fuel dilution of the crankcase oil due to pipe leakage.

A one-piece, forged and case hardened, nickel steel, camshaft is supported in five plain bearings in the Vee of the crankcase and driven by helical gearing from an integral pinion on the crankshaft. Straight lift, mushroom headed chilled iron tappets operate the valve gear. The tappets are phosphate treated and lubricated by individual oil jets.

Aluminium alloy pistons have a cast-in austenitic iron insert to carry the top ring, which has a molybdenum-inlaid face. There are two additional compression and one oil-control rings. Piston cooling is also by a metered supply of oil through jets. The piston crown is machined out to provide an open combustion chamber of "Mexican-hat" form.

The side by side connecting rods are forged from chrome molybdenum steel and are split diagonally at the big-end to permit passage through the cylinder bore. Joint faces at the bearing caps are serrated for positive location and the cap is retained by two set bolts.

The gear train, housed in a gear case integral with the crankcase, provides drives to the camshaft, fuel injection pump, oil and coolant pumps, tachometer generator and engine service counter. Helical gearing is employed throughout and gear end-thrust is taken by phosphor bronze thrust bearings.

Engine lubricating oil is carried in the sump whose capacity is some 31·5 gallons (143 litres). This sump may be emptied via drain plugs or by means of a semi-rotary evacuation pump, dependent on the installation. Provision is made for priming the oil system by means of an electrical or manual pump—either of these being used to draw oil from the sump and deliver it, via a non-return valve, to the engine's lubrication system. The engine's main oil pump is mounted externally for ready access, and delivers oil to a heat exchanger and hence to a triple element full flow filter. This filter head casting incorporates a relief valve (55–65 p.s.i.; 3·87–4·57 kg/cm^2) and also an element by-pass valve. The three filter elements are arranged in parallel and utilize replaceable, impregnated paper elements.

After filtration, the oil continues its path to supply pressure lubrication to all main bearings, crankpins, piston pin bushes, camshaft bearings and valve operating gear. The use of oil transfer bobbins between interfaces is to be noted.

The coolant employed in the engine water jacket may be either an inhibited glycol/water mixture or inhibited plain water. Normally the system is pressurized, a relief valve venting pressures in excess of some 7 p.s.i. (0·5 kg/cm^2) and relieving a depression greater than 1 p.s.i. (0·07 kg/cm^2) below atmospheric pressure. A centrifugal water pump, driven in tandem with the lubricating oil pump circulates coolant through the engine mounted coolant/sea water heat exchanger, lubricating oil cooler, cylinder jackets and cylinder heads. The system incorporates a multi-element wax-type thermostat. As in the lubricating oil system, the use of bobbin type transfer between interfaces is noted in the coolant circuit.

Where applicable an engine mounted and driven sea water pump is provided for raw water supply to both the engine water jacket heat exchanger and also to the intercooler applied to the turbocharged marine engines.

An eight element CAV fuel injection pump is mounted within the engine Vee and driven from the gear case. Twin fuel lift pumps are fitted to fuel pump body and operated from the pump camshaft. Injectors, fed by high pressure pipes from the fuel pump, are multihole type and can be withdrawn for service after the release of a simple clamping device. Dual fuel filters are fitted with easily removable impregnated paper filter elements. A small reservoir above the filters ensures that the fuel system remains primed after long periods of inactivity. For marine propulsion the CAV hydraulic governor is normally used whilst alternative governors such as the Woodward PSG type would be applied to any diesel electric or pumping application calling for close control governing.

An inward flow four entry turbocharger is mounted centrally at the drive end of the engine. Intercoolers are normally sea water cooled on marine propulsion engines, but air cooled on certain non-propulsion applications when radiator cooling is sometimes required. Exhaust manifolds may be air or water cooled.

Engine starting is by 24 volt electrical equipment, with air or hydraulic starters as options. A range of engine mounted alternators is available. In addition to charging starter batteries they also provide small amounts of power for auxiliary circuits on board ship.

17.7. *Volvo Penta TAMD70B*

Volvo Penta TAMD70B engines are made by AB Volvo Penta, Gothenburg, Sweden, who have kindly provided the material for this section.

The Volvo Penta TAMD70B engine is a turbocharged, after cooled, 6 cylinder marine diesel engine designed for inboard mounting and also to operate in conjunction with the Volvo 750 inboard-outboard drive for the propulsion of large leisure craft and commercial boats of the planing type. The engine is compact and has been designed to give a high output in relation to its weight and bulk. Figure 17–15 shows the 6 cylinder engine. Some technical details are given in the following table:

Bore	104·77 mm	4·13 inches
Stroke	130·00 mm	5·12 inches
Continuous output at:		
2,200 rev/min	185 h.p.	
2,000 rev/min	170 h.p.	
1,800 rev/min	155 h.p.	
Maximum output 500 hours/year		
2,500 rev/min	250 h.p.	
2,200 rev/min	230 h.p.	
2,000 rev/min	215 h.p.	
1,800 rev/min	200 h.p.	

FIG. 17–15—*Volvo Penta TAMD70B 6-cylinder engine.*

The cylinder block and crankcase are an integral casting of high strength alloy cast iron in which the crankshaft is supported in underslung fashion. Access to the crankshaft and bearings is by removing the light sump bolted to the underside of the crankcase. The crankshaft is made of drop forged steel, surface hardened on the bearing journals and is carried in seven main bearings, the centre bearing serving as a location. Main and big end bearings are steel shells lined with lead bronze and indium plated on the running surface. The cylinder block is fitted with replaceable wet type liners which are positioned by a flange at the upper end which is gripped between the cylinder head gasket and the block. The lower ends are sealed by compressible rubber joint rings. The cylinder heads are both single piece castings manufactured in cast iron, each covering three cylinders. They are fitted with replaceable valve seats and guides. The gaskets are made entirely of steel and have sealing ring inserts for oil and water.

The valve camshaft, lubricating oil pump, fuel injection pump and sea water pump are driven from the crankshaft by a train of helical gears at the forward end. The camshaft is made from a steel drop forging and is carried in seven bearings of steel with white metal linings. The valves are operated by tappets and push rods actuating rocker levers mounted on top of the cylinder head and enclosed by two rocker box covers. Their stems are fitted with hardened, replaceable wear caps.

The pistons are made of light alloy and have three compression rings and one oil scraper ring situated above the pin. The top compression ring is chromed and is fitted in an alloy cast iron ring carrier cast into the piston. Cooling is by splash. The connecting rods, which are of I section, are of drop forged, toughened steel and are fitted with bronze bushes at the small end. They are drilled through for pressure lubrication of the gudgeon pins.

The lubricating oil is drawn from the sump through a strainer and pumped through the oil cooler and two full flow filters in parallel before passing to the engine. A relief valve built in the oil pump prevents the pressure from reaching excessive values. The whole of the engine is pressure lubricated.

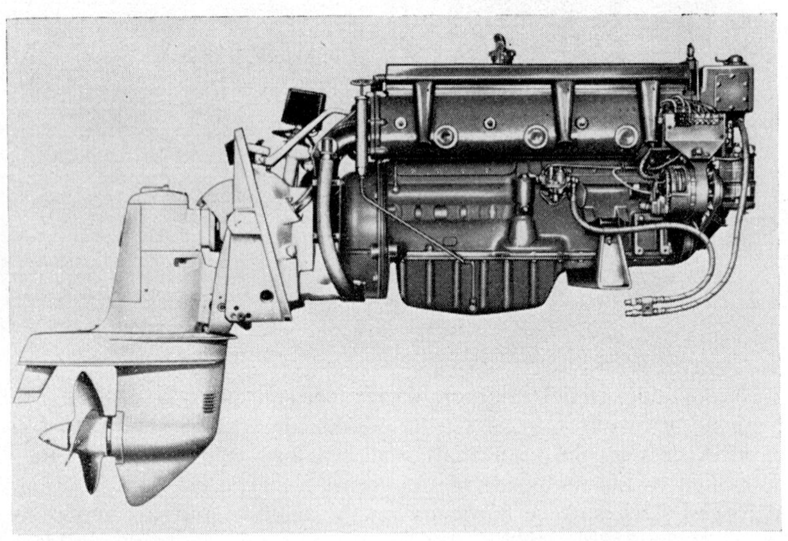

FIG. 17–16—*Volvo Penta TAMD70B 6-cylinder engine with outboard propulsion unit.*

The fuel system consists of a feed pump with pre-filter, fine filter, fuel injection pump and injectors as well as fuel lines. The injection pump is directly lubricated from the engine lubricating system.

An exhaust gas turbocharger is connected directly to the exhaust manifold. The compressor supplies the pressurized air to the engine through an after-cooler. The turbocharger bearings are lubricated by oil from the engine pressure lubrication system and the turbine housing is cooled by water from the engine fresh water system. The aftercooler is fed with sea water to cool the charging air.

The engine is fresh water cooled, the fresh water system being circulated by a pump mounted at the forward end of the engine. This pump draws coolant from the lower part of a heat exchanger which is also mounted on the starboard side of the engine, and forces it into the cylinder block, and to the cylinder heads. The coolant flows through two thermostats to the upper part of the heat exchanger and then passes down through a system of tubes whilst being cooled by sea water. The exhaust manifold is completely fresh water cooled. When the coolant is at low temperatures the thermostat keeps the passage to the heat exchanger closed and the coolant passes through a bypass line directly to the suction side of the pump. This enables the engine to reach operating temperature quickly and to prevent running at too low temperatures during cold weather. The fresh water system is pressurized to raise the boiling point of the water and reduce evaporation. An overpressure safety valve is fitted.

The sea water pump forces cooling water through the oil cooler before forcing it under pressure through the after cooler and the fresh water heat exchanger.

The engine is started by electric starter motor.

The engine is usually sold for propulsion purposes in combination with the 750 outboard drive with which it is combined to form an integral unit rubber mounted to the hull at three points. The arrangement is shown in Figure 17–16.

CHAPTER EIGHTEEN

Air Cooled Engines

18.1. A number of high speed engines for marine use are designed in air cooled form. This fact may be considered remarkable for an application in which there is an abundant supply of cold water close at hand, but air cooling has a number of advantages and developments in this field have given rise to a growing demand.

It is an obvious and well known fact that large bore engines are more difficult to cool than small bore engines because of the adverse ratio of surface area to volume and, as air cooling requires precise design if it is to be effective, it is more commonly found in smaller engines than in larger ones. However, development continues to apply air cooling to larger cylinder sizes and it is difficult to say that a limit has been reached.

In several instances both air and water-cooled versions of an engine are offered by a manufacturer, many components being common to both, but it should be appreciated that the construction of air cooled engines is usually quite different from their water cooled counterparts; they are not just re-designed water cooled engines. A single cylinder air cooled engine has a cylinder head and barrel with fine finning designed to provide cooling at the top of the barrel and the base of the cylinder head. The design has the advantage that the joint between these two components is simple as there are no water passages to seal. For multi-cylinder engines the general practice is to use the cylinder head and barrel of the single cylinder eliminating the complicated castings for the cylinder block and cylinder head of the small water-cooled engine.

The overall design of the engine must be conditioned by the requirement of an adequate flow of cooling air directed to the vital components in proportion to their needs. Provided these points have been taken into account air cooling has advantages to offer in the following areas:

a) Simplicity, leading to enhanced reliability.
b) Less skilled maintenance, as the cooling system does not require high pressure joints to be maintained leakproof.
c) No frost risk and no elaborate precautions to be taken in this respect.
d) In small boats, which may be left in a harbour that dries out, there is no reason why air cooled engines should not be run at any time, provided the drive to the propeller shaft is disengaged.
e) When air cooled engines are used in lifeboats, the required check that all is in working order can be carried out, including running the engine while the lifeboat is still in the davits without any fear of seizure, provided again that the drive to the propeller shaft is disengaged.
f) Air cooled engines are quicker to warm up to running temperature than water cooled engines.

g) The cooling air can be used for warm air ventilation, a particularly useful feature for small craft.
h) Amongst the manufacturing considerations the castings for air cooled engines are usually simpler than those for water cooled engines as they do not involve many cored passages and compartments.

18.2. *Lister JA Engines*

Lister JA (air cooled) and JW (water cooled) engines are manufactured by R. A. Lister & Co. Ltd., Dursley, Gloucestershire—a Hawker Siddeley Company. This section is based on material kindly provided by them in the form of an article by their Technical Director, Mr. A. J. Morris.

R. A. Lister & Co. Ltd. have a long experience in the design, development and manufacture of both air and water-cooled engines. A new range of direct injection engines has been introduced, suitable for marine propulsion or power generation as well as a wide range of industrial duties. It is significant of the growing demand both for air cooling and increased power, together with the developing techniques to meet these demands, that this new range of engines is offered in both air and water-cooled versions, although it is of larger bore than the older designs it is intended to replace.

The JA engine has six cylinders of 5 inches bore and $5\frac{1}{2}$ inches stroke. It is offered in either naturally aspirated or turbocharged versions. The engines all meet Lloyd's requirements. Figure 18–1 shows the air cooled engine and Figure 18–2 the water cooled version, both arranged as propulsion engines with hydraulic reverse-reduction gears.

The total cylinder capacity of the six cylinders is 647·9 cu. in. The engine has been developed to run continuously at any speed between 1,000 and 2,000 rev/min and Figure 18–3 shows the performance in naturally

FIG. 18–1—*JA6 MGR engine 138 bhp at 2000 rev/min air-cooled.*

FIG. 18–2—*JW6 MGR engine 138 bhp at 2000 rev/min water-cooled.*

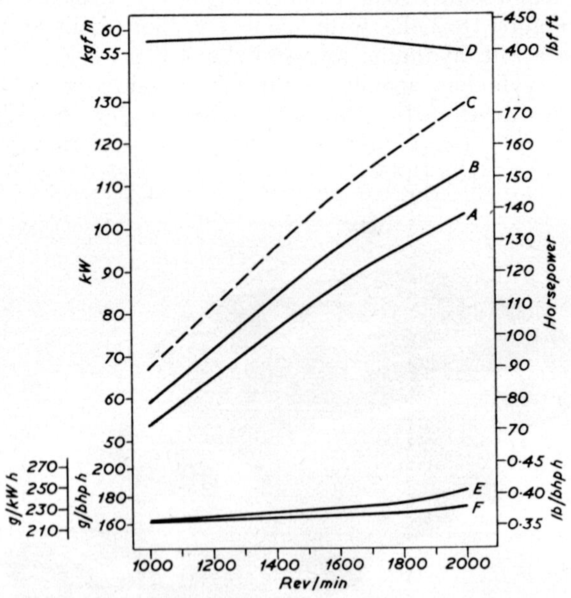

FIG. 18–3—*Performance of naturally-aspirated JA6 MGR engine.*

aspirated form. The continuous British Standard rating for this engine is 138 h.p. at 2,000 rev/min and the table below gives leading particulars of it. The turbocharged version of the six-cylinder air cooled engine will, in the first instance, develop 180 h.p. at 2,000 rev/min. The design follows the well-tried lines of the Lister air cooled engines, but with some additional features which are necessary for marketing these larger engines

Bore and stroke. in (mm)	5 × 5·5 (127·0 × 139·7)
Total cylinder capacity. in^3 (litre) ...	647·9 (10·617)
BHP continuous BS or Din A. (kW) ...	138 (102·91)
Rev/min	2,000
B.m.e.p. at 1,500 rev/min (lb/in^2) (bars)	89·2 (6·15)
Weight with 220 lb (99·8 kg) flywheel and electric starting. lb (kg)	2,094 (950)
Overall dimensions. in (mm)	
Length including $2\frac{1}{2}$ in (63 mm) shaft extension	58·5 (1,485)
Width	30 (762)
Height	40 (1,016)

The cooling air blowers were the object of separate development. The design finally decided upon is 10 inches in diameter, runs at 7,900 rev/min and gives an installed flow of 23 cu ft per h.p. at a pressure of $4\frac{1}{2}$ inches WG. The stator blades have a mean angle of 20° and the impeller blades 35·5°.

The cast iron crankcase extends well below the crankshaft centre line to form a rigid frame carrying the underslung main bearings.

The crankshaft is a steel forging having a minimum tensile strength of 55 t.s.i. and all the journals and crankpins are induction hardened to a minimum of 53 Rockwell C and lapped after grinding to 12 micro inches CLA. The drillings for the lubricating oil have cross holes, so that the loaded part of the crankpins are plain, with no oil holes breaking through them.

A 0·25 inch oil hole passes longitudinally through the centre of the connecting rod H section. This hole is drilled from the small end towards the big end, but does not break through into the latter as this would weaken the bearing shell. To feed oil to the small end and to assist in cooling the piston, oil is fed from the crankshaft through a hole in the unloaded part of the top half of the big end, and then through a cross hole into the main longitudinal hole previously mentioned.

One of the important considerations in the developing and running of an air-cooled engine of this size, with very high reliability, is the cooling of the lubricating oil. Figure 18–4 shows a picture of the very large oil cooler which is used and which gives a sump oil temperature rise of 55°C over ambient, at full engine speed and power. The cooler is connected to the main oil system by means of two flexible hoses so as to prevent pipe fractures and a by-pass valve is included to deal with cold ambient operations.

The lubricating oil pump is driven through an idler at the free end of the engine crankshaft, which is steadied by means of an 11 inch viscous crankshaft damper. The oil is fed under pressure to all moving parts, with the exception of the pistons and valve stems, which are splash lubricated. Special attention has been given to the lubrication of the valve stems. One of the problems in lubricating these components is that the oil splash is

FIG. 18–4—*Oil cooler.*

equally distributed between the inlet and exhaust valve and in naturally aspirated engines it is often necessary to allow more lubrication for the exhaust than for the inlet, since in general there is a suction on the inlet port and a pressure in the exhaust one. In the J range of engines the top plate of the cylinder head, which carries the valve gear, has been designed to retain more oil under the springs of the exhaust side than the inlet regardless of the angle of the operation of the engine. The crankcase is drilled for the distribution of the oil and there are only two main oil pipes —the main oil suction, and the internal delivery from the pump to the crankcase.

A pressed steel sump is used and this is clamped to the crankcase, the sealing being made by a round section strip of special rubber. The clamps are bolted solid to the crankcase and the rubber compressed a predetermined amount, so that once the sump is tightened it does not require re-torqueing, as it would if a flat joint under compression were used. The rubber strip sealing system with metal to metal bolting is used in various other parts of the engine, for example, the rocker covers and the oil filter mounting, in order to make a good seal without re-torqueing.

The cylinders are closely finned, and special attention has been given to the finning between the cylinders, to maintain an adequate air flow area.

The pistons are effectively cooled by the oil splash from the high pressure oil feed to the small end. The combustion chamber is machined in the crown and has an effective compression ratio of 14·5:1 Five rings are used: two scraper rings, one above and one below the gudgeon pin and a specially honed chromium plated top ring. This arrangement gives satisfactory continuous no-load running and a full load oil consumption in the region of 0·25% to 0·75% of the fuel consumption.

A two-part cylinder head is used, with an open simple die casting of heat treated aluminium alloy forming the lower part, and a heavy duty cast iron deep top plate above it, housing the valve gear, and taking the load from the four holding down studs, which go right down to the crankcase. This system of cylinder head construction permits the use of cylinder head finning designed almost without production limitations, the only aim being to use the available air in the most effective way. As shown in Figure 18–5 the exhaust side of the head is divided from the air intake side by a fin running the whole length of the air path; the air that cools the exhaust part of the head is passed right through, whilst the air that goes over the relatively cool inlet port is turned round and used to cool the injector, before it is discharged. The air velocities through the head are maintained more constant with this system than generally are achieved with some other designs, which have casting limitations.

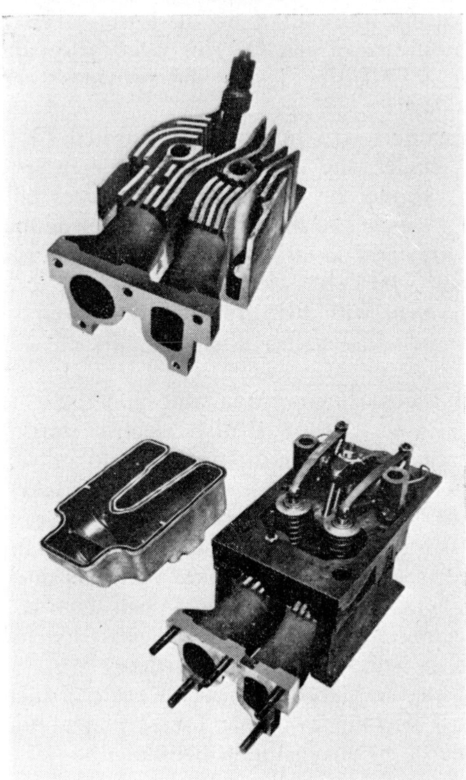

FIG. 18–5—*Cylinder head, JA6 MGR engine.*

The joint between the cylinder and crankcase is solid copper and that between the cylinder and the cylinder head is solid steel. Here again, no re-torqueing is necessary, as all the joints are solid metal to metal. The same system is used in the water-cooled version of the engine, except that in this case the additional water passages are circular in section and sealed by heavy section silicone rubber "O" rings, prevented from collapsing internally by short stainless steel ferrules.

Two valves are used per cylinder with individual cast iron rocker brackets retained by the cylinder head holding down studs, as has been normal Lister practice for 18 years.

The injectors are positioned between the rockers but are externally mounted, with the rocker cover protecting them from dirt; they are held by a single bolt clamp which ensures even pressures and good sealing. Bryce injection equipment is used, consisting of interchangeable, individual, high flange, 9 mm pumps, one for each cylinder symmetrically and externally mounted, and four hole nozzle injectors. A simple and quick system for balancing the calibrated output of the pumps has been devised and each pump can be timed by means of split shims without lifting the pump from its seating more than about 1 mm. The pumps are sealed against oil leaks by means of an "O" ring fitted below the flange.

The governor is of Lister design and provides speed regulation well within British Standards.

Crankcase breathers are internally connected to the inlet ports in every individual cylinder, the gases passing through special oil separating chambers in each cylinder head cover. This reduces pollution, and maintains a vacuum of $1\frac{1}{2}$ inches to three inches WG inside the crankcase.

Double belts are used to drive the cooling air fan. The belts have an automatic tensioner which has adjustment for the tension range and a damping device to deal with belts of slightly uneven section. In the unlikely event of one belt failing the other has sufficient capacity to maintain the drive.

A number of accessories are available and these include SAE No. 1 Flywheel housing, American or British electric starting equipment and other interchangeable starting equipment, single or double plate power take-off clutches at the flywheel end, built in power take-off at the free end of the engine up to 70 h.p.—and more in special cases, provision for gear driving a compressor, water pumps and hydraulic pumps facing towards the flywheel or the free end of the engine, air outlet ducting, Donaclone or Cyclopac air cleaners, flexible couplings, belt guards, oil pressure and temperature protecting devices, tachometer, shaft extensions at either end of the engine with or without pedestal bearings.

The standard engines can be run at an angle of inclination of 15° but engines shortly will become available for special applications capable of running at an angle of 30° from the horizontal.

18.3. *Petter Marine Diesel Engine*

Petter marine diesel engines are manufactured by Petters Limited, Staines, Middlesex who have kindly provided the material for this section.

Petters manufacture air cooled diesel engines in the range from 3 brake horse power to 45 brake horse power for marine use. They are suitable for both propulsion and auxiliary duties. The smaller engines are made in single cylinder form only, whilst the larger engines of the range are offered in 1, 2, 3 or 4 cylinder versions, according to the power. Water cooled versions are also available.

Engines of small horse power are offered in a very competitive market for many kinds of duty. In consequence the marine version is necessarily based on adaptation of successful industrial engines. Fortunately, the major application of this type of engine, which is found in civil engineering construction work, demands an extremely robust design capable of standing considerable abuse, which can be operated over a wide range of speed and load in all weathers and can be maintained with a minimum of skill. These are exactly the qualities required in marine engines for small boats and the ready acceptance of these engines in marine engineering applications such as emergency marine auxiliary sets and lifeboat propulsion units, is proof of the reliable and successful adaptation.

This section describes the 'A' series, the smallest series of the range which is manufactured in single cylinder form only; sectional views are shown in Figure 18–6. Technical data for the series is given in the following table:

Engine type		*AA1*	*AB1*	*AC1*
Bore	inches	$2\frac{3}{4}$	3	3
Stroke	inches	$2\frac{1}{4}$	$2\frac{1}{4}$	$2\frac{5}{8}$
Power	b.h.p.	$1\frac{1}{2}$–$3\frac{1}{2}$	$2\frac{1}{4}$–5	$2\frac{3}{4}$–$6\frac{1}{2}$
Rev/min		1,500–3,600	1,500–3,600	1,500–3,600
Weight	lbs	95	98	104
Size:				
Height	inches	17	17	$18\frac{1}{8}$
Length	inches	$14\frac{1}{4}$	$14\frac{1}{4}$	$14\frac{1}{4}$
Width	inches	13	13	$14\frac{1}{2}$
Compression ratio		17:1	16·25:1	17:1
Piston speed	ft/min	1,125	1,125	1,312
Mech. efficiency	%	59	63	68
Brake thermal efficiency	%	25	28	32

The engine type AA1 is the basic engine, whilst the AB1 and the AC1 are consecutive developments to higher powers and increased efficiency. A performance curve is shown in Figure 18–7. The combustion chamber is a Lanova type air cell, as shown in Figure 18–8. The main combustion chamber is formed in the cylinder head and the air cell, which is in connexion with it, is situated opposite the pintle type single hole injector nozzle. The fuel is sprayed across the main combustion chamber, the centre of the spray being directed just below the centre line of the air cell. Combustion starts separately in both the chamber and the cell, the combustion in the cell causing a discharge into the main chamber as burning proceeds which promotes turbulence in the gases already in process of combustion in the main chamber. The result is a very thorough mixing of fuel and air giving complete combustion in spite of the difficulties of distributing the very small volume of fuel which is required to be injected for each cycle. The characteristics of this type of combustion chamber are low maximum pressure and a smooth rate of pressure rise. It is also noted for its good starting characteristics, being practically as good as the

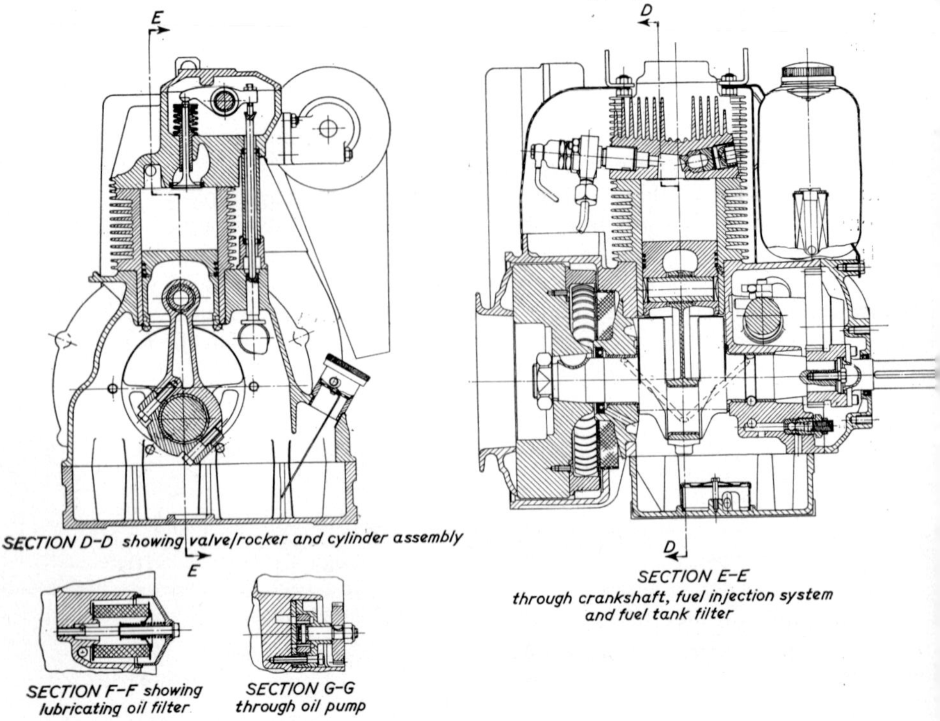

E
D
SECTION D–D showing valve/rocker and cylinder assembly
E
D
SECTION E–E
through crankshaft, fuel injection system
and fuel tank filter
SECTION F–F showing
lubricating oil filter
SECTION G–G
through oil pump

FIG. 18–6—*Petter Marine Diesel engine, "A" series.*

Continuous rating, bhp
kW
ACI
ABI
AAI
1
2
3
4
5
6
7
Specific fuel consumption, lb/bhp h
0·60
0·55
0·50
0·45
g/kWh
375
350
325
300
275
AA1
ABI
ACI
1000
1500
1800
2100
2500
3000
3600
Rev/min

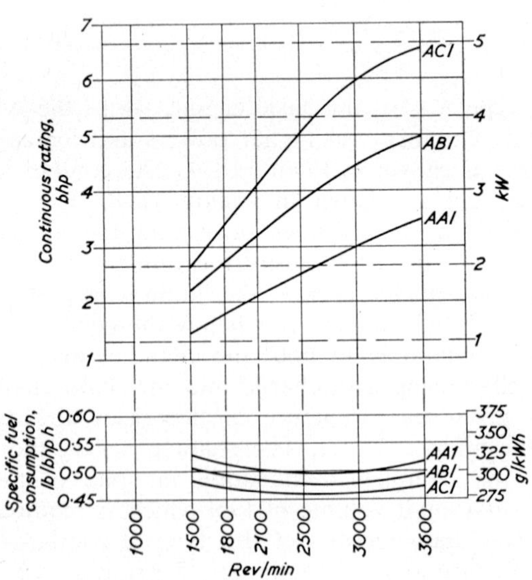

FIG. 18–7—*Performance curves, Petter Marine Diesel "A" series.*

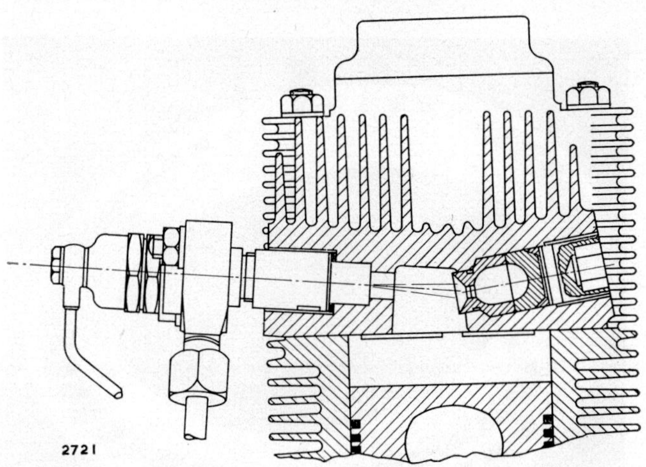

FIG. 18–8—*Combustion chamber, Petter engine.*

traditional best design for starting, the direct injection combustion chamber.

In small engines of this type, which may be used as lifeboat engines, propulsion units for small craft or as emergency generator sets, there is in many cases no source of electrical supply and for reasons of cost only hand starting may be provided. In the case of this small engine it is provided with a rope start from a pulley fitted on the crankshaft. Hand cranks and electric starting are alternatives which are available if required. To improve starting at low temperatures a permissible aid is the use of lubricating oil in the combustion chamber and this is normally introduced into the inlet port from a cylinder with a hand operated plunger. The object of this priming is to make a seal round the piston to ensure reaching the required compression pressure.

The cylinder head is a die cast aluminium alloy component with the Lanova air cell made of heat resistant alloy steel fitting directly into it. The valve guides and seats are of austenitic steel and are shrunk into the head. The combustion chamber is cast into the head and finning is carefully arranged on the outside to provide adequate cooling.

The cylinder barrel is of cast iron and has dense finning especially at the top. It will be noted that the fins are cast completely round the barrel and the holding studs pass through holes drilled in the fins. This design results in the minimum variation of temperature round the barrel so keeping distortion to the smallest possible amount.

The cooling air is carefully ducted round the cylinder barrel and head; a large proportion being allotted to cooling the head. This cooling air is supplied by a centrifugal fan mounted on the flywheel and shown in Figure 18–9. It can be seen that the blades are of good aerodynamic design, accurate shape and smooth finish. The material is polypropylene which is the lightest of the commercial plastics (S.G. 0·905). The air flow is 185 cu ft/min at 3,600 rev/min.

The crankcase is an aluminium alloy die casting. It incorporates the

FIG. 18–9—*Air cooling fan, flywheel mounted, Petter engine, "A" series.*

flywheel housing, the cooling fan volute and the gear end casing. This design minimizes assembly cost and obviates many possible sources of leakage. It also ensures concentricity without time consuming assembly problems. The oil sump is detachable.

The crankshaft is a steel forging and runs in aluminium tin shell bearings. The connecting rod is a forged steel component, the large end bearing being aluminium tin and an aluminium alloy piston is used.

The lubricating oil system is of conventional form with pressure feeds to the main bearings, crankpin bearings and valve rockers; a gear type lubricating oil pump is the source of supply. The camshaft operates in bearings which are steel backed bushes with PTFE lead impregnated sintered bronze linings and is lubricated by splash.

A paper element type air cleaner with a muffler tube and wire gauze pre-cleaner is a standard fitting for the combustion air intake together with a pepper pot type exhaust silencer. Both oil bath air cleaners and acoustic exhaust silencers are available, if required. The fuel tank is mounted on the engine.

Adaptation of the engine to suit a wide range of applications has been achieved by the development of components which may virtually be bolted on to produce different variants. Figure 18–10 shows an AB1 air cooled marine propulsion engine fitted with hot air ducting and reverse-reduction gearbox. Figure 18–11 shows the water cooled version of the same engine AB1W arranged as a marine auxiliary generator set fitted with raised hand start, water circulating pump and anti-vibration mountings. Obviously,

FIG. 18–10—*ABI air cooled marine propulsion engine.*

FIG. 18–11—*Water cooled ABI "W" engine.*

many combinations of such parts are possible to suit individual circumstances.

The complete engine has very good tilt characteristics, being capable of normal operation when inclined from the vertical in any direction up to an angle of $22\frac{1}{2}°$ permanently and 27° intermittently.

CHAPTER NINETEEN

Some Turbochargers

19.1. *Brown Boveri turbochargers*

Brown Boveri turbochargers are manufactured by Brown Boveri & Co. Ltd. Baden, Switzerland. They are marketed in Great Britain by British Brown Boveri Limited, Glen House, Stag Place, London S.W.1 who have kindly supplied the material for this section.

Brown Boveri turbochargers are suitable for a wide range of engines covering all powers from about 300 h.p. upwards. The basic design of the turbocharger is shown in Figures 19–1 and 19–2. It comprises a single stage, axial flow gas turbine driving a radial flow blower. The turbine and compressor wheels are mounted inboard of the bearings which are normally of the ball or ball and roller type and are lubricated by self-contained lubrication systems. The exhaust inlet and outlet casings are water cooled, the supply being provided from the engine jacket system. Air entry to the blower may be by inlet duct and suction branch or via a combined silencer filter. After leaving the turbine outlet casing the exhaust may go straight to atmosphere or it may pass through a waste heat boiler.

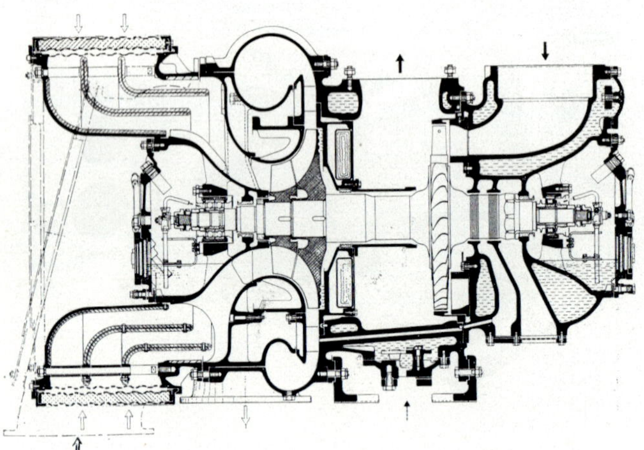

FIG. 19–1—*Brown Boveri turbocharger.*

The exhaust outlet casing, which also encloses the turbine wheel, forms the basic frame of the turboblower. It may be mounted on the engine in any one of twelve angular positions at 30° intervals round the axis of the shaft. The turbine entry casing assembly, including the bearing for the shaft at that end, is bolted on to the frame at one end and the compressor assembly with the other bearing is bolted to the other end.

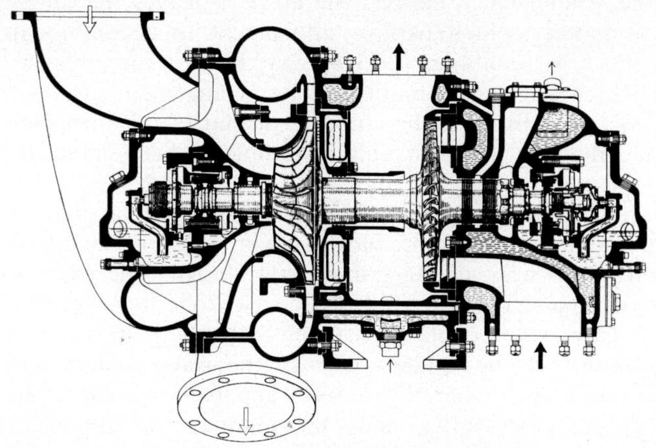

FIG. 19–2—*Brown Boveri turbocharger.*

The turbine wheel and shaft are formed from a single piece steel forging. The rotating parts of the compressor are comprised of the inducer and the impeller. They are keyed to the shaft and held in position by a threaded nut. The inducer is manufactured from a light alloy forging and consists of a ring of rotating guide vanes designed to conduct the air from an axial direction of flow at entry to a radial direction of flow at the impeller. The impeller is a single disk manufactured from a light alloy forging. It performs the task of compressing and imparting kinetic energy to the air flowing through it. On one side of the disk are the blades and on the other side, towards the outer edge, grooves are machined to form labyrinth glands; the slight flow of air which passes through these glands into the exhaust casing is used to cool the turbine shaft. The cooling air is kept close to the shaft by a sleeve and the labyrinth glands are positioned so that the end thrust on the shaft remains within admissible limits. The partition wall, which separates the compressor from the exhaust gas casing, embodies thermal insulation to prevent heat being transferred to the air with a resultant deterioration in compressor efficiency.

Air may be fed to the blower either from the atmosphere through a suction duct and branch or it may be drawn in from the engine room through a silencer filter. The suction noise of the impeller is reduced by the silencer and a cover plate, as shown in Figure 19–1, absorbs the radiated structure borne noise which would otherwise escape from the central part of the silencer. The increasing levels of pressure charging and higher speeds of turboblowers have made attention to noise reduction of greater importance. When an air suction branch is used it may be attached to the compressor air outlet casing at any angle of 30° interval.

The air leaving the impeller at a high velocity enters the diffuser. It is the duty of the diffuser, which is made from light alloy, to slow down the air and convert its kinetic energy into pressure energy. It will be seen from Figures 19–1 and 19–2 that the components forming the compressor casing, including the volute which collects the air as it leaves the diffuser, are designed in a manner which permits adjustment to accommodate different sizes of diffuser so enabling it to be matched to the pressure and flow conditions which the engine requires. The arrangement also permits withdrawal of the rotor through the compressor end of the turbocharger and it will be noted that the partition wall, including the heat shield, is withdrawn with the rotor.

The whole compressor assembly can be located on the frame at any angle of 30° interval. With some blowers twin outlets are available.

The turbine is a single stage axial flow design. The gas inlet and outlet casings are made of high grade grey cast iron; they are double walled to accommodate water cooling passages to protect the metal against the high temperature of the exhaust gas. In the water spaces anti-corrosion anodes are mounted on corehole covers and baffles in order to safeguard against corrosion. Depending upon the number and arrangement of the engine cylinders and its firing order, the gas inlet casing may have 1, 2, 3 or 4 inlet openings and can be positioned on the frame at any angle of 30° interval.

The turbine nozzle which admits the gas to the turbine blades has passages formed from high grade, heat resistant sheet metal which is cast-in between an inner and outer ring of cast iron. The maximum admissible gas inlet temperature to the turbocharger is normally between 600 and 650°C depending on the particular design, but special metals can be used for the nozzle rings and turbine blades if temperatures of 700°C or more are required. The turbine nozzle is secured to the inboard side of the turbine entry casing and it will be observed from Figures 19–1 and 19–2 that the construction is such that this component can be changed in order to match the boost conditions required by the engine.

The bearings are situated at the ends of the rotor shaft. In this position they are easily accessible and can be inspected without dismantling the turbocharger. They are normally ball, or ball and roller, bearings which are spring mounted within the casings. Their situation gives a stable arrangement so that uneven fouling of the turbine blades or breaking away of deposits which may cause imbalance of the turbine wheel does not result in high out-of-balance forces on the bearings. The bearing at the compressor end provides axial location for the rotor shaft and takes any resultant end thrust from the pressures upon the two wheels. The frictionless characteristics of the bearings simplifies the starting up of the engine, its response to change of load and, in the case of propulsion engines, assists in maintaining lower dead slow speeds. If required by the customer, sleeve bearings are available as an alternative.

The ball and roller bearings at each end have their own lubrication system. Oil is contained in a small sump from which it is drawn by a pump and sprayed onto the bearing or alternatively, it is distributed by a rotating disk. After flowing through the bearing it falls down again into the sump. It will be observed from Figures 19–1 and 19–2 that at the turbine end this sump is contained within the water jacket so that the oil is maintained in a cool condition away from the heat of the exhaust gases. Oil level inspection windows, together with filling and drain plugs, are provided at each end of the turboblower.

In order to prevent exhaust gases entering the bearings at the turbine end, a bleed of air is taken from the diffuser and conducted via passages, which can be seen in Figures 19–1 and 19–2, to labyrinth glands situated between the turbine disk and the bearing. This sealing air leaks along the shaft both towards the turbine disk to keep exhaust gases away from the bearing and also in the other direction to a vent space between the gland and the bearing.

19.2. *Hispano-Suiza turbochargers*

Hispano-Suiza turbochargers are manufactured by Hispano-Suiza, 92 Bois-Colombes, Hauts de Seine, France, who have kindly provided the material for this section.

This range of turbochargers falls into three groups, the HS300, HS400 and HS500. Each group covers a band of air flow and pressure ratio as shown in Figure 19–3 and the members of each group differ from each other both in their characteristics and internal design which are adapted to suit the requirements of the engines to which they are fitted.

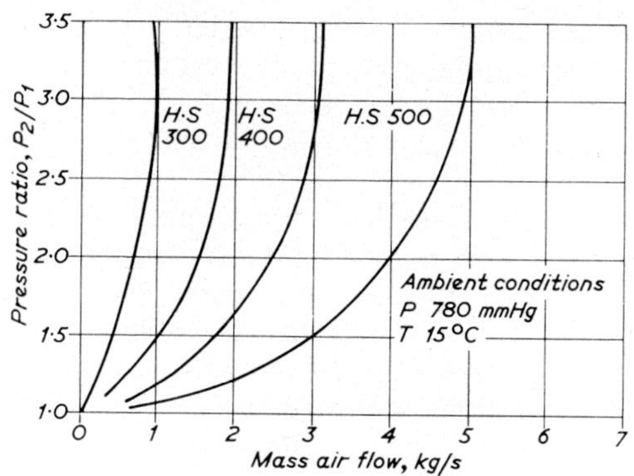

FIG. 19–3—*Hispano-Suiza turbochargers: airflow-pressure ratio.*

All the turbochargers have the following features in common. An axial flow turbine drives a centrifugal compressor, the compressor and turbine rotors are overhung, the shaft on which they are mounted being carried in two plain bearings inboard of the rotors. Each of the wheels is separately balanced and can be replaced during maintenance, if necessary. The journals are pressure lubricated by oil from the engine lubrication system and the exhaust gas casing is cooled by water from the engine jacket system. Turbine inlet temperatures up to 700°C may be used.

Figure 19–4 shows a section through an HS400 turbocharger from which the construction may be understood. The exhaust casing forms the main structure; it is carried on the mounting adaptor which holds the turbocharger on the engine and to it are fastened the other components of

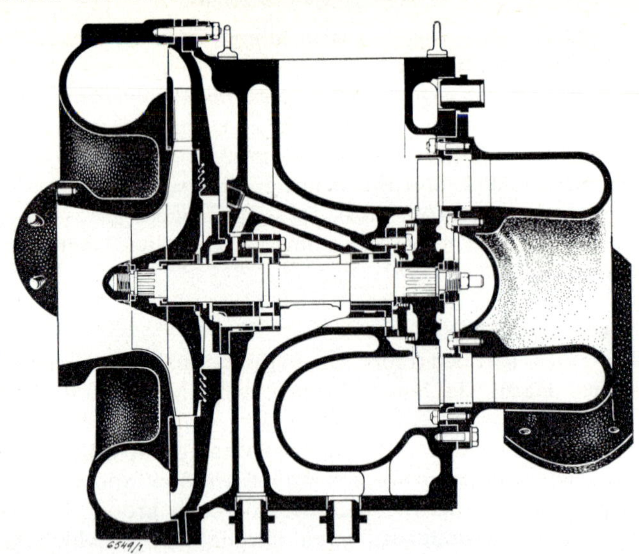

FIG. 19-4—*Section through Hispano-Suiza HS.400 turbocharger.*

the turbocharger. The lubricating oil inlet and outlet holes, cooling inlet and outlet holes and the exhaust gas outlet port are all formed in this casing as well as the internal passages for oil circulation and rear bearing sealing air. A hole, normally blanked off by a plug, is provided at the bottom of the casing so that any rain water that enters via the exhaust may be drained.

The shaft is supported in lead-bronze plain sleeve bearings at front and rear and is positioned axially by a pressure lubricated double acting thrust bearing assembly fitted to the rear face of the front bearing. Labyrinth seals are fitted to the front face of the front bearing and the rear face of the rear bearing to prevent oil leaking out of the bearings and air or exhaust gas leaking into them.

The stationary parts of the compressor are the impeller casing, the diffuser cover and the diffuser. The diffuser is secured to the diffuser cover and this assembly is positioned and secured to the exhaust casing by three countersunk screws. The impeller casing is machined in the form of a volute with a central air inlet opening and one, or if required, two tangential air outlets. It is secured to the exhaust casing by studs and encloses the diffuser and cover assembly.

The impeller consists of two rotating members, the steel rotating guide vanes and the light alloy impeller itself, assembled together in correct angular position. This impeller assembly is positioned on the shaft by splines and locating diameters. It is held by a blind nut at the air entering side. The diffuser cover has circular grooves machined in it which match circular ridges on the back of the impeller to form a labyrinth gland seal.

The turbine consists of a disk and blades made of high heat resisting steel. The blades are secured to the disk by fir tree type roots and are locked in position. The turbine disk is splined to the shaft, located radially by close fitting diameters and is retained by a nut and lock washer. The turbine nozzle is made up of the exhaust gas intake casing and two concentric nozzle guide vane rings which hold the nozzle guide vanes. The vanes fit in the slots of these inner and outer rings, some clearance being provided at assembly to allow for thermal expansion. The inner and outer rings are secured to the intake casing with bolts and lock washers and the whole turbine nozzle assembly is secured to the exhaust casing, a gasket being interposed to seal.

The turbine blading and nozzle guide vanes can be selected to give the desired supercharging ratio and the corresponding conditions of air pressure and flow are obtained by choosing the appropriate impeller and diffuser vane angle.

Lubricating oil is fed to the journal bearings and thrust bearing through internal passages machined in the exhaust casing. The oil returns to the engine sump under gravity. Cooling is achieved by feeding water from the engine coolant system, the circulation being in an upward direction entering the bottom of the exhaust casing and leaving at the top. The turbocharger cooling system can handle pressures up to a maximum of 6 bars.

There are two sealing systems in the turbocharger, one to seal air leaks at the impeller and one to seal oil leaks at the front and rear bearings. The first is achieved by the concentric grooves and disks in the impeller rear face and the diffuser cover. Leakage of air is used to pressurize the front and rear bearings and also for turbine disk cooling. The annular space between the diffuser cover and the exhaust casing forms a pressure chamber in which leaking air from the impeller periphery is discharged. Oil leakage from the outer faces of front and rear bearings at the impeller and turbine ends is prevented by the combined use of labyrinth seals and air pressurizing. The front bearing air pressurizing is direct as the labyrinth seal is situated in the pressurizing chamber. Rear bearing pressurizing is achieved through a passage provided in the casing and connecting the pressurizing chamber with the rear face of the labyrinth seal. At the rear bearing, owing to the close proximity of exhaust gases, two sealing devices are used, one on the outside of the turbine disk sealing lip and the other on the inside at the turbine disk hub.

The exhaust casing may be set on the engine in the axial direction at any angle not exceeding 15° on either side of the horizontal. In the transverse direction the exhaust casing may be set at an angle which does not exceed 60° on either side of the vertical axis. The impeller casing may be mounted on the exhaust casing at any 15° position over the complete circle. Similarly, the gas inlet casing may be mounted on the exhaust casing at any 30° position over the complete circle.

19.3 *Holset Turbochargers*

Holset turbochargers are manufactured by Holset Engineering Co. Ltd. of Turnbridge, Huddersfield, England who have kindly provided the material for this section.

These turbochargers are compact, lightweight, units that can be mounted directly on the engine. They are suitable for high-speed engines in the power range from about 70 bhp to about 750 bhp. Various frame sizes are available, fitted with compressor wheels of between 3 in and $6\frac{1}{2}$ in diameter. All are of similar basic design, but in detail vary from the very simple small (3 in) unit shown in Figure 19–5 to the more complex ($6\frac{1}{2}$ in) unit shown in Figure 19–6. A sectioned model of the former appears in Figure 19–7.

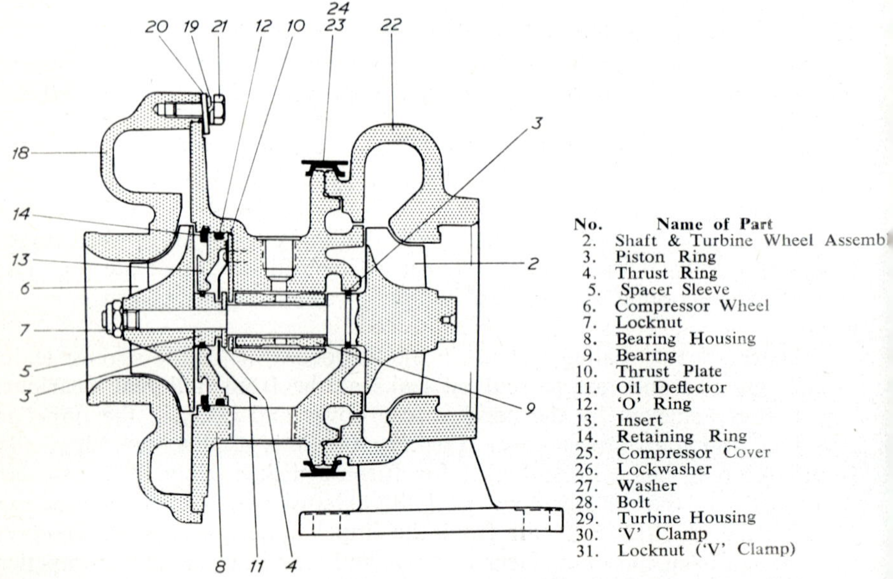

FIG. 19–5—*Holset Model 3LD turbocharger.*

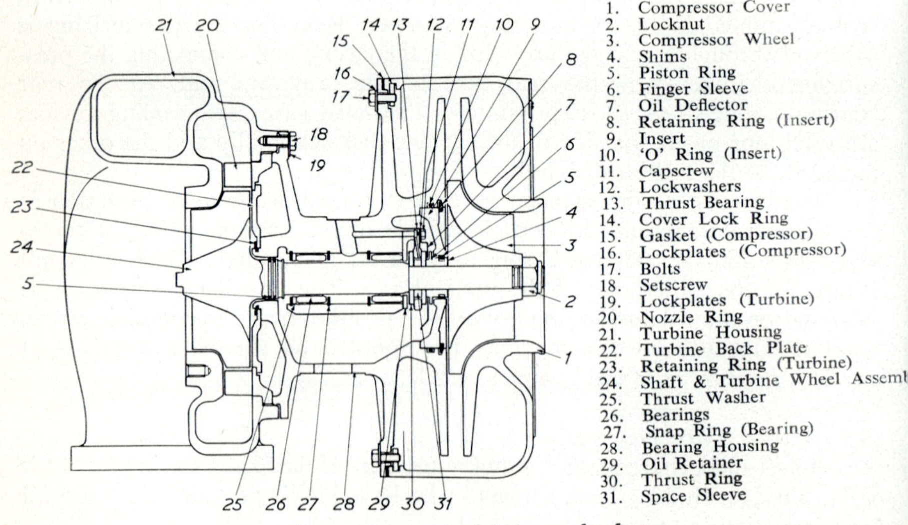

FIG. 19–6—*Holset Model 6D turbocharger.*

FIG. 19–7—*Holset Type 34 turbocharger.*

A centrifugal compressor is driven by a radial flow turbine. The rotor consists of an integral turbine wheel and shaft, to which the compressor wheel is secured by a nut. This rotor runs in fully floating sleeve bearings located in a bearing housing which, with a few other components, forms a central "core assembly" of the turbocharger. At one end of the core assembly is a turbine housing, secured either by a V-clamp or bolts; in the larger, low quantity units, turbine flow is controlled by a separate nozzle ring clamped between the turbine housing and the core assembly; in the smaller, high quantity, units turbine flow is controlled by the area of the shaped volute casing and different casings are provided for different applications.

To the other end of the core assembly is attached the compressor cover, which is a die-cast volute type in the case of the smaller units; this cover and the compressor wheel are machined from standard castings to provide flow passages that meet the air requirements of the engine. Diffuser vanes may be fitted in the larger model compressor to provide the highest possible peak efficiencies, but vaneless compressor diffusers are normally used in order to provide a wide flow range at acceptable efficiencies.

A tapping from the engine lubricating system supplies oil to the bearing housing, the oil serving the double purpose of lubrication and cooling. The oil enters the bearing housing at full pressure, flow being metered by the bearing clearances themselves. A separate thrust bearing is provided in the larger turbochargers, but in the smaller units the single journal bearing itself functions as part of the thrust bearing arrangement.

In all cases, the bearings are fully floating, to suppress oil whirl at the high speed, light load, conditions at which they operate. After leaving the bearings the oil passes to the drain cavities and back to the engine through the drain connexion.

As with all high speed machinery, adequate oil filtering is essential. All oil supplied to the turbocharger should first pass through a filter that will hold back all particles above 15 microns. If the engine oil filter will not clean the oil to this standard, a separate oil filter of adequate capacity should be provided in the turbocharger oil supply line.

Piston ring type oil seals are fitted outboard of each bearing. These are not positive seals, which would be impractical at the operating speeds of the turbochargers, but in fact act as restrictive labyrinths to limit the amount of air or gas leaking into the core assembly. It is this controlled leakage which holds back the oil and the process is assisted by generous sizing of the oil drain cavities and provision of deflector seals in the bearing housing.

The oil leaving the turbocharger may be in a foamy condition as a result of the small amount of air that passes the shaft seals. It is necessary, therefore, to have a relatively larger drain line which should slope downwards for its entire length without restrictions, sharp bends, or unnecessary turns, and connect into the engine crankcase at a point above the oil level.

The smaller sizes of Holset turbochargers are mounted by the turbine inlet flange being bolted rigidly to the engine exhaust manifold. This mounting should be the only rigid connexion to the turbocharger. All other piping should be sufficiently flexible in all directions so that no significant forces can be applied to the turbocharger as a result of thermal expansion or loads due to weight of piping. The largest size of Holset turbocharger (model 6D) is provided with mounting feet. If these feet are not used, the foregoing remarks also apply to this size of turbocharger. However, if the turbocharger is supported on the mounting feet, adequate leak-tight expansion joints may be provided in the exhaust pipe leading to the turbocharger.

The compressor cover can be rotated relative to the bearing housing, by loosening the fixing screws or V-clamp, to locate the outlet in any position required. A V-clamp mounted turbine housing can similarly be rotated to any required position, whilst on the larger units either 8 or 12 positions are available, depending on the number of turbine housing bolts.

The performance of the turbocharger will be improved if a divergent diffuser section as illustrated in Figure 19–8 can be used at the turbine outlet to recover the kinetic energy of the gas leaving the turbine wheel. The divergent section should have a diameter exactly equal to the turbine outlet diameter at one end, and a diameter equal to the exhaust pipe diameter at the other. The total angle divergence between the sides of the

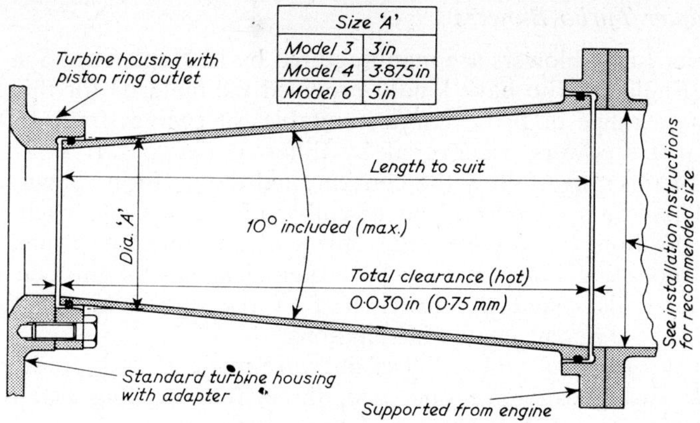

FIG. 19–8—*Holset diffuser section.*

diffuser should not exceed 10°. Such a diffuser section may often be impractical because of space limitations but is well worthwhile if it can be included.

With all turbocharged engines attention to the cleanliness of air entering the turbocharger is important. The air should be as free as possible from both oil and dust which cause fouling of the compressor flow passages, and rapid loss of efficiency. (If this in fact happens, it may be a very useful indicator of imminent damage to the engine cylinder bores by the same dust).

Figure 19–9 shows a typical illustration of a turbocharger on a high speed engine.

FIG. 19–9—*High speed engine fitted with Holset turbocharger.*

19.4 *Napier Turbo Blowers*

Napier turbo blowers are manufactured by D. Napier and Son Limited, Lincoln, England, who have kindly provided the material for this section.

A wide range of turbochargers suitable for engines from 160 bhp up to the largest powers, is covered by different design forms. The largest blowers of the range follow the conventional design form shown in Figure 19–10. The frame consists of an assembly of four casings each of which may be positioned relative to the others at 30° intervals round the axis and the mounting brackets may be mounted at any of thirteen positions at 15° intervals round the lower half of the turbine outlet casing for adaptation to different engine installations.

The turbine inlet and outlet casings are made of high grade cast iron. The inlet casing can have one, two, three, four or "big and little" gas entries with internal ducts arranged to keep the exhaust streams separate up to the nozzle ring. In the "big and little" casing the larger port serves $\frac{2}{3}$rds of the area of the nozzle and the smaller serves $\frac{1}{3}$rd. Internal passages accommodate the cooling water which is normally supplied from the engine jacket system; unions and blanking plugs fitted to the inspection plates can be interchanged to suit any arrangement of casing positions.

The compressor casings are cast in corrosion resisting aluminium, the design of the components permitting the accommodation of various widths of impeller and diffuser to enable the air flow to be matched to the requirements of the engine. The compressor inlet casing is available with either a side entry to connect to a ducted air supply or with a filter silencer drawing from the engine room.

The steel rotor shaft is in two parts spigotted and bolted to the turbine disk to form the assembly. Turbine blades are mounted in fir tree roots in the disk, secured by tab washers and wire-laced to counter vibration from the pulsating gas flow. The turbine wheel and blades are made from high temperature resistant alloy steel.

The compressor consists of an inducer and an impeller section, both machined from aluminium forgings. The whole rotating assembly is carried in bearings at each end. These may be ball and roller or plain sleeve bearings, the latter usually being recommended for the higher pressure ratios demanded for the newly designed constant pressure engines now being introduced.

The ball and roller type has a two row angular contact ball bearing at the compressor end of the shaft, incorporating controlled end-float to absorb the thrust. Roller bearings are fitted at the turbine end and both bearings are carried in resilient mountings to ensure smooth running and to protect the bearings when the rotor is stationary. Each bearing is lubricated by a separate gear type oil pump or by a suitably filtered supply taken from the engine.

The sleeve type has plain bearings which include hardened steel sleeves fitted to the shaft, enabling both fixed and rotating surfaces to be extracted and replaced easily and quickly. A separate thrust bearing at the compressor end permits end-float adjustment before assembly of the bearing on the shaft. Lubrication of the sleeve bearings may be provided from any

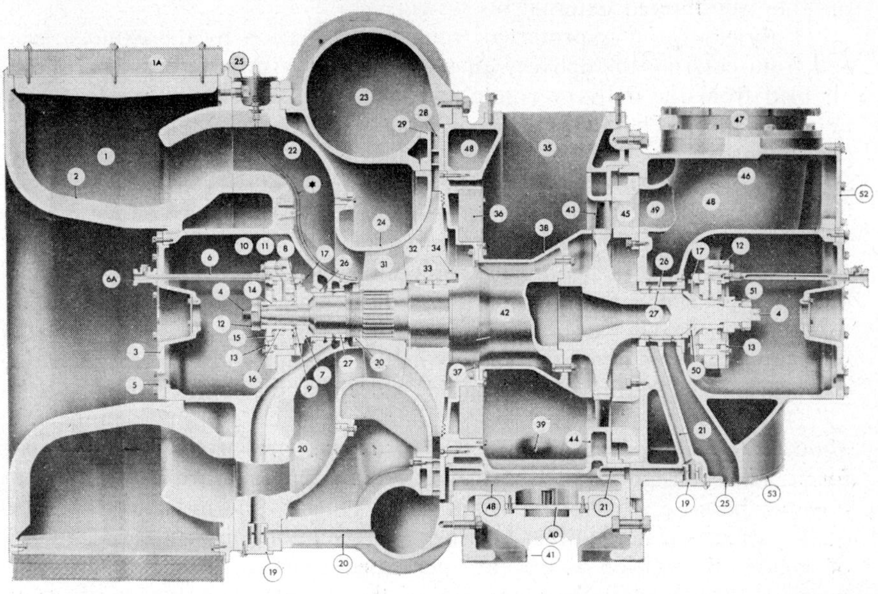

FIG. 19–10—*Napier turbocharger type S610.*

1. Compressor inlet casing (air filter/silencer shown; side entry version alternative)
1a. Filter element
2. Polyurethane foam lining
3. End cover
4. Tachometer mounting positions
5. Oil drain connexion
6. Oil feed pipe
6a. Oil supply connexion
7. Thrust block and spacer
8. Major thrust plate
9. Thrust clearance adjustment shim
10. Bearing housing
11. Turbine wheel clearance adjustment shim
12. Shaft end nut
13. Bearing housing locking plate
14. Compressor end bearing sleeve
15. Minor thrust plate
16. Bearing bush
17. Oil seal plate
19. Air seal passage restrictor plug
20. Air seal feed passages (compressor end)
21. Air seal feed passages (turbine end)
22. Sump portion of compressor outlet casing
23. Volute portion of compressor outlet casing
24. Compressor casing insert
25. Breather plate
26. Oil seal bush
28. Diffuser
29. Packing ring
30. Inducer ring nut
31. Inducer
32. Impeller
33. Impeller key
34. Impeller clearance adjustment shim
35. Turbine outlet casing
36. Heat shield (asbestos packing)
37. Air seal plate
38. Shaft cooling shroud
39. Drain plug
40. Water connexion cover plate (incorporating water deflector)
41. Mounting feet
42. Rotor shaft
42. Rotor shaft (incorporating turbine disc)
43. Turbine blades
44. Shroud ring
45. Nozzle assembly
46. Turbine inlet casing
47. Transition piece
48. Cooling water jackets
49. Engine exhaust gas passage
50. Turbine end bearing sleeve
51. Locking tool bush (with keys attached)
52. Core plate
53. Blanking cover plate (with anti-corrosion plug,

suitable well-filtered external oil system.

Lubricating oil is protected from contamination by the exhaust gases and from entering the delivery air stream, by labyrinth seals pressurized by air bled from the delivery volute and led to the seals by internal passages in the casings. The lubricating oil system will operate efficiently with a permanent tilt of 15° from the horizontal or a temporary tilt of 22½°, such as may arise in a vessel which develops a list or is rolling heavily. An initial installation of a blower in an inclined position may necessitate the use of an external oil supply.

For servicing the complete rotor assembly can be removed through the compressor end of the unit without disturbing the air or gas connexions to the casings. The air and oil seal labyrinths are sufficiently robust to withstand the weight of the rotor without damage during the removal of the bearings; a feature which enables simplification of the blower tool kit.

The turbochargers for the smaller powers covering the range from about 160 bhp up to about 3,000 horsepower are of light and compact form. Their design is based on a different approach which has become possible because of two important developments; the availability of high quality precision made plain bearings and the adoption of higher standards of engine oil filtration which have permitted a change from rolling to plain bearings. These plain bearings are of the double shell type with robust fully floating steel bushes. They are mounted inboard of the turbine wheel and impeller, reducing the distance between the bearing centres considerably, giving a short rigid shaft, with the thrust taken on conventional thrust pads adjacent to the journal bearings. The overhung turbine wheel and impeller are much more accessible for maintenance and the simpler construction results in easy access to the bearings even though they are inboard.

The smallest machines of the range have centripetal turbines in air cooled turbine casings. Those for somewhat larger power engines have axial flow turbines in water cooled casings, the arrangement being as shown in Figure 19–11. Turbochargers of this series, designated "SA" type, are suitable for pressure ratios from 1·5 to 3·5.

The air enters by a drum form combined filter and silencer or by side entry duct if desired, those components being of generally conventional form. The centrifugal compressor has a simple impeller held on to the rotor shaft by splines and a nut. The casing carries the diffuser and has a scroll form of volute.

The overhung construction of the rotor enables the gas entry to the turbine nozzle to be axial, minimizing flow losses. The nozzle ring directs the gas into the axial flow blades which are carried on a turbine wheel of conventional form.

The turbine outlet casing forms the main structure and embodies the mounting foot. The turbine inlet casing is bolted to it. The centre casing contains the bearings and the lubricating oil feeds. With the rotating assembly and compressor casing fitted to it, it is inserted as a cartridge assembly into the turbine outlet casing, an arrangement which provides a convenient means of dismantling for maintenance purposes.

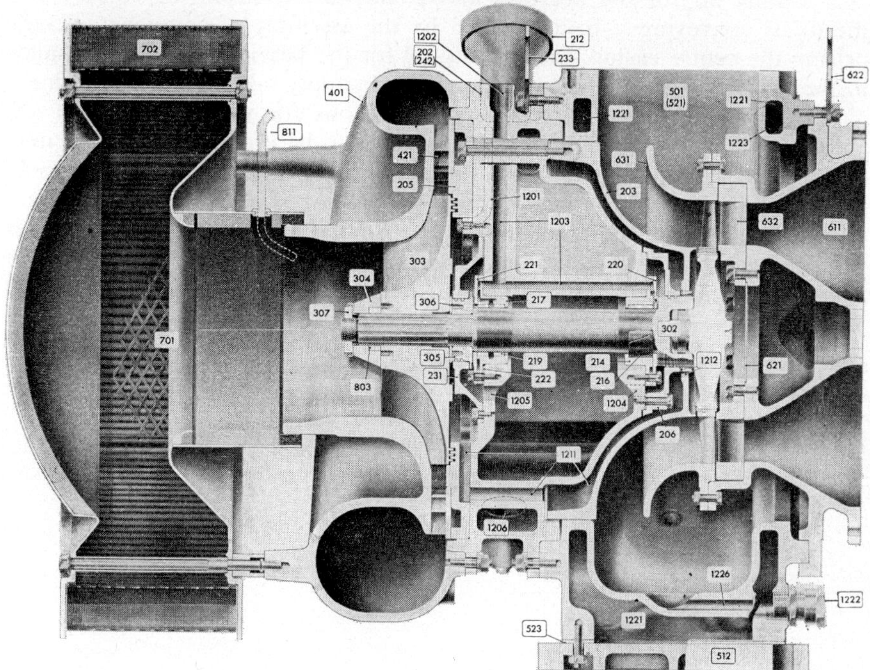

FIG. 19–11—*Napier turbocharger type SA.*

The item numbers in the following annotation have been selected to agree with those in the respective sections of the Spare Parts List for this turboblower:

202 Main (centre) casing (cast iron
(242) or aluminium)
203 Cone portion of main (centre) casing
205 Seal plate (labyrinth form)
206 Sea ring bush (and oil drain chamber), Turbine end
212 Breather (cowl form) to oil chamber
214 Bearing housing, Turbine end
216 Sleeve bearing, Turbine end
217 Bearing housing, Compressor end
219 Sleeve bearing, Compressor end
220 Thrust washer, Compressor end
222 Thrust block clearance (end float adjustment shim, Compressor end
231 Seal ring bush (and oil drain chamber), Compressor end
233 Lifting bracket
302 Rotor shaft, turbine disc and blades
303 Impeller (with labyrinth seal)
304 Spacer
305 Thrust block and seal ring carrier, Compressor end
306 Impeller clearance adjustment shim
307 Shaft end nut
401 Compressor outlet (volute) casing
421 Diffuser
501 Turbine outlet casing
(521) (cast iron or aluminium)
512 Mounting foot (and coolant jacket cover plate)
523 Sandwich plate (only for aluminium casing)
611 Turbine inlet casing
621 Gas barrier plate
622 Lifting bracket
631 Shroud ring
632 Nozzle ring
701 Air filter/silencer (sleeve-attached duct alternative)
702 Filter element
803 Speed-measuring magnetic washer (optional)
811 Compressor washing equipment (optional)
1201 Breather passage for seal plate
(normally blanked off)
1202 Lubricating oil inlet
1203 Lubricating oil passages
1204 Lubricating oil drain slots, Turbine end
1205 Lubricating oil drain slots, Compressor end
1206 Lubricating oil drain
1211 Sealing and cooling air passages
1212 Sealing and cooling air holes
1221 Coolant passages
1222 Coolant inlet
1223 Coolant outlet
1226 Drain to gas passage

Sealing air for the bearings and glands is bled from the compressor, the ducts conveying it being formed by the assembly of components comprising the centre casing. Lubricating oil for the bearings is fed to passages in the centre casing from a suitably filtered supply taken from the engine.

With all Napier turbochargers connexions for water washing of the impeller are provided and connexions may also be provided in the gas inlet ducting for washing of the turbine if heavy fuel is used.

In the design of the intake filters and passages attention has been paid to silencing, sound absorbent linings being fitted at effective positions. This requirement has become more important as the pressure ratios and air flows have increased demanding still higher speeds of these fast rotating machines.

INDEX